Weather Map Handbook

Fourth Edition

A guide to the Internet, modern forecasting,
and weather technology

Tim Vasquez

Copyright © 2003-2023 Weather Graphics Technologies

No part of this publication may be reproduced, stored in a retrieval system, or transmitted by any means without the express written permission of the publisher.

For information about permission to reproduce selections from this book, write to Weather Graphics Technologies, P.O. Box 170, Palestine TX 75802, or to servicedesk@weathergraphics.com .

Weather Map Handbook
Fourth Edition
April 2023

ISBN 978 0 9969423 5 5

Weather Graphics Technologies
P.O. Box 170, Palestine TX 75802
Web site: www.weathergraphics.com
servicedesk@weathergraphics.com

Contents

- 1 The forecast process

- 8 Analysis charts
    - 12 SURFACE
    - 14 850 MB
    - 16 925 MB
    - 18 700 MB
    - 20 500 MB
    - 22 300/250/200 MB
    - 24 TERRAIN FOLLOWING CHARTS
    - 26 THICKNESS
    - 28 ISENTROPIC ANALYSIS
    - 30 VORTICITY DIAGNOSTICS
    - 32 VERTICAL VELOCITY)
    - 34 Q VECTOR DIAGNOSTICS
    - 36 PRECIPITABLE WATER
    - 38 POTENTIAL VORTICITY
    - 40 CROSS SECTION DIAGRAM
    - 42 THERMODYNAMIC DIAGRAM
    - 44 WIND PROFILERS
    - 46 LIGHTNING DETECTION

- 50 Satellite imagery
    - 54 VISIBLE IMAGERY
    - 56 INFRARED IMAGERY
    - 58 WATER VAPOR IMAGERY
    - 60 GOES ABI BANDS
    - 62 MULTISPECTRAL IMAGERY

- 66 Radar
    - 68 REFLECTIVITY
    - 70 COMPOSITE REFLECTIVITY
    - 72 VELOCITY
    - 74 SPECTRUM WIDTH
    - 76 DIFFERENTIAL REFLECTIVITY
    - 78 CORRELATION COEFFICIENT
    - 80 SPECIFIC DIFFERENTIAL PHASE
    - 82 HYDROMETEOR CLASSIFICATION
    - 84 PRECIPITATION TOTAL
    - 86 VERTICALLY INTEGRATED LIQUID
    - 88 ECHO TOPS
    - 90 STORM TRACKING INFORMATION
    - 92 HAIL ALGORITHM
    - 94 MESOCYCLONE DETECTION
    - 96 TORNADO DETECTION
    - 98 VAD WIND PROFILE
    - 100 FREE TEXT MESSAGE

- 103 Human forecasts
    - 104 STORM PREDICTION CENTER
    - 106 WEATHER PREDICTION CENTER
    - 108 AREA FORECAST DISCUSSIONS

- 112 Numerical weather prediction
    - 116 GFS MODEL
    - 118 NAM MODEL
    - 120 RAP & HRRR MODELS
    - 122 ECMWF MODEL
    - 124 OTHER IMPORTANT MODELS
    - 126 ENSEMBLE PREDICTIONS
    - 128 MODEL SOUNDINGS
    - 130 HISTORICAL MODELS

- 135 Raw data
    - 136 SYNOP SURFACE OBSERVATIONS
    - 138 METAR SURFACE OBSERVATIONS
    - 140 TERMINAL AERODROME FORECAST
    - 142 RADIOSONDE OBSERVATIONS

- 145 Appendix
    - 146 1A — SURFACE PLOT SCHEMATIC
    - 147 1B — UPPER AIR PLOT SCHEMATIC
    - 148 2A — ICAO REGIONS
    - 148 2B — WMO REGIONS
    - 149 3 — DESCRIPTORS
    - 150 4 — PRESENT WEATHER
    - 151 5 — CLOUD GROUPS
    - 152 6 — ISOPLETHS
    - 153 7 — CHART ANALYSIS SYMBOLOGY
    - 156 8 — STABILITY INDICES
    - 160 9 — MILLER'S SEVERE PARAMETERS
    - 161 10 — SATELLITE GALLERY
    - 164 11 — POLARIMETRIC PRECIPITATION
    - 169 12 — WSR-88D PRODUCT LIST

- 171 Glossary

- 181 References and Further Reading

- 184 Index

Introduction

The previous version of Weather Map Handbook, the Third Edition, was written between 2013 and 2015, adding content like polarimetric radar products, newer modeling information, and updated sources. This newer Fourth Edition is intended to ensure relevancy with the state of forecast systems, techniques, and procedures of the mid-2020s. This includes greater use of AWIPS, a computer system which has been used by National Weather Service offices for over 20 years and is now available to users in CentOS environments. I have also phased out the NCEP DIFAX charts which were finally discontinued in 2017 after 50 years of distribution to the field. They are no longer used by today's forecast offices.

The previous edition included a number of historical and thought-provoking passages and graphics which I have decided to leave intact. After all, there is no need to rewrite history. Some articles and paragraphs also remain factually correct and there's no need to change essential material for the sake of change. Graphics however do deserve to be updated with new case studies, so I have added newer counterparts. With regard to the text, I have revisited every sentence in this book and made numerous changes throughout the text where new information is required or where I think a fresh perspective or better detail is useful. There are many changes, which makes this Fourth Edition an extensive update. A lot of outdated Internet links also had to be fixed or removed.

There has been some consideration of changing the units of pressure from millibars (mb) to hectopascals (hPa). These units are equivalent. The millibar is not a SI (International System) unit, but was originally introduced by the synoptic meteorologist Vilhelm Bjerknes. It still remains highly influential, particularly in American forecasting operations. I conducted an audit of internal National Weather Service discussions between 2002 and 2022 and found only a very minor shift toward hPa. Since this book is geared toward operational use, it has been decided to leave the units as millibars in this edition. Comments on this are welcome.

Thanks to Rob Dale, Dustin Pruett, Michael Galindo, Scott Nordstrom, Dan Young, Valerie Smith, and Jarrod Schoenecker for ensuring radar apps, software, and websites are up to date in these pages. I also thank John Monteverdi for this and other inputs, and Greg Stumpf for providing some clarification on algorithms used by the WSR-88D.

I welcome all suggestions, corrections, and ideas for future editions. I ask that you let us know of any errors you might find so that we can track it and make sure future editions are corrected. You may send these along to servicedesk@weathergraphics.com. Feel free to write us at any time to request an errata sheet.

TIM VASQUEZ
April 11, 2023

The forecast process

Meteorology has undergone a revolutionary shift over the past century. Instead of there being too little data to work with, now there is too much! The forecast process has become increasingly reliant on time management techniques to get the job done, and to be most effective it requires a thorough understanding of each one of the products that are available and which ones are most appropriate in a given weather situation.

The saturation of weather data has been made possible in part due to rapid advances in small computers and modem technology during the late 1970s and early 1980s. The modems in particular are a noteworthy technology, as they allow any available phone line to transmit data. The appearance of the Internet in the 1990s dramatically expanded the availability of data to anyone connected to the network. Both of these advances helped deliver more data to the forecast center and also allowed weather data to flood into the hands of hobbyists and students. The median Internet speed in the United States in 2022 was reported to be 192 Mbps. A complete NCEP model run can be downloaded and stored on an average family computer in minutes.

Home computing power is thousands of times faster than what it was when Windows 95 made its debut. This has brought surprising advances like numerical modeling to into offices and homes. A home computer equipped with Linux is easily capable of running the WRF model, one of the most sophisticated prediction models in the world.

Starting the forecast process

The forecast process does not involve looking at every single product. With the plethora of products and data available today, the forecaster must now carefully manage their time in selecting what to look at. Forecasters must continuously ask themselves *what am I trying to accomplish, and what is the correct tool for the job?* The answer, combined with a rough understanding of meteorology, will help you seek out the process that's right for the forecast situation. This book in turn will help you figure out what charts you should concentrate on.

For example, if you are sailing, are you trying to find the best wind patterns? If so, you'll be focusing on surface charts, and will also examine upper-level charts for features that could locally strengthen the surface pressure gradient. And if time allows, you'll work on tasks like reconciling models with observed data like satellite images and SYNOP surface observations to keep tabs on the accuracy of model details.

Are you a chaser expecting storms to fire any minute? The ingredients that would favor initiation in a specific location would include convergence and the presence of cumulus towers. So you will be looking at things like mesoscale surface plots for convergence and fine lines and enhanced cumulus in the visible satellite pictures. And where time allows, you'll look at model fields supportive of destabilization, and perhaps evaluate forecast initiation in model precipitation fields.

Weather essentials

Although a complete discussion about forecasting is far beyond the scope of this book, a brief overview of forecasting process is warranted. The overwhelming question that most forecasters try to solve daily is, "Where will there be clouds and precipitation?"

Obviously the clouds and precipitation may not have formed yet, so looking for the underlying causes, rather than what's there already, is the crux of the forecast process. In either case, the clouds and precipitation are caused by some type of lift, often called *ascent* by meteorologists. Stratiform rain is caused by slow, large-scale ascent of a very humid air mass. Convective showers and rain can be aided by this process, but depend much more on the presence of instability (very warm air underlying cold air). The ascent occurs in the form of very small pockets of rising motion, the size of a cumuliform cloud.

Sources of ascent

Slow, large-scale (synoptic-scale) ascent is one of the easiest problems for amateur forecasters, professionals, and numerical weather models to tackle. It occurs over such a large area that the air mass characteristics are sampled quite well by surface and radiosonde stations. In most cases, large-scale ascent is revealed in many different ways, including:

■ **Upper-level forcing**, often called "dynamics". This occurs when air is no longer in geostrophic balance, typically because of thermal contrasts in the air mass below. The air is forced to seek out sinking or rising motion to try to compensate for the lack of geostrophic balance. When the response is divergence in the upper atmosphere, which removes mass from the column and lowers surface pressures, air tends to rise to "fill the void". The famous "four-cell concept" of vertical motion and Q-vector divergence are all indicators of upper-level forcing.

■ **Convective lift**, which occurs in an unstable air mass when air ascends freely due to buoyancy. Any cold-over-warm situation is inherently unstable and favors convective lift. Sources of increased instability include insolation, advection over a warm surface, and upper level cooling.

■ **Isentropic lift**, which is most prominent when air parcels are travelling over rapidly varying air mass temperatures. The parcels must rise or sink in order to conserve their potential temperature.

■ **Surface convergence**, due to a clash in low-level wind direction or a low pressure area, causes ascent. Air converges and is forced to rise. This is closely related to orographic lift.

■ **Orographic lift**, where air is forced higher and higher as it travels along ascending terrain. This can occur on scales anywhere from ascent up a mountain to long runs of ascent measuring 1000 miles or more on the Great Plains. The longer runs of ascent are often referred to as upslope flow.

Finding areas of ascent

Each of these mechanisms has indicators, patterns, and characteristics which show up on standard weather charts. For example, warm air advection (WAA) occurs where winds and pressure gradient are bringing in warmer air. This tends to imply the presence of isentropic lift. At the 500 mb level, cyclonic vorticity advection (CVA), also known as positive vorticity advection (PVA) in the Northern Hemisphere, is often associated with upper-level divergence, which implies upward motion.

Although there are model charts that depict omega, which is the value of vertical velocity, these do not provide an easy, cookie-cutter solution. The omega fields tend to be noisy and difficult to scrutinize. Furthermore, they say nothing about why the lift or subsidence is occurring, its character, and what implications it has for the forecast area. A wise weather forecaster will look past these nondescript "headlines" and open up the charts to get a full understanding of what is happening.

Unfortunately this is where the discussion must end. To jump into these issues in much greater detail, see titles like the author's *Weather Analysis & Forecasting Handbook*, Gary Lackmann's *Midlatitude Synoptic Meteorology*, and *Meteorology Today* by C. Donald Ahrens and Robert Henson.

Picking a good web site

Whether you are a professional or a hobbyist, you'll need to get familiar with all of your products and know how to get to them. If you are a professional, this means knowing the key

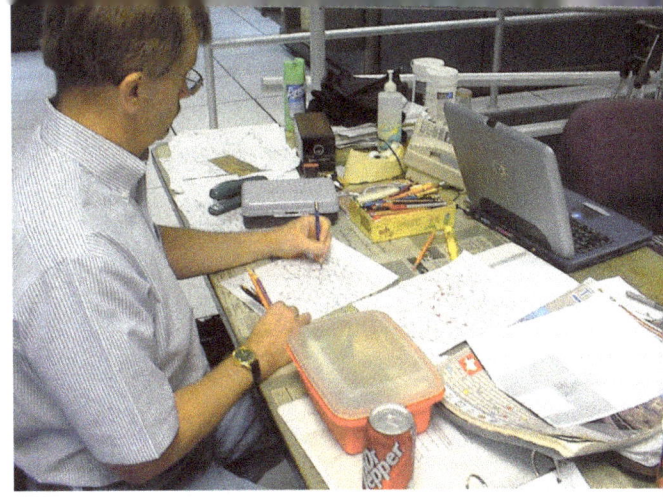

Preceding page: Forecasters at the Storm Prediction Center discussing a risk area. *(NOAA)*

Right: Alan R. Moller (1950-2014), forecaster at WSFO Fort Worth. He was a passionate advocate of hand analysis and an influence on the author. He is pictured here during a May 2003 severe weather outbreak. One of the books he endorsed was Robert Pirsig's *Zen And The Art Of Motorcycle Maintenance*, an exploration of quality. *(Tim Vasquez)*

product displays used at your location. If it involves the Internet, you'll need to set up your bookmarks and organize them in a useful manner. It's always best to research your options on a fair-weather day, since an unfolding severe weather outbreak is not the best time to be looking for new sources.

Some criteria to ask before including a web site in your forecast process includes:

■ **How timely are the updates?** Your favorite site might have the best graphics, but if 06Z rolls around and the HRRR isn't yet posted, it's a good idea to have a backup source in mind.

■ **Do the charts print well?** If you love making printouts and posting them on your bulletin board, charts with black backgrounds will print illegibly and exhaust your toner See if the web site offers a "printer friendly" option for each map. This provides the same graphic with a white background. You may also want to explore browser extensions or utilities that can provide a "negative" image of a graphic.

■ **Will software products do a better job?** Depending on what kind of chart analysis you are doing, Digital Atmosphere (the author's plotting software), RAOB, and other standalone programs may be more appropriate for the task at hand. Experiment with tools that might work and explore how they might help.

■ **Are there newer sources?** Weather charts on the web change monthly. Listen in on social media sites like Twitter and Reddit, and check discussion forums such as stormtrack.org and storm2k.org. Examine other people's bookmarks to see what they are using. Check in on sites you don't typically use. Many online sites often go dormant after a few years, and links rarely last more than ten years. If you get a "file not found" error you can often find new charts by shortening the URL. If it's still no-go, try a web search on the URL or the resource name in case it moved.

■ **Bookmark your favorite sites.** A sound forecast process means that you have a list of bookmarks or favorites set up in advance. Add to it as you find new charts you like.

Other valuable resources

Though book knowledge will take you far, there are a few other things you need to know before you get started. Practice often, and do get familiar with the UTC time zone.

■ **Pencils and crayons**. If you are planning to tackle hand analysis, a good supply of HB or B pencils (Staedtler Lumographs are top quality), colored pencils, and crayons are a great idea. Grease pencils or Vis-A-Vis transparency markers will give especially bold results when you want fronts, lows, and highs to leap out of the map visually. Keep a box with these pens and pencils near your analysis table.

And don't settle for the rubber eraser on your pencil. A good-quality vinyl ("plastic") eraser is amazingly effective at wiping away a badly placed isobar without smudging. Finally, a divider is handy for measuring map distances. Most of these items can be purchased online and at art and hobby shops, office supply stores, and architectural supply companies.

■ **Atlases**. Topography and geography plays tremendously into all aspects of meteorology, from orographic lift to storm reports from small towns. A good knowledge of topography is absolutely essential for forecasting. National Weather Service analysis desks have long been known for their excellent assortment of beat-up atlases.

Some online sources of standardized, regional-scale topographic maps are the aeronautical charts at skyvector.com. These provide elevations of all airports (and by extension most METAR stations and major facilities). The University of Texas Perry Castaneda library offers worldwide sets of Operational Navigational Charts (ONC) and Tactical Pilotage Charts (TPC) online that provide a similar function. They are available at *maps.lib.utexas.edu/maps/onc* and *maps.lib.utexas.edu/maps/tpc*.

Paper references do seem to be slowly disappearing. Still, though, you may want to consider keeping a few atlases near your desk. For worldwide topographic views the *Planet Earth Macmillan World Atlas* maps out the world at two sets of equivalent scales. For road networks in the United States and tracking severe weather, the *Michelin Road Atlas and Travel Planner* edges out ahead of the standard Rand McNally and AAA atlases since it offers full map coverage at a constant scale and has maps which cross state lines. This makes it very easy to follow weather systems and visualize their size as they progress across state and national borders.

It may seem strange in the Internet era to recommend paper atlases, considering resources like Google Maps, Open Street Map, and Google Earth are available. A very significant drawback of Google Earth is elevation contours and tint bands are neglected. Most paper world atlases portray this type of elevation data with great detail, so even in the 2020s and 2030s a good paper atlas still offers value to forecasters.

■ **Practice, practice, practice!** Skills only come with experience, and experience cannot be built if you are taking a casual, uninvolved approach to the weather. Supplement your experience by building a good weather library and using it.

Below: Forecasters at work at the Chicago/Joliet WSFO in March 2023. *(NWS)*

Time zone conversions. The "code" is the military designator, typically expressed as the phonetic word for each letter (Alpha, Bravo, etc). For further information see <*www.timeanddate.com*>.

When standard time is in effect (no local daylight saving or summer time)

Code	If your time zone is	To convert from UTC to local	Examples
Y	International Date Line West (IDLW)	subtract 12 hours	(none)
X	Samoa Standard Time (SST)	subtract 11 hours	Apia, Niue, Midway, Pago Pago
W	Hawaii Standard Time (HST)	subtract 10 hours	Honolulu, Hilo, Tahiti
V	Alaska Standard Time (AKST)	subtract 9 hours	Anchorage, Fairbanks, Barrow
U	Pacific Standard Time (PST)	subtract 8 hours	Los Angeles, Seattle, Vancouver
T	Mountain Standard Time (MST)	subtract 7 hours	Phoenix, Salt Lake City, Denver, Calgary
S	Central Standard Time (CST)	subtract 6 hours	Dallas, Chicago, Minneapolis, Winnipeg
R	Eastern Standard Time (EST)	subtract 5 hour	New York City, Boston, Toronto
Q	Atlantic Standard (AST), Western Brazil (WST)	subtract 4 hours	Halifax, Moncton, Bermuda, Thule
-	Newfoundland Standard Time (NST)	subtract 3.5 hours	Stephenville, St. Johns
P	Brazil Time (BRT), Western Greenland Time (WGT)	subtract 3 hours	Nuuk, Buenos Aires, Rio de Janeiro
O	Fernando de Noronha Time (FNT)	subtract 2 hours	Noronha
N	Azores Time (AZOT), Eastern Greenland (EGT)	subtract 1 hour	Scoresbysund, Cape Verde
Z	Greenwich Mean Time (GMT)	no change	London, Glasgow, Dublin, Lisbon, Dakar
A	Central European Time (CET), W. Africa Time (WAT)	add 1 hour	Paris, Frankfurt, Stockholm
B	E. European Time (EET), Central Africa Time (CAT)	add 2 hours	Helsinki, Kiev, Sofia, Athens
C	Arabian Standard Time (AST), E. Africa Time (EAT)	add 3 hours	Moscow, Baghdad, Aden, Nairobi
D	Gulf Standard Time (GST), Russia Zone 4 (ZP4)	add 4 hours	Samara, Dubai, Muscat
E	Pakistan Time (PKT), Russia Zone 5 (ZP5)	add 5 hours	Yekaterinburg
-	India Standard Time (IST)	add 5.5 hours	Calcutta, New Delhi, Bombay
F	Bangladesh Time (BDT), Russia Zone 6 (ZP6)	add 6 hours	Omsk, Dhaka
G	Indochina Time (ICT)	add 7 hours	Krasnoyarsk, Bangkok, Jakarta, Hanoi
H	Aust. Western Std. Time (AWST), China Std. (CST)	add 8 hours	Perth (WA), Manila, Taipei, Beijing
I	Japan Standard Time (JST), Korea Standard (KST)	add 9 hours	Tokyo, Seoul, Yakutsk
-	Australia Central Standard Time (ACST)	add 9.5 hours	Darwin, SA/NT
K	Australia Eastern Standard Time (AEST)	add 10 hours	Sydney, QL/ACT/NSW/VIC/TAS
L	Magadan Time (MAGT), New Caledonia (NCT)	add 11 hours	Magadan, Noumea, Ponape
M	New Zealand Standard Time (NZST)	add 12 hours	Kamchatka, Auckland, Kwajalein, Fiji
-	Tonga Time (TOT), Phoenix Isl. Time (PHOT)	add 13 hours	Enderbury, Tongatapu
-	Line Island Time (LINT)	add 14 hours	Christmas Island

When local daylight saving or summer time is in effect

Code	If your time zone is	To convert from UTC to local	Examples
U	Alaska Daylight Time (AKDT)	subtract 8 hours	Anchorage, Fairbanks, Barrow
T	Pacific Daylight Time (PDT)	subtract 7 hours	Los Angeles, San Francisco, Vancouver
S	Mountain Daylight Time (MDT)	subtract 6 hours	Denver, Salt Lake City, Calgary
R	Central Daylight Time (CDT)	subtract 5 hours	Dallas, Chicago, Minneapolis, Winnipeg
Q	Eastern Daylight Time (EDT)	subtract 4 hours	New York City, Boston, Toronto
P	Atlantic Daylight Time (ADT)	subtract 3 hours	Halifax, Moncton, Bermuda
A	British/Irish Summer Time (BST)	add 1 hour	London, Belfast, Glasgow
B	Central European Summer Time (CEST)	add 2 hours	Frankfurt, Paris, Oslo, Zurich, Warsaw
C	Eastern European Summer Time (EEST)	add 3 hours	Helsinki, Athens, Sofia, Kiev
D	Moscow Summer Time (MSD)	add 4 hours	Moscow
K	Japan Daylight Time (JDT)	add 10 hours	Tokyo
-	Australian Central Daylight Time (ACDT)	add 10.5 hours	Adelaide, SA
L	Australian Eastern Daylight Time (AEDT)	add 11 hours	Sydney, ACT/NSW/VIC/TAS

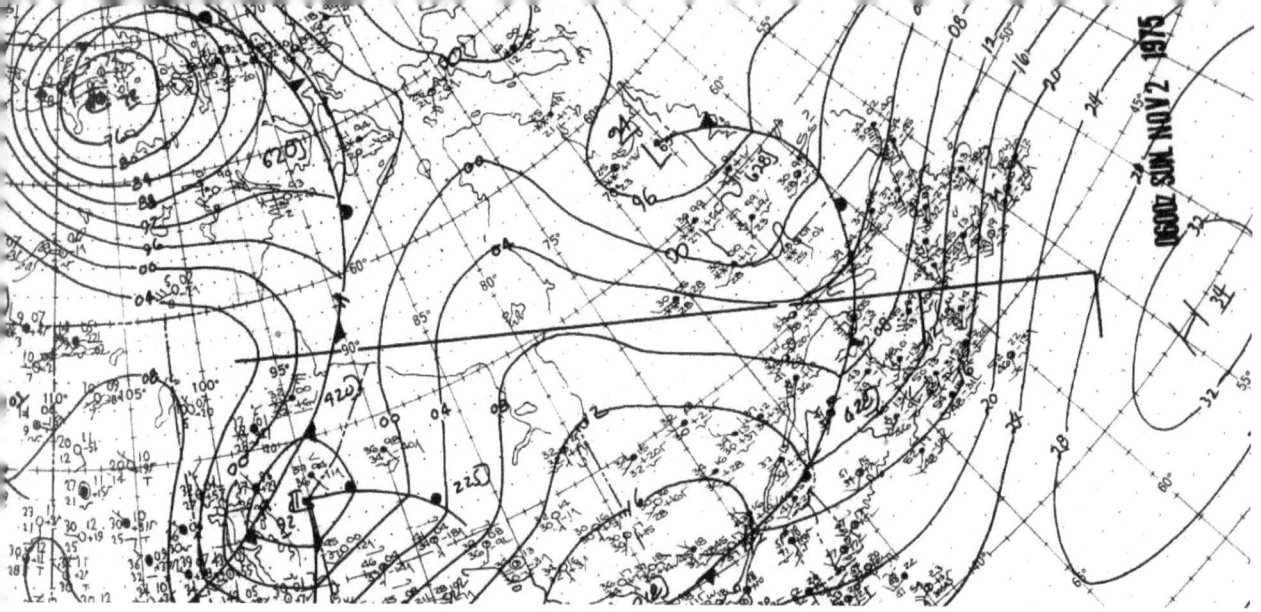

The day computers took over station plots. November 1975 brought a major change to the analysis procedures at the National Meteorological Center (now NCEP).

Above: One of the very last charts to feature hand-drawn station plots on November 2, 1975.

Below: Six hours later we see the introduction of automated station plots, the culmination of a significant amount of FORTRAN code development to translate the coded teletype reports to a graphical format. This eliminates drudgery that even today is not considered worthwhile, because station plots are a part of the chart preparation process, not the analysis process. This automated station plot upgrade required serious memory power at the time just to display the maps: the 4000 x 2500 pixel raster field required 10 million bits (about 1.2 megabytes) to store. At the time, NMC computers were limited to 4 MB. Buying this much memory for a homebrew system at the time would have cost $60,000. This chart was printed instead of displayed, since high-resolution digital displays were not widely available until the late 1980s. Instead, these maps were directly offloaded to 36-inch to 60-inch wide electrolytic radiofax printers for hand analysis. This paper was noteworthy for leaving the machine wet and carrying a strong ammonia smell until it dried. The National Weather Services AFOS (Automation of Field Operations and Services) system was the first major move away from fax charts to digital charts. It was deployed nationwide in the early 1980s.

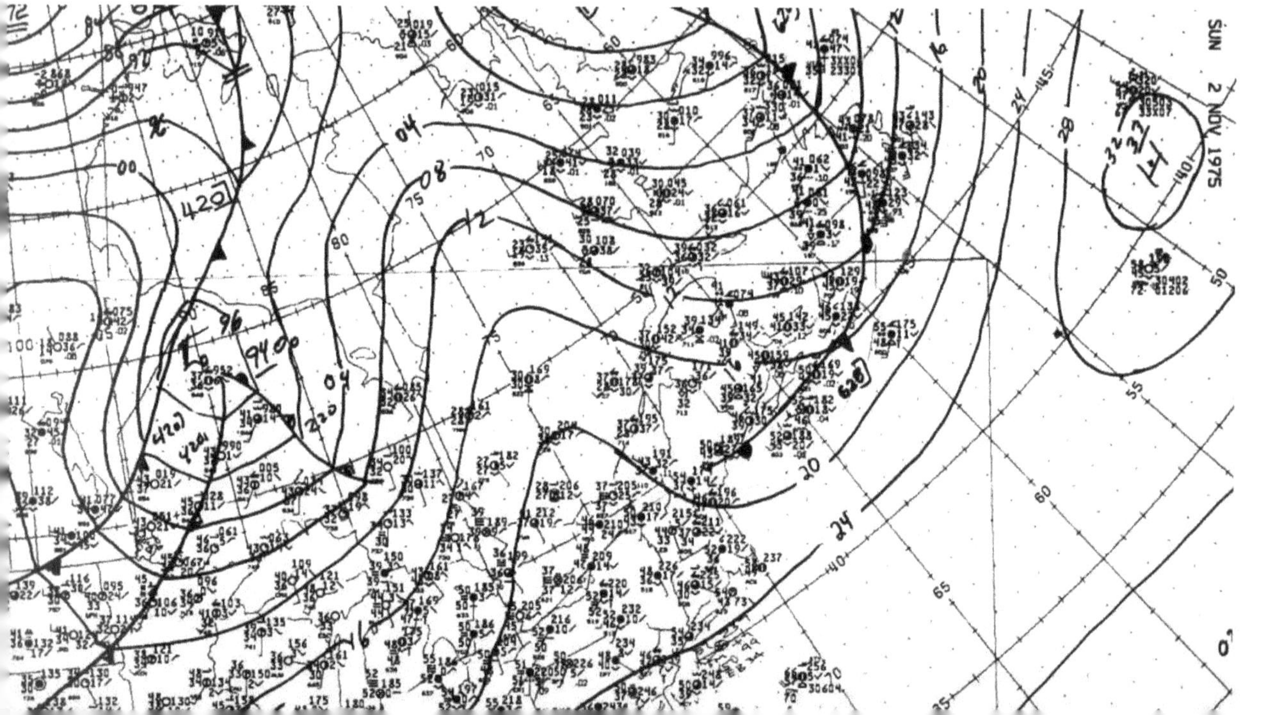

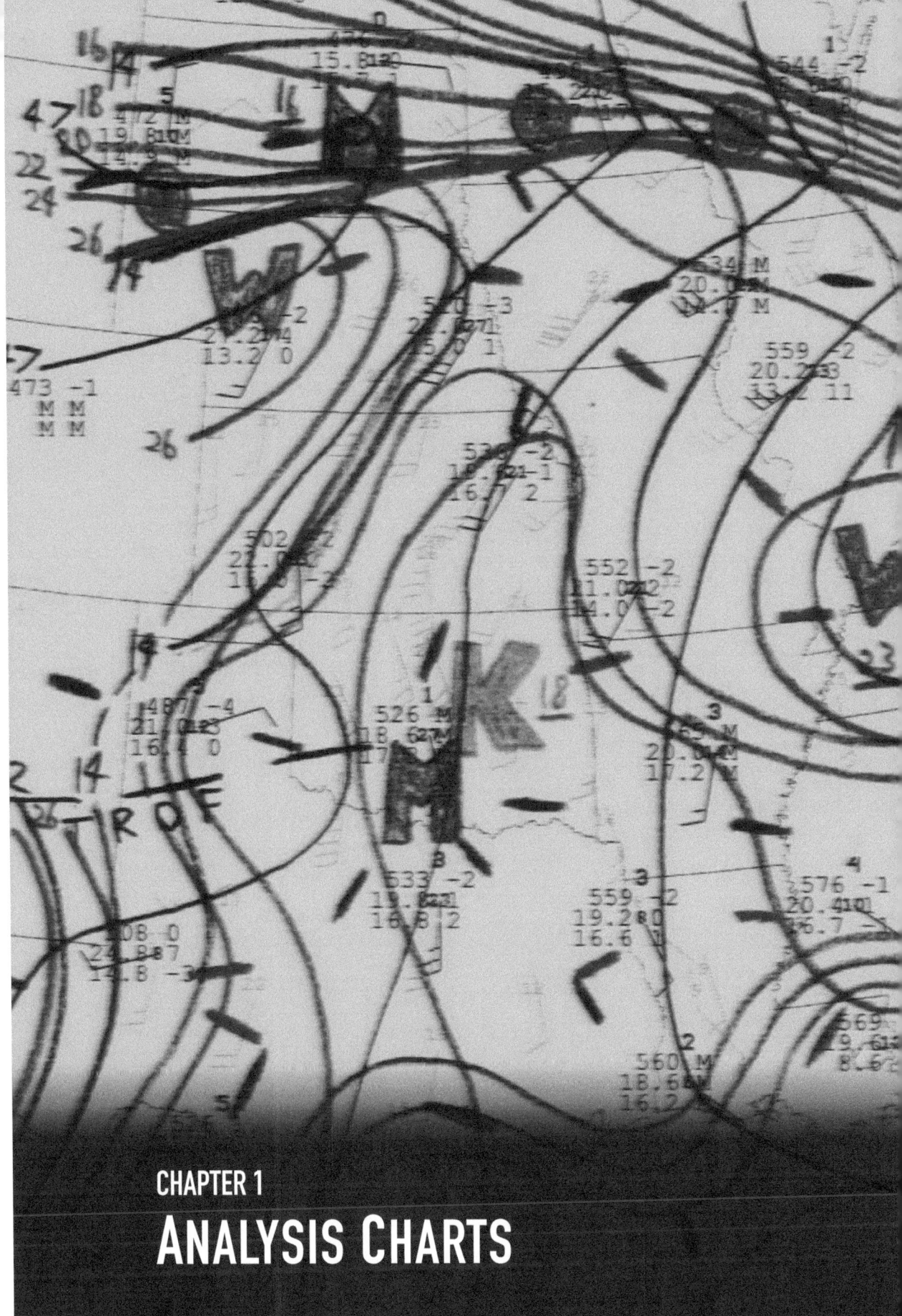

CHAPTER 1
ANALYSIS CHARTS

Analysis charts

A weather product is often used merely to obtain the basic information that it shows. What is the temperature? Is it raining over Pittsburgh? Is there a threat of clouds for tonight's star party? Will the flight crew have a good tailwind?

In a more organized forecast setting, meteorologists are not as interested in exact numbers as in being able to visualize processes in the atmosphere and look for clues. Therefore this book attempts to explain not only how to interpret maps but also to find the subtle signals, patterns, and trends that may have a much greater impact on the forecast. This is the science and art of meteorological diagnosis. Developing the talent for diagnosis requires a little philosophical insight and the willing participation of the forecaster.

The ancient days of analysis

Systematic gathering of weather data over a wide area began in the 18th century with the logbooks of scientific institutes and naval . This led to the understanding of the Hadley Cell which drives the trade winds and weather in the tropical regions.

However there was no effective way to transmit weather data in realtime. One of the earliest inventions that attempted to solve the problem of long-distance communication was the optical signal system. Interestingly the first system was driven by British gamblers in 1767 who sent the results of horse races ahead of the news.

The first government system was deployed in France in 1792 in the midst of the turbulent French Revolution. It was a great success and by 1794 it was carrying coded diplomatic and military messages at an effective speed of 300 mph, and was used by other European countries. However due to the very slow signaling speed it was not capable of transmitting weather data.

The real data revolution came in 1837 with the development of the electric telegraph. Surprisingly this was slow to gain acceptance, but by the early 1840s its potential was realized and a rapid expansion of telegraph links followed during the decade. This is perhaps the single most significant contribution that would make modern meteorology possible.

By 1848, the telegraph was being used in Britain to gather weather observations for publication, but there was not yet any effort to produce centralized weather maps or investigate the forecast potential of the system. It would be 1857 before the British Board of Trade began creating daily weather maps.

The stormy month of October 1859 witnessed the sinking of 200 vessels in British waters. The worst was the foundering of the steam clipper *Royal Charter* off the Welsh coast, which claimed over 450 crew and passengers. This prompted renewed interest in synoptic analysis, and led to an expansion of the observation network. By 1861, lighthouses were displaying storm warnings.

In the United States, the Smithsonian Institution in 1849 was the first to exploit the telegraph for a national weather network. Some of the first weather charts were created over the following years, but the logistical and military demands of the Civil War would sideline meteorology during much of the early 1860s.

In 1870, a Congressional resolution restarted the weather network for the purpose of protecting shipping along the coasts and in the Great Lakes. Responsibility was delegated to the War Department, who placed the program under the Army Signal Service.

The Johnstown, Pennsylvania flood of 1889 underscored the need to place weather forecasting under civilian control to help meet the needs of the general public. This led to the creation of the U.S. Weather Bureau in 1890.

For the most part, the U. S. Weather Bureau focused on general weather prediction and agricultural forecasts. By the 1920s the U. S. Weather Bureau shifted to teletype, which dramatically increased the ability of regional offices to process many different kinds of weather data. Radio facsimile machines were also developed, in which a source image was scanned, the information transmitted via shortwave radio, and duplicated by a mechanical printer which simulated the same scanning process. This system was used for a limited time during the 1920s, but the cost was prohibitive.

As a stopgap solution, during the 1930s the Weather Bureau devised the Sectional Teletype Map, in which an analyzed pressure field was transmitted graphically on teletype machines, similar to the "ASCII art" of the computer era. The receiving weather office simply drew lines along the margins of the "shaded" areas, then underlaid a map transparency and traced the state borders. Since this only transmits a very coarse representation of the weather field, it was soon abandoned.

Modernization and computers

Aviation, instrument flying, and commercial passenger flights all sparked the next big revolution. These became common in the years leading up to WWII, and by this time airlines had begun setting up their own meteorology departments.

After World War II, facsimile machines were miniaturized and became economical enough to put in all regional forecast offices and dispatch centers. Thy were widely adopted, first in 1946 by U.S. forces in postwar Europe, using a network of 23 radio stations. The Weather Bureau took note and deployed them at its forecast offices during the 1950s.

Chart analysis continued much the same through the 1960s and 1970s. However late in the 1970s the National Weather Service recognized the potential of the "mini-computer", the predecessor to the personal computer. This led to the development of AFOS (Automation of Field Operations and Services). The AFOS system was deployed in the early 1980s nationwide and formed the backbone of weather charts during the 1980s and 1990s.

The AFOS system was quickly made obsolete, outpaced by the rapid developments in

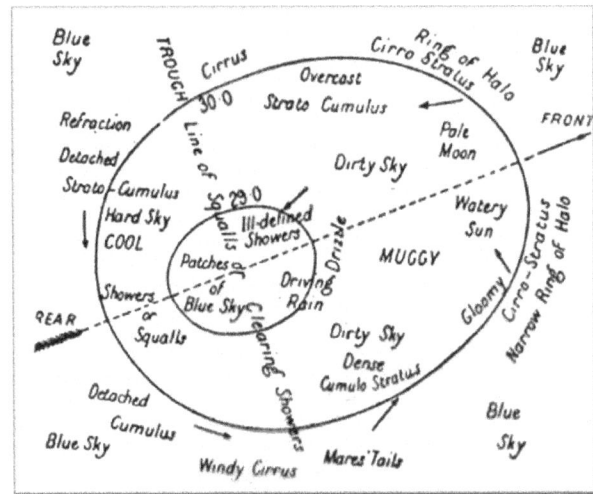

Above. During the 19th century, cyclones and fronts could be observed on weather charts, but the techniques for forecasting them were not understood. Forecasters used rules of thumb such as these, which relates locations of typical weather relative to the surface cyclone, and with respect to its movement.

Quick tip

Want professional-looking paper charts? This is possible with an 11x17" (wide format) printer.

In previous editions of this book we recommended looking at online auction sites. By 2019, however, affordable wide format printers were going mainstream. A cheap printer can be bought new for about $200. Keywords to look for are "11x17", "wide format", and "tabloid". Check reviews and the cost of ink.

As far as paper, office supply stores are very hit and miss when it comes to reams of 11x17" paper. A ream is 500 sheets of paper. Buying a ream is far cheaper per sheet than buying smaller quantities. If you're in a city, look for an architectural supply store. If all fails, you can get such paper from online sources.

The very first system of weather symbols, published in 1771 by J. H. Lambert. In order they represent clouds, rain, snow, fog, and thunder.

Lambert's Symbols, 1771.

computing during the 1980s. By 1993, desktop weather programs like the author's own WeatherGraphix software greatly surpassed what was possible on AFOS. The AFOS system was also hampered by monochrome screens, limited data storage, and slow speed.

These shortfalls were recognized early on, and in the early 1980s a vision for a follow-on system was developed as part of the National Weather Service modernization effort. This became known as AWIPS (Advanced Weather Interactive Processing System). A lengthy, difficult development process began in the late 1980s. The first systems were deployed between 1998 and 2001. This was followed by the deployment of an upgraded system, AWIPS-II, in 2011.

Weather symbols

The first system of weather symbols was devised in 1771 by J. H. Lambert. By the early 19th century, many different countries had adopted their own systems of weather symbols, few of which are recognizable today.

The system of International Meteorological Symbols was standardized at the Vienna Meteorological Congress of 1873. Here the modern symbols for rain, snow, the thunderstorm, fog, and blowing snow were published. Many countries, including the United States, continued to use their own national systems, but they were gradually phased out by the early 20th century.

The chart analysis process

Some excellent papers have been written on the philosophy behind using weather charts. Two of them are, "The Role of Diagnosis in Weather Forecasting" (1986) by Charles Doswell III and Robert Maddox; and "The Human Element in Weather Forecasting" (1986) by Charles Doswell III, both online via a web search. In these essays, Doswell and Maddox identified two separate steps that must occur before a prognosis (forecast) can be made: analysis and diagnosis.

The first step, analysis, is the identification of ingredients in the atmosphere. It is the process of drawing lines and identifying fronts, lows, highs, wind shift lines, jets, and air masses. A computer may assume a lot of the analysis procedures. Following this, a process called diagnosis is performed. This is composed of thought, ideas, and conclusions. It is the process of visualizing the completed analysis and relating it to other products. Diagnosis is also the precursor to a coherent, robust forecast.

Unfortunately in today's computer-driven age, some forecasters complete the analysis but fail to perform any diagnosis. Quick glances at the maps and reliance on computer-drawn isopleths do not constitute diagnosis. As a result, the misguided forecast process inevitably leans toward very heavy emphasis on numerical models. The balance in the forecast process is spelled out surprisingly clear in forecast discussions, case studies, and even technical papers.

There are countless situations where important ingredients are resolved not through the models but only through careful analysis and diagnosis. One case in point is a jet max moving out of New Mexico on 3 May 1999, linked to a deadly tornado outbreak near Oklahoma City, and initially detected only by the Tucumcari NM wind profiler. In this case, meteorologists detected the clue and worked it into the forecast.

A debate has sometimes emerged among some of the best operational forecasters: is it absolutely necessary to put pencil to paper and "hand-analyze" the chart? The consensus is generally "yes". Either way, there is nothing to be lost by forcing yourself to scrutinize the data in better detail and find a relationships in the patterns.

Even so, hand analysis only works when the forecaster makes an effort to think about the data. Drawing lines with a closed or distracted mind accomplishes nothing more than drawing lines. Make an effort to visualize the meaning of the data as you put pencil to paper. Glance at the station plots as you go, picturing the weather and contrasting it to conditions you see nearby. The hand analysis is not a picture to be drawn; it is a canvas for thought.

Dedicated analysts will want to invest in a wide format printer, as conventional letter-size sheets offer only a small workspace for a detailed analysis field. Wide format refers to the tabloid paper size, equaling 11x17" in the United States, or A3 or 297x420 mm internationally. At press time, economical wide format printers were available for about $200 in the U.S.

Frontal placement

Fronts are one of the most basic components of a weather chart, as they highlight *baroclinic zones*, where temperature advection is taking place and atmospheric energy is at work. A front is always drawn with its barbs or pips facing the direction of movement. The figure at right details the styles used for depicting a front.

Fronts are always placed on the *warm side of a temperature gradient*. This makes perfect logical sense, as when we picture a cold air mass moving into a region, the front occurs when the temperature first begins falling, not when the temperature drop is complete. When warm air is invading, it is not quite as intuitive, but the most basic weather books demonstrate that the warm front passes after the temperature has finished climbing (not counting the effects of diurnal heating, of course).

The occluded front is a little more challenging. It is covered at length in some of the author's other book titles, such as *Weather Analysis & Forecasting Handbook*. The occluded front at the surface will take on the characteristics of a warm front or cold front, depending on what type of occlusion it is, and there will also be a thermal trough associated with the deepest warm air aloft. This trough will be parallel to and ahead of the front in the case of a warm occlusion, and behind or overlapping the front if there is a cold occlusion.

Drylines are sharp boundaries between a moist tropical air mass and a very dry continental airmass, usually originating from higher terrain. They are common in the southern Great Plains of the United States but may also be found in India, the Sahel, and Australia. The dryline is always located on the *moist side of the dewpoint gradient*. The barbs point toward the direction of the dryline's movement.

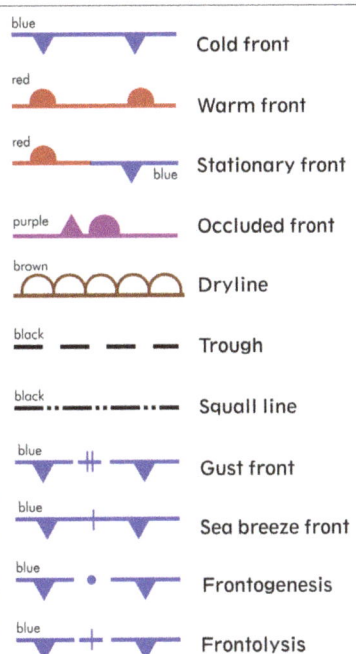

Above. Conventions for drawing fronts and boundaries. The scheme for a gust front and sea breeze front is adopted from Young and Fritsch (1989).

Finally, troughs and other features can be sketched in. However it should be noted that it is important to spend more time refining your position of the feature than attempting to categorize it. The feature can be revisited later in the diagnosis process and may be much more meaningful then.

The symbols for fronts and boundaries are shown in the table above. Throughout this book, various styles and colors for other parameters are recommended. Most of these assume you are marking on white charts. For black (on-screen) graphics, these colors are all still valid, except that white and black are reversed.

MILLIBARS VERSUS HECTOPASCALS

Millibars (mb) and hectopascals (hPa) are exactly the same. One hectopascal (hPa) equals 100 Pa, which equals one millibar. Millibars remains the conventional unit used in weather map analysis in the United States and is still used extensively in operational meteorology. In this edition we continue to use millibars.

Surface chart

The surface chart is the backbone of weather forecasting and is the most familiar map to both meteorologists and the general public. It provides a display of parameters such as temperature, dew point, and pressure at the Earth's surface.

The biggest advantage of the surface chart is that data is available quite frequently: as often as six hours in remote regions and 20 minutes in North America. A less obvious but extremely important use of the chart is to search for imbalances that reflect processes occurring aloft, which may not be reflected by the coarse radiosonde data that is only available every twelve hours. For example, pressure fall centers or very gusty winds may be closely linked to areas of strong upper-level forcing moving with the upper-level flow.

Although winds closely follow the pressure (height) lines on upper-level charts, this is not necessarily the case on the surface chart. The winds usually turn more directly toward low pressure. This is because the effect of friction is much more pronounced, which diminishes the Coriolis effect and allows air to move more directly toward low pressure. The effect is not so strong on ocean surfaces due to weaker friction.

Fronts are frequently found on the surface chart. The primary indicator of a frontal boundary is the temperature field. In mountainous areas it can be helpful to analyze using theta, or potential temperature, to normalize the effects of elevation.

A recommended sequence for analyzing the surface chart is to look for obvious fronts and boundaries and sketch them in lightly. Then sketch in isobars lightly with the pencil. Use the isobars and other indicators, and even other tools such as satellite and radar, to further refine front and boundary positions. Then "harden in" the final position of the front. Following this, "harden in" the isobars, forcing them to kink along the front. While this process is underway, never forget to spend time visualizing what is seen on the chart, feeling the data as you go. This will help make your diagnosis complete.

Starting in the 2000s and continuing through the 2010s there has been an effort to integrate mesonet data from a wide variety of sources into the surface analysis process. The NOAA MADIS program (madis.ncep.noaa.gov) channels this data into forecast systems. This data may be seen on some sources of analysis charts. As it depends on a large number of private networks, not all of this data is releasable outside the federal government.

key details

The surface chart depicts conditions at ground level. It shows processes in the lowest layer of the troposphere, taking advantage of the very dense observation network with frequent observation times.

The chart is drawn as follows:
- Data plots are given in Appendix Table 1A.
- Isobars are drawn in solid black every 4 mb or 2 mb (hPa) or 0.05 in Hg.
- Isotherms (optional) are drawn every 2 C° or 5 F°. For detailed analysis, 1 C° / 2 F°
- Add other base fields as needed.
- High and low pressure centers are marked using, respectively, a large blue "H" or a large red "L". The center pressure value may be labelled below the H or L, typically in tens and units of a millibar.
- Fronts are drawn in blue and red using standard notation.
- Trough axes are drawn as a thick black dashed line.
- Drylines are drawn as a thick brown line with hollow pips connected to one another, facing toward the moisture.

Observations of U.S. surface weather have been made primarily by automated sensors since the mid-1990s. Human weather observations are common outside North America and Europe.

The vast majority of official surface data flows through FAA and ICAO (AFTN) and WMO data circuits in METAR and SYNOP format. The rise of mesonets has created additional formats and data sources. Some popular formats for secondary data include BUFR (WMO), XML, and CSV. Unlike METAR and SYNOP, these formats are designed for computers rather than for human readability.

real-time Internet sources

www.aviationweather.gov/metar
weather.gc.ca/analysis/
www.wpc.ncep.noaa.gov
weather.ral.ucar.edu/surface/
weather.cod.edu/analysis
www.spc.noaa.gov/exper/mesoanalysis
www.wetter3.de

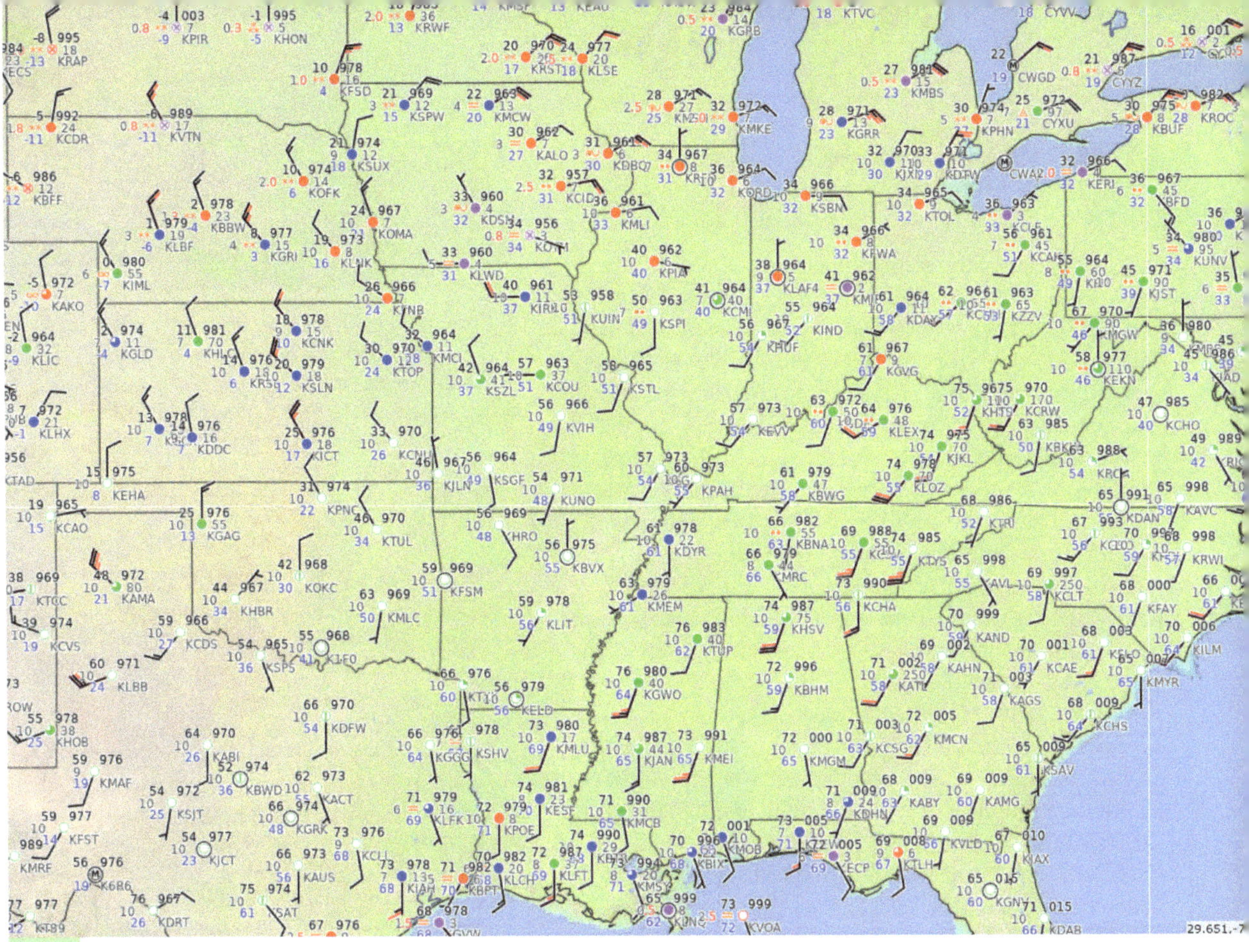

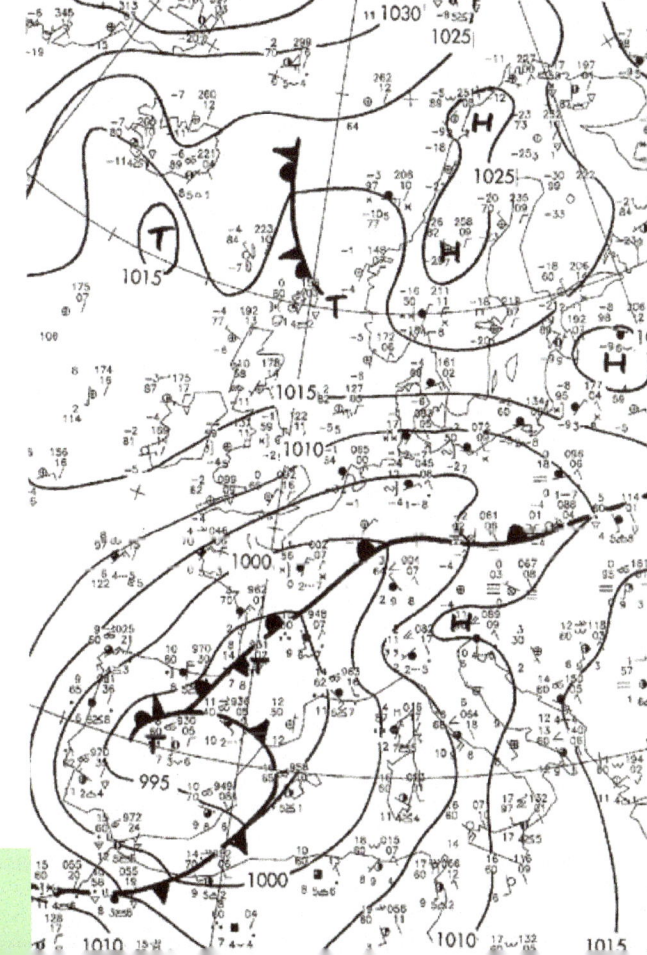

Above: Raw surface chart (surface plot) for 22 February 2023 showing a powerful weather system moving through the Midwest states. This chart originates from the National Weather Service Aviation Weather Center and is built from METAR observations. This tool is essential for following fast-breaking mesoscale developments. Note that this is not referred to as an "analysis" because nothing has been analyzed. No attempt has been made to diagnose weather fields and features. This is simply a display of the original observations. It serves as a starting point for an analysis, either through automated or manual means. *(NOAA/NWS Aviation Weather Center)*

Right: German surface analysis for 23 December 2010 during one of the most severe cold air outbreaks to hit the British Isles in modern times. The temperature fell to -18.7 °C (-1.7 °F) in Northern Ireland. The easterly gradient across northern Europe was instrumental in advecting cold air from Scandinavia westward. A blanket of fresh snow across much of the region helped insulate the air mass from the effects of conductive heating. *(Deutscher Wetterdienst)*

925 mb chart

The 925 mb level is located at about 2500 ft (760 m) MSL, assuming a standard atmosphere. In practice, it can vary by a few hundred feet or more. Since this is the lowest part of the troposphere, it is likely to intersect the ground in mountainous terrain and plateau regions, so it must be used with caution. In the Great Plains, the 925 mb level will typically intersect the ground along a line from western South Dakota to western Kansas and into west Texas along the Caprock. It often clips the top of the larger ridges in the Appalachian Mountains.

The chart was unknown in operational forecasting until late 1991 when international coding standards began prescribing 925 mb as a standard radiosonde level. Given the rapidly expanding interest in mesoscale forecasting during the 1990s, the 925 mb chart was quickly adopted as an essential chart at forecast offices for obtaining data on frontal locations, moisture depth during severe weather events, and finding low-level jets, particularly in the eastern half of the United States and Canada.

The 925 mb level is close to the 1 km AGL altitude for stations close to sea level and is equivalent with the 1 km AGL altitude in many parts of the Midwest and lower Great Plains. Thus it can be used as a proxy for 1 km conditions. This provides information on the top of the 0-1 km AGL layer in which parameters such as mixed layer CAPE and as well as shear parameters are evaluated.

At elevations under 1000-2000 ft MSL, the 925 mb level commonly intersects the lower portions of the low-level jet, and can be used to keep track of its behavior leading up to a severe weather event. The 925 mb chart at such locations is usually found within or above the nocturnal inversion, but as surface heating takes hold, the level becomes firmly rooted in the lower part of the planetary boundary layer where deep mixing takes place.

During quiet weather patterns, especially during the summer, this chart is useful for establishing the prevailing low-level flow through the use of streamline analysis. By being located just above the ground, frictional effects and terrain interactions are reduced and the winds more closely approximate geostrophic flow.

It may be more appropriate for some forecasters to use a terrain-following chart (q.v.) such as the 1 km AGL analysis when forecasting severe weather events on sloped terrain, such as that in the Great Plains and the Canadian Prairies. This prevents problems with terrain intersection and interpretation of patterns at irregular AGL heights.

key details

The 925 mb chart is usually near 2500 ft MSL. It's useful for assessing severe thunderstorm moisture quality and inflow wind fields, for evaluating the depth of tropical and polar air masses alike, and for finding low-level jets. It usually lies below convective cloud bases but may intersect stratocumulus layers and low level jets at night.

The 925 mb chart intersects the ground in mountainous terrain, high plains, and plateau regions and cannot effectively be used there.

The chart is drawn as follows:
- Data plots are given in Appendix Table 1B.
- Geopotential height contours are drawn as solid black lines every 30 m (3 dam) using 60 dam as a base value (60, 63, 66, etc).
- Isotherms are drawn as dashed red lines every 2 C°.
- The type of moisture analysis (optional) is contingent on the type of weather regime:
  -- In stratiform situations, a relative humidity parameter is recommended. Shade all areas in green with dewpoint depressions below 5 C° (this is roughly a 75% relative humidity).
  -- In convective situations, use an absolute humidity parameter. Use the dewpoint value to draw isodrosotherms, or outline specific humidity or mixing ratio contours and shade the highest values.

Features are drawn as follows:
- Low level jet axes (narrow bands of strong winds exceeding 40 kt) should be drawn as a thick red arrow.
- Moisture axes may be drawn as thick, wavy green lines.
- Fronts should be drawn in standard blue and red colors. Since this is an upper-air chart, do not shade the barbs and pips.

real-time Internet sources

weather.ral.ucar.edu/upper
weather.cod.edu/analysis
www.spc.noaa.gov/exper/mesoanalysis
www.wetter3.de

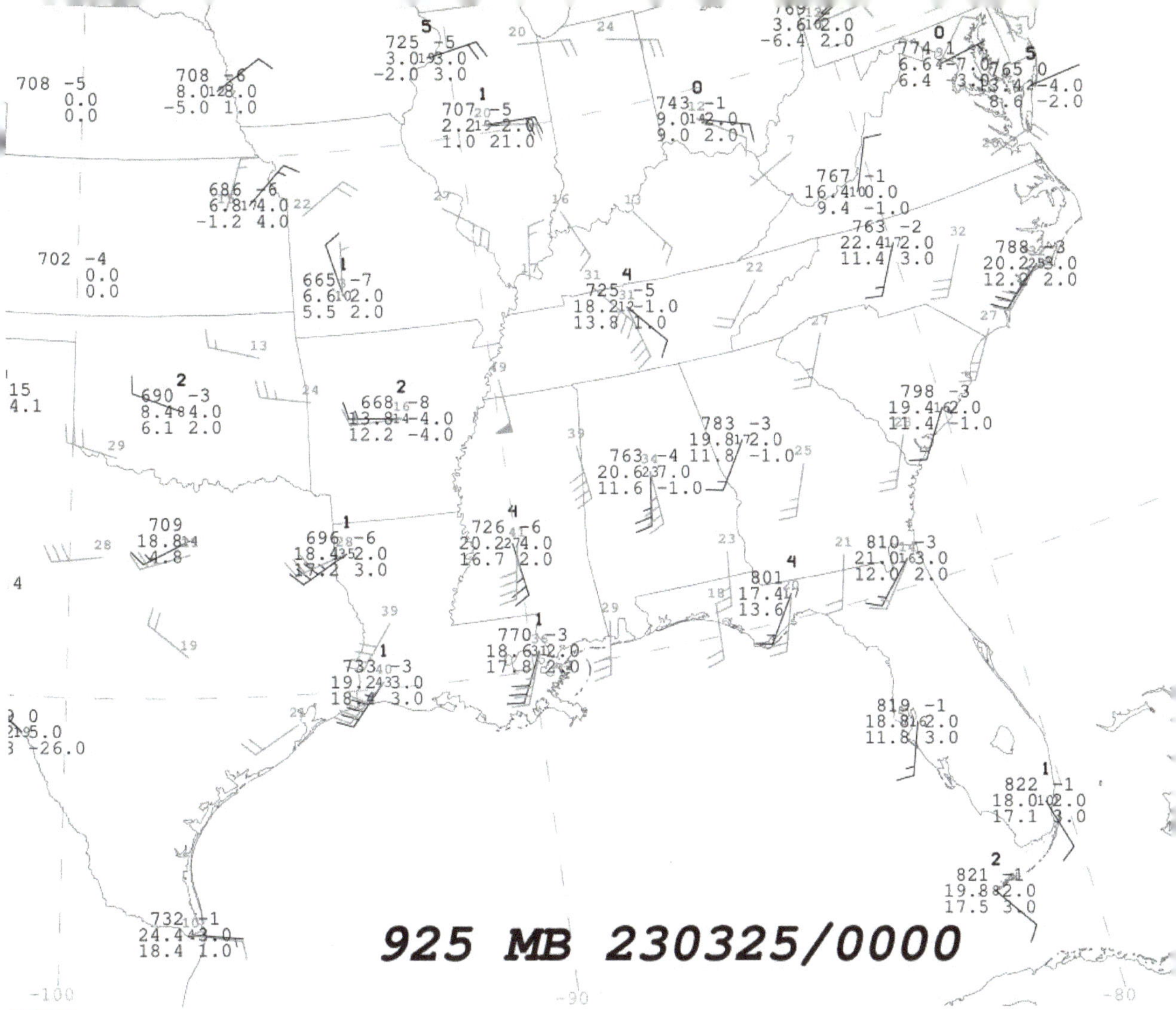

925 MB 230325/0000

Above: Station plot of 925 mb data from the Storm Prediction Center about three hours before the Missisippi tornado outbreak of 24 March 2023. This shows an extended system of upper-air plots which include (left, top to bottom) height (m), temperature (C), and dewpoint (C), and (right, top to bottom) corresponding 12-hour change values with the top right figure in decameters. Bold figure at top is dewpoint depression (C) when 5 or below. Number at center is wind speed in wind speed in knots. This plot also includes VAD wind data from WSR-88D radars in gray. *(AWIPS-II)*

Right: Forecast plot of 925 mb conditions on the evening of 25 March 2023. Dewpoint is highlighted in color bands. *(AWIPS-II)*

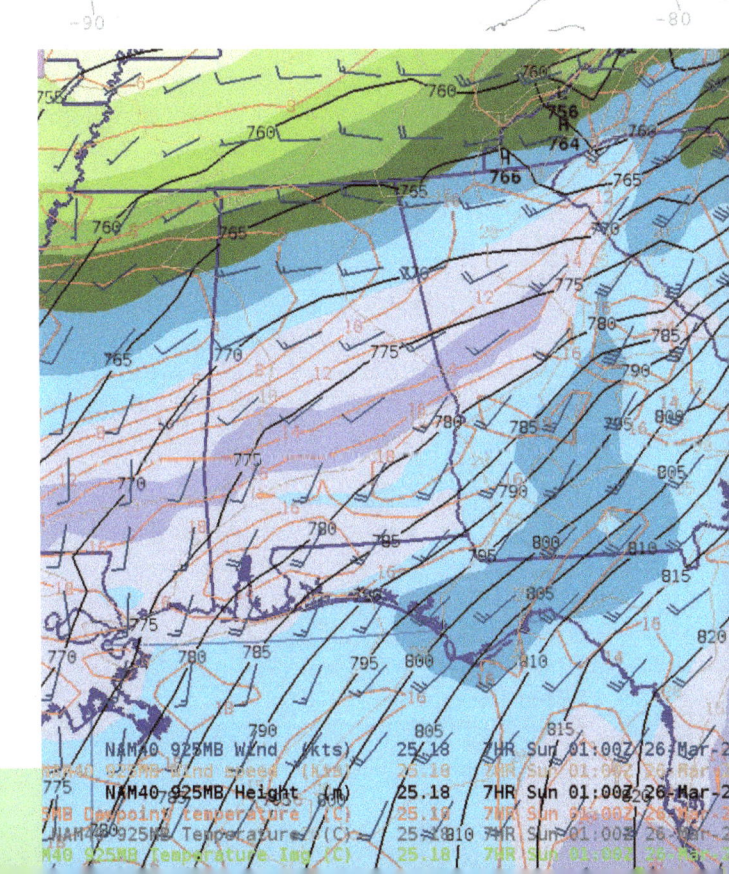

850 mb chart

The 850 mb level is located at about 5000 ft MSL, assuming a standard atmosphere. In practice, it can vary by several hundred feet or more. Since this is the lowest part of the troposphere, it is likely to intersect the ground in mountainous terrain. In the United States, the radiosonde stations at Denver, Flagstaff, Reno, and other locations in the Rockies and Great Basin area are typically below the surface. As a result, the 850 mb level comprises a mix of locations in the free atmosphere, within the boundary layer, and underground. Because of this, the forecaster must have a thorough knowledge of the terrain in the forecast area.

The planetary boundary layer (PBL) is that part of the troposphere in contact with the ground. A typical depth is about 2000 ft. The PBL is strongly influenced by the earth's surface through heating, drag, and evapotranspiration. The top of the PBL can range from the Earth's surface on a clear, cold night to 10,000 ft AGL or more given a windy, unstable, uncapped air mass with very strong heating.

Therefore depending on the surface elevation, the time of day, the season, and the ongoing weather regime, the 850 mb level, if not underground, will be within the PBL or within the free atmosphere. It is useful to a sense of which layer the radiosonde is representing and get familiar with all of the radiosonde station elevations in your region to see how high the 850 mb level usually is at each site. It is also important to bear in mind that the PBL and free atmosphere can seem to merge, especially during the summer in weak wind patterns.

When it comes time to put the pencil to the maps, the most popular use for the 850 mb chart is to locate frontal systems, particularly when their positions are not clear on the surface chart. Remember that frontal surfaces slope up and into cold air, therefore the 850 mb front will almost always be found poleward of the surface front. There are some exceptions, particularly on the Plains, where cold fronts will move much faster aloft than near the ground, and this may result in a cold front being placed equatorward of the surface front.

During the spring months, the 850 mb chart helps paint out the configuration of the low-level jet (LLJ). This feature is common before severe weather outbreaks and typically stretches from the Gulf of Mexico coastline into the central United States. The depth of tropical air masses can be assessed by determining the dewpoint at 850 mb and relating that to surface dewpoints.

key details

The 850 mb chart is usually near 5000 ft MSL. It is useful for finding the structure of frontal systems aloft, for assessing the depth of tropical and polar air masses alike, and for finding low-level jets. It intersects with the typical level of cumulus and stratocumulus clouds.

The chart is drawn as follows:
- Data plots are given in Appendix Table 1B.
- Geopotential height contours are drawn as solid black lines every 30 m (3 dam) using 150 dam as a base value (147, 150, 153, etc).
- Isotherms are drawn as dashed red lines every 5 C°, however 2 C° and 4 C° intervals are common.
- The type of moisture analysis (optional) is contingent on the type of weather regime:
  -- In stratiform situations, a relative humidity parameter is recommended. Shade all areas in green with dewpoint depressions below 5 C° (this is roughly a 75% relative humidity).
  -- In convective situations, use an absolute humidity parameter. Use the dewpoint value to draw isodrosotherms, or outline specific humidity or mixing ratio contours and shade the highest values.

Features are drawn as follows:
- Low level jet axes (narrow bands of strong winds exceeding 40 kt) should be drawn as a thick red arrow.
- Moisture axes may be drawn as thick, wavy green lines.
- Fronts should be drawn in standard blue and red colors. Since this is an upper-air chart, do not shade the barbs and pips.

real-time Internet sources

weather.ral.ucar.edu/upper
weather.gc.ca/analysis
weather.cod.edu/analysis
www.spc.noaa.gov/exper/mesoanalysis
www.ametsoc.org/amsedu/dstreme
www.wetter3.de

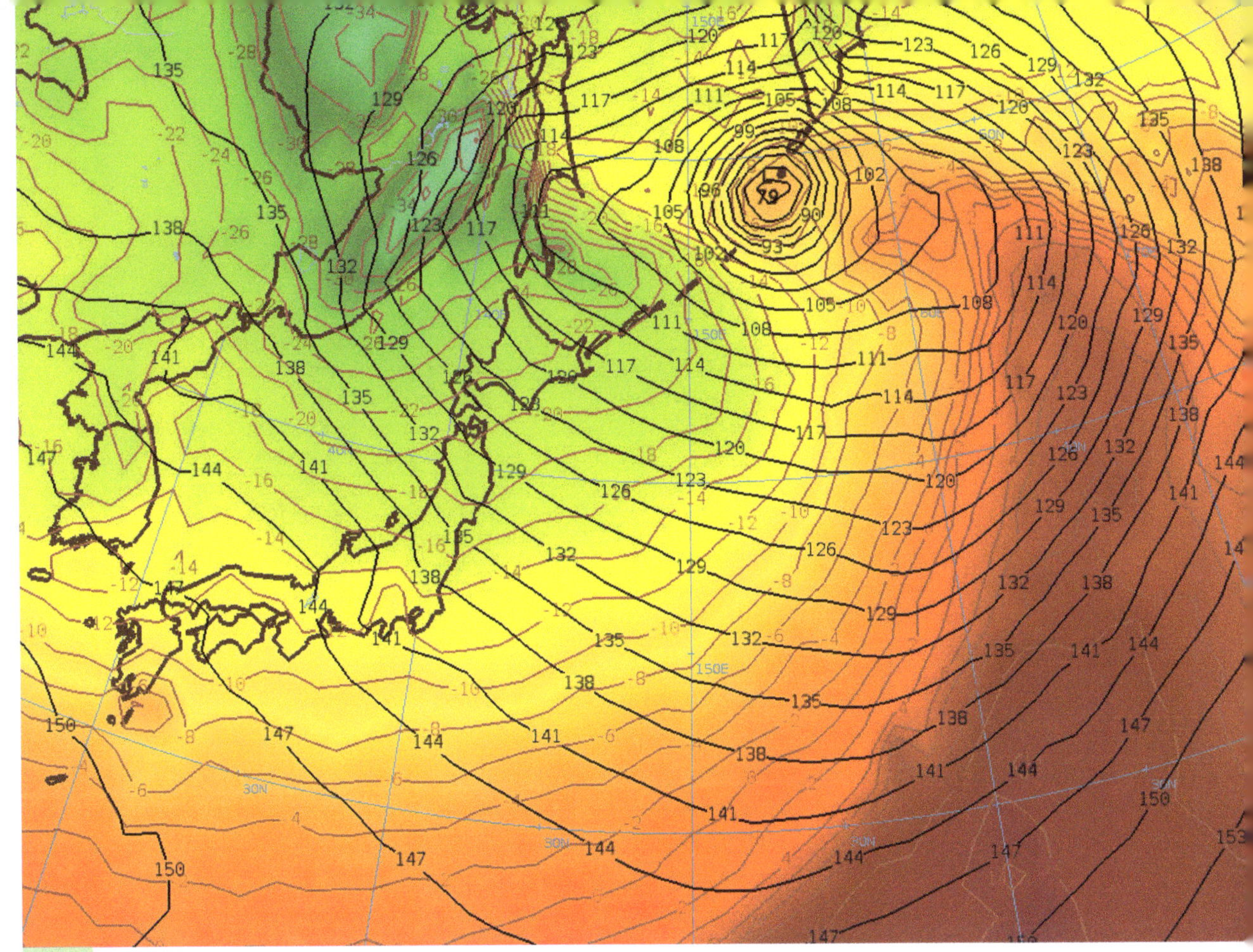

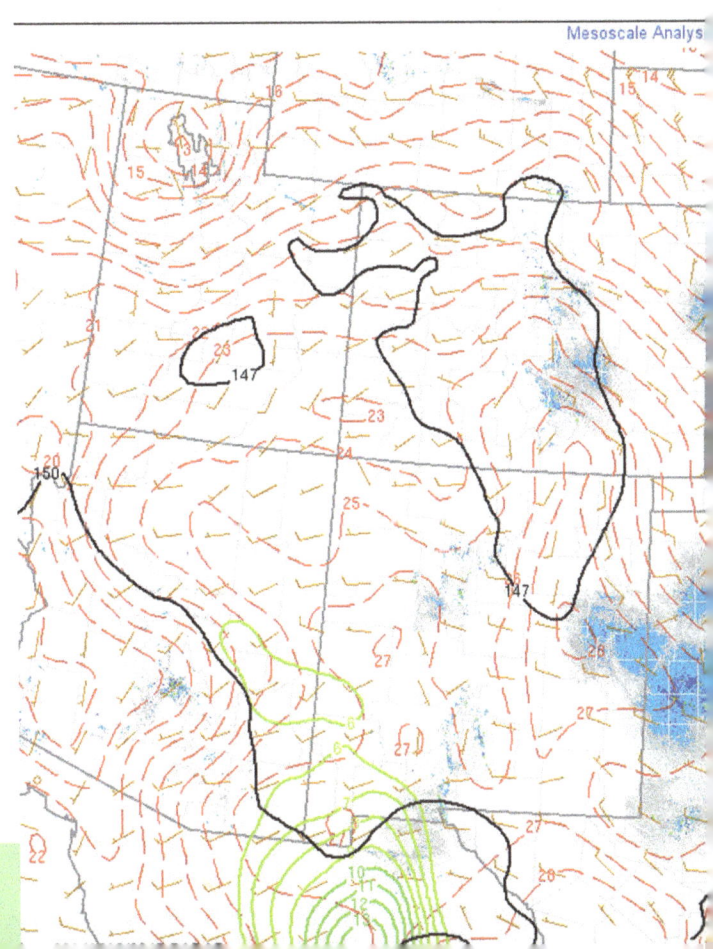

Above: An example of an 850 mb GFS forecast, with heights (black lines) and temperature (red lines). This is the wake of a powerful cold wave that swept through Japan with rare snowstorms in January 2023 after producing the coldest temperature ever observed in China (-53°C, -63°F). Note the -14°C temperatures at 850 mb over Tokyo. *(AWIPS-II)*

Right: 850 mb analysis from SPC. Note that some analysis products like this are actually generated from a mix of observed and model data, and may vary from actual observed data. For example, the plot shown here shows an unbroken grid of wind plots, but in many places here the 850 mb level is actually found underground. It is always important to be aware of where the data is coming from so it can be used properly. *(NOAA/SPC)*

700 mb chart

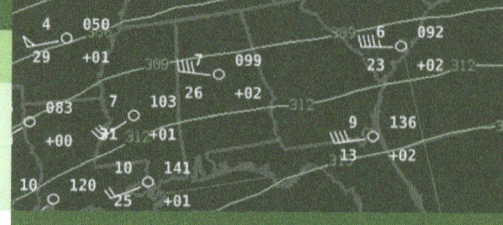

The 700 mb level is roughly at 10,000 ft MSL. This is considered to be somewhere between the lower and middle troposphere. At desert stations and higher elevations (such as above 5000 ft MSL), this level is often within the planetary boundary layer during the warm season. In regions like the Rocky Mountains, the 700 mb level intersects the tops of many of the mountain ranges.

Weather systems at 700 mb take on an open, broad appearance compared to the patterns at lower levels. The fronts are usually found further poleward compared to the 850 mb and surface charts, owing to the slope of fronts upward and into the cold air. This relationship can help forecasters locate the low-level front when the 700 mb chart shows a thermal gradient and the 850 mb and surface charts don't have a clear position.

The orientation of 700 mb streamlines with respect to the surface cold front helps indicate whether it is a katafront or anafront. A katafront is associated with a streamline that crosses from the cold to the warm side. A component that is parallel or flows from warm to cold air at 700 mb suggests an anafront.

Considering that the vast majority of cloud cover in a developing frontal weather system occurs in the 5,000 to 15,000 ft MSL range, humidity is a favorite quantity for measuring the extent of synoptic-scale lift and moisture. Relative humidity values of 70% or dewpoint depressions of 5 C° or less at 700 mb are considered synonymous with overcast cloud cover. Model output charts tend to give relative humidity, while upper air plots provide dewpoint depression, which bears an inverse correlation with relative humidity. When areas more humid than the threshold values are shaded in green, a definition of the area and shape of strongest upper-level forcing emerges. A "wrapped" structure may be seen on the charts, matching quite well with the cloud bands observed on satellite imagery.

In springtime thunderstorm situations, the 700 mb level is usually within the heart of the elevated mixed layer (EML), a broad area of warm, dry air originating from the southwestern United States that is lofted eastward into the central United States. The air at this level tends to be warmer than that below it. This forms an inversion, or "cap", that prevents thunderstorms altogether or suppresses them until afternoon when heating and instability are maximized. Therefore the 700 mb isotherm pattern can help define characteristics of the cap.

key details

The 700 mb chart is usually at about 10,000 ft MSL. It is used to find mid-level wind cores, short waves, capping inversions, and areas of mid-level clouds and moisture. It intersects with the level where mid-level clouds such as altocumulus and altostratus are found.

The chart is drawn as follows:
- Data plots are given in Appendix Table 1B.
- Geopotential height contours are drawn as solid black lines every 30 meters (3 dam) using 300 dam as a base value (297, 300, 303, etc).
- Isotherms are dashed red lines every 5 C°. Intervals of 2 C° or 4 C° are common.
- Areas of significant moisture (dewpoint depression of 5 C° or less, or relative humidity of 70% or greater) may be shaded in green. This emphasizes areas of possible dense cloud cover and precipitation.
- Jet axes are represented by a thick red arrow.
- Fronts should be drawn in standard blue and red colors. Since this is an upper-air chart, do not shade the barbs and pips.
- Short wave trough axes are drawn as thick black straight lines.
- Short wave ridge axes are drawn as thick black zig-zag lines.
- Col. Robert Miller's Severe Weather Analysis Notes suggests coloring jets and moisture in brown for this level to signify that they apply to 700 mb.

Surface lows tend to move at about 70% of the 700 mb wind speed above the low.

real-time Internet sources

weather.ral.ucar.edu/upper
weather.gc.ca/analysis
weather.cod.edu/analysis
www.spc.noaa.gov/exper/mesoanalysis
www.ametsoc.org/amsedu/dstreme
www.wetter3.de

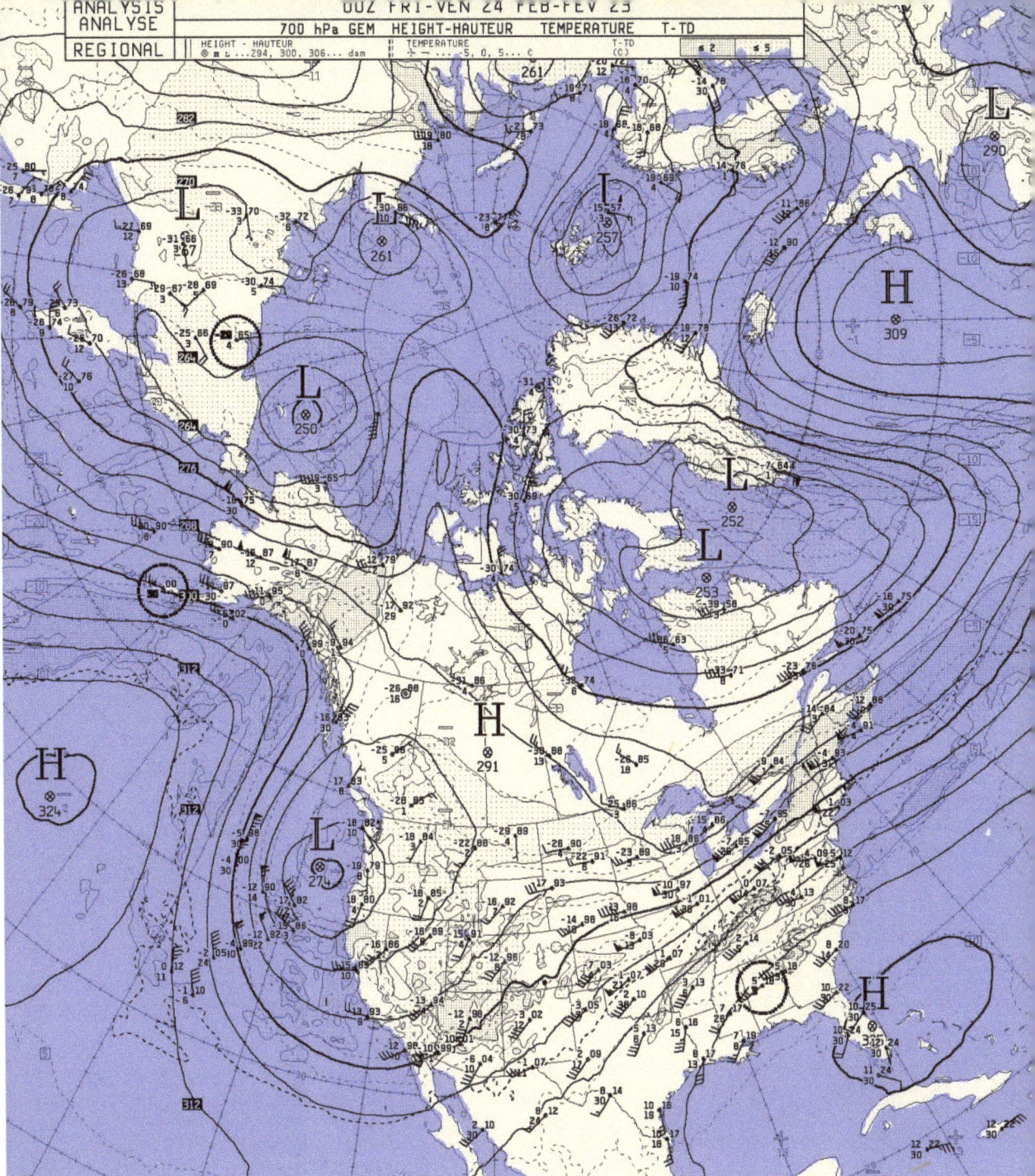

Above: Environment Canada produces excellent high-resolution chart sets, available at <weather.gc.ca/analysis/>. They provide the only surviving counterpart to the NWS DIFAX charts which were discontinued in 2017. Shading corresponds to dewpoint depression: that is, the difference between temperature and dewpoint in Celsius degrees. This may surprise forecasters accustomed to schemes in which isotachs are shaded. It's a reminder to always inspect new charts before using them. Light shading corresponds to 5 C° or less, which is roughly a 70% humidity or higher and suggests cloud cover. Heavy shading corresponds to 2 C° or less, which is approximately a 90% relative humidity and indicates thick cloud cover with the possibility of precipitation. The chart is useful for showing the distribution of available radiosonde observations. This is a blend of the original chart with a color outline of land and ocean developed by the author; readers may download it at <www.weathergraphics.com/dl/canadaoverlay.jpg>. *(Environment Canada)*

500 mb chart

The 500 mb level is considered the middle of the troposphere, except in polar regions. It exists at a height of about 18,000 ft MSL (±2000 ft typically), and so is nearly always within the free atmosphere and not part of a planetary boundary layer.

At this level, forecasters see an excellent mix of small-scale and large-scale systems. At the large scale, the upper-level jet pattern begins emerging, painting out the regions of strongest baroclinicity (energy available to the atmosphere). Superimposed on this are a series of large troughs and ridges. The troughs correspond to very cold air masses, while the ridges exist above areas of warmth.

At even smaller scales we find the notorious short waves. These are small-scale troughs and ridges embedded in the large-scale flow which are just barely resolved in the existing radiosonde network. Short waves are reflections of important thermal perturbations in the air mass beneath, whose influence easily translates to higher levels.

These perturbations, if a strong thermal gradient is present in the lower troposphere, often go on to amplify the short wave which in turn deepens the short wave, which in turn imparts more energy to the system beneath. This is a chain reaction called "self development" and is broken only when the system finally occludes and washes out the thermal gradient. The short wave trough, by this time, has often deepened into an upper-level low.

Short waves are located using either the absolute vorticity field or by looking for wind shifts. On standard upper-air analyses, vorticity overlays are sometimes not available. The vorticity fields are largely a product of numerical weather prediction output, and the short waves tend to lie along elongations of vorticity axes, particularly those that cross the flow in a perpendicular orientation. The short wave will tend to be located along a boundary oriented in a perpendicular fashion across the 500 mb flow which separates two stations with a sharp wind shift. A cyclonic wind shift defines a short wave trough, while an anticyclonic wind shift indicates the presence of a short wave ridge.

Finally, it is crucial to make that distinction between a "short wave" and a "short wave trough", because the term short wave applies to both. Also short waves can sometimes be resolved at lower levels, such at the 700 mb level. Short wave troughs stack downward toward the warmer air (upward toward colder air).

key details

The 500 mb level is usually at a height of about 18,000 ft MSL. It is used for finding mid-level jet, for seeing the general overview of the tropospheric pattern, and is the primary chart for finding short waves. It is near the level of non-divergence (the center of mass in the troposphere).

Upper air plots are explained in Table 1B in the Appendix.

Surface lows tend to move at about 50% of the 500 mb wind speed above the low.

The chart is drawn as follows:
- Geopotential height contours are drawn as solid black lines every 60 m (6 dam) using 570 dam as a base value (564, 570, 576, etc).
- Absolute vorticity contours are machine-produced and are typically drawn as dashed black lines every $2*10^{-5}$ sec.
- Isotherms (optional) are dashed red lines every 5 C°. Intervals of 2 C° or 4 C° are common.
- High and low height centers are plotted as a large black "H" or "L", with the decameter value below it.
- Areas of PVA are shaded red (optional) while areas of NVA are shaded blue (optional).
- Short wave trough axes are drawn as thick black straight lines; short wave ridge axes are drawn as thick black zig-zag lines.
- Jet axes are drawn as a thick red line.
- Jet maxes may be identified with an X and a value in knots. This is done on SPC's SREF charts.
- Col. Robert Miller's Severe Weather Analysis Notes suggests a blue color for any jet depiction to signify that the markings are for 500 mb.

real-time Internet sources

weather.ral.ucar.edu/upper
weather.gc.ca/analysis
weather.cod.edu/analysis
www.spc.noaa.gov/exper/mesoanalysis
www.ametsoc.org/amsedu/dstreme
www.wetter3.de

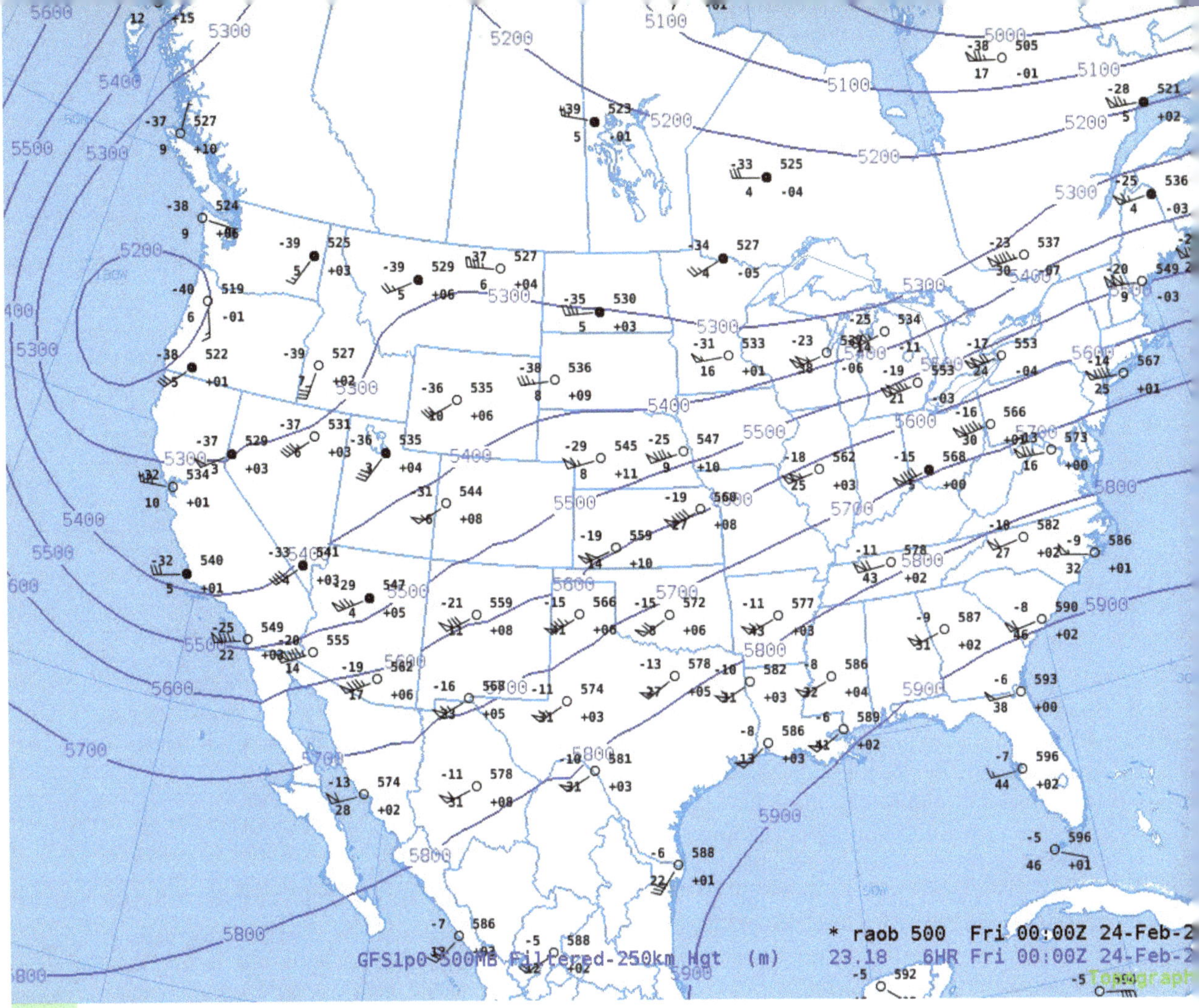

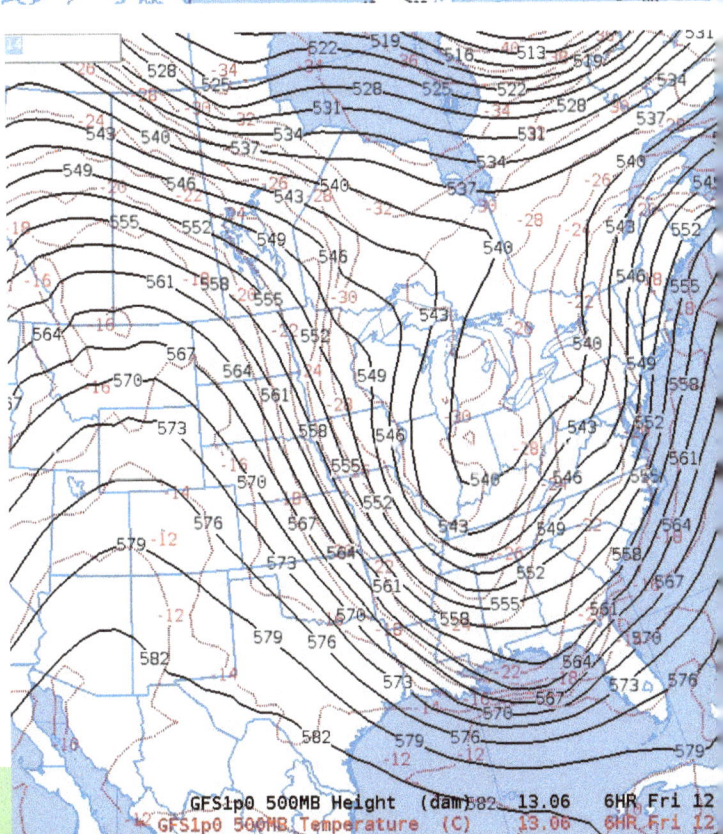

Above: Typical 500 mb station plot and contour analysis from AWIPS. This has been combined with a GFS geopotential height field (solid blue line). Significant differences between the plotted values and the contour field may arise from smoothing of the height field, but may also occur due to rejection of data due to instrumentation errors or contamination. *(AWIPS-II)*

Right: 500 mb GFS plot for 13 January 2023 using a 3 dam height interval (black lines) and temperature (thin red dotted lines). It showing deep troughing across the eastern United States, with a large ridge across the Rocky Mountains and Great Basin region. This produces a strong northerly component across the Midwest and Mississippi River valley, bringing unseasonably cold weather to these areas. *(AWIPS-II)*

300/250/200 mb chart

The practicing meteorologist always wants to have a glance at the upper tropospheric conditions, as the polar front jet lies in its topmost portions. Unfortunately picking a level is complicated, because the top of the troposphere (the tropopause) in temperate latitudes ranges in height from 30,000 ft during the winter to 45,000 ft during the summer.

Therefore, given the standard levels of 300 mb (30,000 ft MSL), 250 mb (34,000 ft MSL), and 200 mb (39,000 ft MSL), it is necessary to choose different charts depending on the season to find most representative "upper tropospheric chart". During the winter, 300 mb is used. In the transition seasons of autumn and spring, the 250 mb chart is preferred. In the summer, the 200 mb is selected. The others are discarded as they are either too low or tap into the stratosphere where the polar jet is rapidly weakened with height. Though there is little use for stratospheric charts, some studies have been published on stratospheric warm sinks and cold domes, which have ties to areas of upper divergence and convergence, respectively.

Overall, the axis of strongest winds paints out the jet stream. This pattern is by far the highlight of the upper tropospheric chart, and defines the weather regime that is in place. Long waves are formed by the very broad troughs and ridges that ring the hemisphere. The long wave troughs are caused by large masses of cold air, and the ridges by warm air.

Jet maxes, sometimes called jet streaks particularly when referring to smaller scales, are very important features. The winds are frequently out of balance around them, resulting in strong vertical motions. A careful isotach field that provides the correct shape of the jet streak can be very helpful in inferring areas of vertical motion.

A conceptual model exists which suggests the type of vertical motion that may exist around a jet max. It is usually referred to as the "4-cell jet max concept". The coordinate system specifies that "left" is the poleward direction, "right" is equatorward, "rear" is upstream, and "front" is downstream. Upward motion should be found in the left front quadrant (LFQ) and right rear quadrant (RRQ), with downward motion in the right front quadrant (RFQ) and left rear quadrant (LRQ). In cyclonic flow the left quadrants are enhanced with the right quadrants nullified, with the opposite true in anticyclonic flow. This concept is highly subjective and is subject to assumptions, but demonstrates an excellent use for the upper tropospheric chart.

key details

The 300 mb chart is usually at a height of about 30,000 ft. **The 250 mb chart** is usually at a height of about 34,000 ft. **The 200 mb chart** is typically near 39,000 ft. All three charts are used primarily to establish the pattern of upper-level jets, to locate areas of shear, and to assess the nature of the hemispheric circulation and identify any blocking patterns. All levels intersect with the height regime where cirriform clouds are found.

The chart is drawn as follows:
- Geopotential height contours are drawn as solid black lines every 120 m (12 dam) using 900 dam as a base value (888, 900, 912, etc.).
- Isotachs are drawn as purple lines every 20 kt starting at 30 kt as a base (30, 50, 70, 90, etc.).
- Isotherms (optional) are dashed red lines every 5 C°, but may clutter the chart.
- High and low height centers are plotted as a large black "H" or "L", with the decameter value below it.
- Jet axes. The axis of jets should be drawn as a heavy red arrow. Col. Robert Miller's Severe Weather Analysis Notes suggests a purple color for this level.
- Jet cores. Draw a red ellipse to mark the jet core.

Isotach bands can be shaded using colored pencils.
- One possible shading spectrum is green to yellow to red.
- Another spectrum is green to red to blue to purple to yellow. This scheme is used by Environment Canada.

real-time Internet sources

weather.ral.ucar.edu/upper
weather.gc.ca/analysis
weather.cod.edu/analysis
www.twisterdata.com
www.spc.noaa.gov/exper/mesoanalysis
www.ametsoc.org/amsedu/dstreme
www.wetter3.de

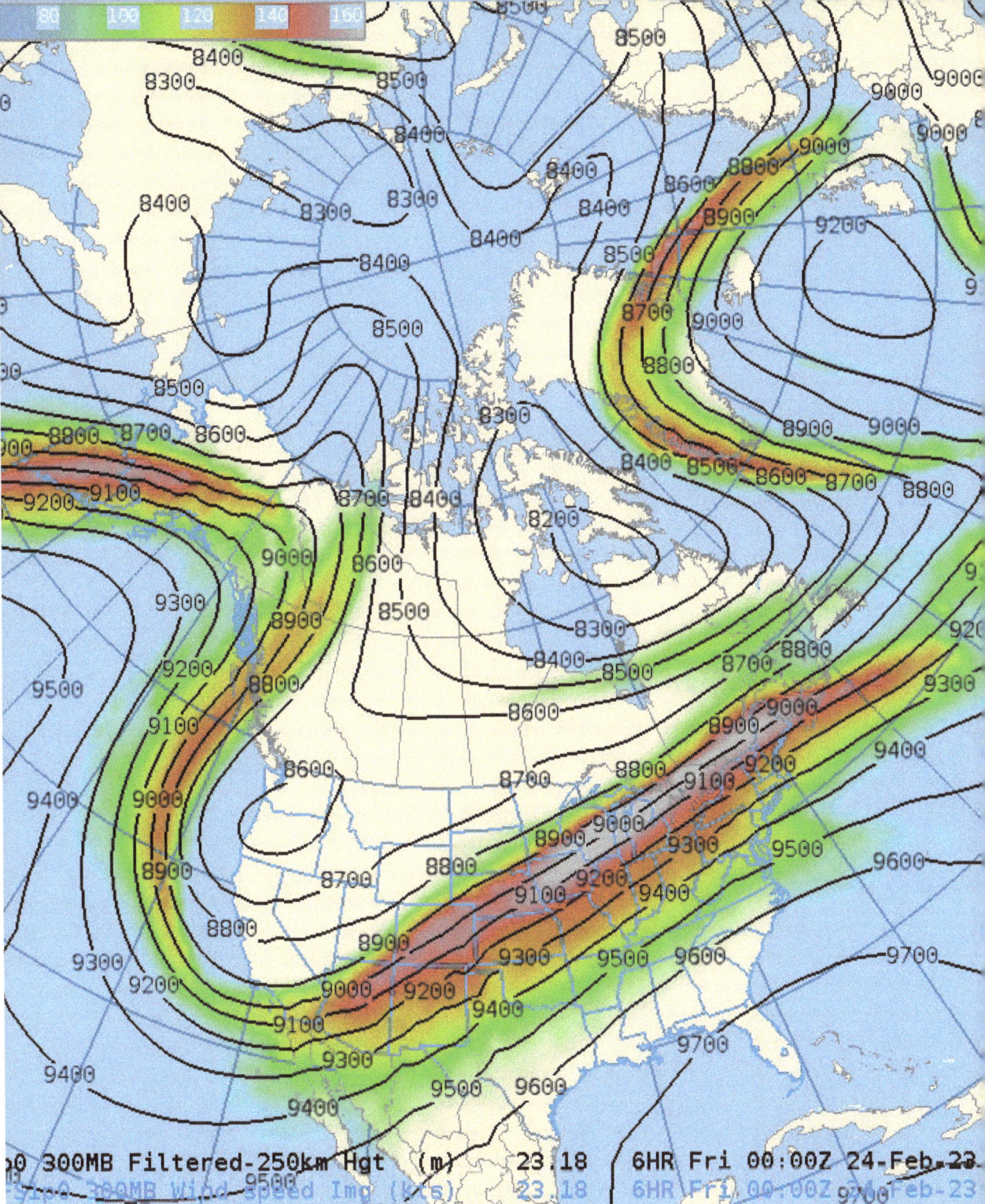

Above: 300 mb chart for North America showing heights (black) and wind velocity (colored shading, with scale at top in knots). This outlines a polar front jet arcing from Alaska south into the East Pacific Ocean, then flowing from Arizona to New England. The strongest winds are found over the Great Lakes, showing as an elongated 170 kt jet max. We also find a large upper low over Washington and Oregon, and a Hudson Bay Low representing part of the polar vortex. *(AWIPS-II)*

Terrain-following charts

Severe thunderstorm analysis has traditionally used the 850 mb and 925 mb charts. In flat regions such as the U.S. Gulf Coast and the Great Lakes, these correspond to a relatively uniform height above the terrain. However on the Great Plains and in mountainous regions there can be vast differences in height above ground level (AGL) from place to place. This difference can have substantial implications in the forecast process. For example, an 850 mb chart may intersect both the decoupled nocturnal radiation layer at Dodge City and the free atmosphere above a low level jet over St. Louis. A terrain-following chart removes this ambiguity by maintaining a fixed height AGL, typically in km or ft, so that air masses can be interpreted with a sense of vertical consistency.

Terrain-following charts have not yet come into widespread use. Much of this is due to tradition, with constant-pressure charts remaining dominant in forecasting. A constraining factor has also been a relative scarcity of software to produce such plots, and the lack of bandwidth in prior years to deliver high-resolution model data so that this type of chart can be constructed on a client computer. Another shortfall was the limited vertical resolution of forecast models and centralized objective analyses in earlier years, which resulted in coarse vertical profiles that were not very useful. All of this gradually changed in the late 2000s and into the 2010s. On modern display systems it is now possible to construct detailed terrain-following charts that provide considerable information on the environment of immediate use to the forecaster.

An example of a traditional process that references altitude above ground level is sounding analysis. The bottom of the profile almost always marks the surface. Forecasters are accustomed to interpreting the chart based on the vertical distance from the ground to features occurring aloft, especially features in the lower troposphere. Surface-dependent quantities such as mixed-layer CAPE and 0-1 km shear have also come into extensive use. A terrain-following chart complements these measurements by providing a view of the center or the tops of these layers. This can be very useful on sloped terrain such as the Great Plains where severe weather is common.

Some caution is warranted. Fields will be noisy in and around mountainous areas. Also temperature and dew point fields undergo adiabatic changes with height, and are distorted by a terrain-following surface. Conserved quantities such as potential temperature and specific humidity or mixing ratio should be used instead.

key details

Terrain-following charts are locked to a specific height above ground level (AGL). The 1 km AGL base level is recommended for severe weather forecasting. If resources permit, other charts can also be constructed for 0.5 and 1.5 km.

The chart is drawn as follows:
- Height and pressure are not suitable parameters for display since the pressure and AGL height are largely a function of terrain elevation.
- Temperature is strongly influenced by terrain elevation. A more appropriate quantity for synoptic or mesoscale analysis is potential temperature (theta). Isotherms are drawn as dashed red lines every 2 C°.
- Equivalent potential temperature (theta-e) may be used for display of buoyant energy. Contours may be drawn as blue lines every 2 K.
- Wind plots, wind vectors, or streamlines should be used to visualize the flow and infer where pressure centers are located.

Features are drawn as follows:
- Low level jet axes (narrow bands of strong winds exceeding 40 kt) should be drawn as a thick red arrow.
- Moisture axes may be drawn as thick, wavy green lines.
- Fronts should be drawn in standard blue and red colors. Since this is an upper-air chart, do not shade the barbs and pips.

Where terrain is at sea level, the 1 km AGL chart corresponds to the 899 mb level in a standard atmosphere. The 1 km AGL chart is equivalent to the 925 mb chart where the surface elevation is 780 ft (238 m) MSL in a standard atmosphere, and for everyday general forecast applications for all regions where terrain lies between approximately 500 and 1000 ft (150 to 300 m) MSL.

real-time Internet sources

None were known to exist at press time.

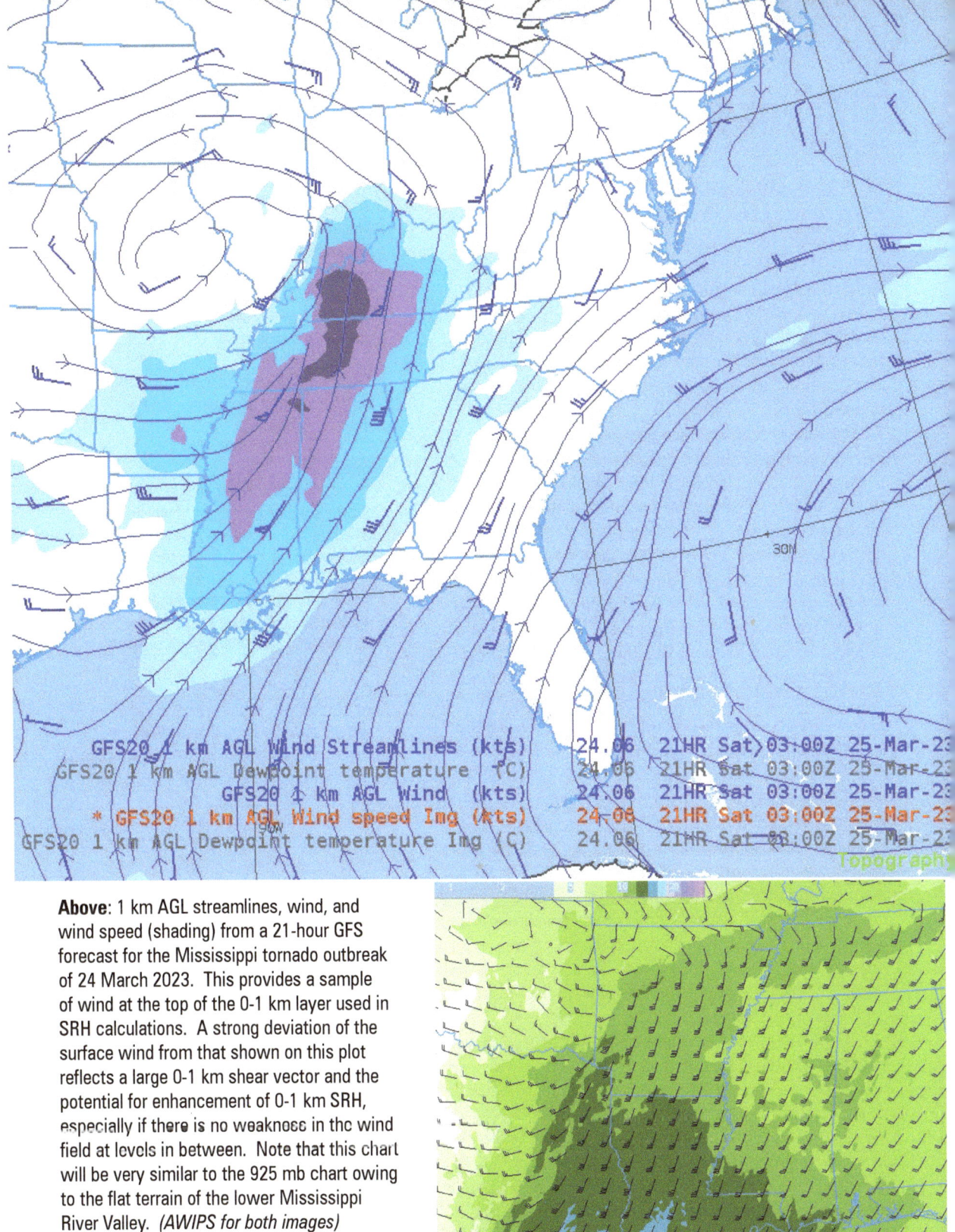

Above: 1 km AGL streamlines, wind, and wind speed (shading) from a 21-hour GFS forecast for the Mississippi tornado outbreak of 24 March 2023. This provides a sample of wind at the top of the 0-1 km layer used in SRH calculations. A strong deviation of the surface wind from that shown on this plot reflects a large 0-1 km shear vector and the potential for enhancement of 0-1 km SRH, especially if there is no weakness in the wind field at levels in between. Note that this chart will be very similar to the 925 mb chart owing to the flat terrain of the lower Mississippi River Valley. *(AWIPS for both images)*

Right: 1 km specific humidity for above event, 7 hr HRRR model forecast valid 2000 UTC. The moisture axis is correlated with depth.

Thickness chart

Thickness measures the vertical distance between one pressure surface and another. It is expressed either in whole meters or in dekameters (tens of meters), abbreviated as "dam". Contours are drawn every 60 meters or 6 dam. Thickness provides a measure of average density or virtual temperature within the given layer. In operational forecasting, thickness provides information on the average temperature within the layer being sampled, helps identify baroclinic zones, and distinguishes different air masses.

Increasing temperature, decreasing density, and increasing moisture content all contribute to an increase in thickness. Likewise, decreasing temperature, increasing density, and decreasing moisture content all contribute to reduction of thickness. Moisture content is determined by dewpoint, mixing ratio, or specific humidity. Temperature is by far the biggest contributor to thickness. In a typical forecast environment, a dew point drop of approximately 5 to 15 C° is needed to counteract a 1 C° temperature rise in order to maintain the same thickness value.

Thickness is always based on a specific layer in the atmosphere. The most common layer used in operational meteorology is the 1000-500 mb layer, which yields the 1000-500 mb thickness. This layer, which occupies the area roughly between sea level and 18,000 ft MSL, represents the bottom half of the troposphere where the majority of air mass contrasts exist. During extremely cold events that involve shallow air masses, the 1000-700 mb or even the 1000-850 mb thickness may be more appropriate for sampling air masses.

On horizontal maps, thickness plots are almost always displayed together with surface isobars, which connect lines of equal pressure. This provides an accurate relationship of the pressure gradient (and wind) to the thermal contrasts and air masses that exist.

Thickness charts have been traditionally used to make quick assessments of thermal advection. Advection occurs where wind is blowing colder or warmer thicknesses into a given location.

Specific thickness lines have been used since the 1960s for highlighting rain-snow transition areas. The most common transition line is the 1000-500 mb 540 dam line (usually ±4 dam, but as low as 526 dam in specific areas like the Pacific Northwest). The corresponding line on the 1000-700 mb chart is 284 dam and on the 1000-850 mb chart is 130 dam.

key details

Thickness approximates the average temperature through a given layer of the atmosphere, usually from 1000 to 500 mb. Thickness lines can be thought of as isotherms for the entire layer. Moisture contributes a small amount to thickness values.

Thickness charts are used to locate and delineate fronts, air masses, and areas of thermal advection.

Values:
* Low thickness corresponds to cold air, and may be enhanced by a lack of moisture.
* High thickness implies warm air, and may be enhanced by the presence of moisture.

Advection: Warm advection is associated with large-scale ascent, clouds, and rain. Cold advection is associated with subsidence and clear skies.

Fronts: Cold fronts tend to be associated with cyclonically curved thickness lines; warm fronts are associated with anticyclonically curved thickness lines.

Bands of thickness lines are thermal gradients which tend to separate different air masses. A surface front usually exists on the warm edge of bands of thickness lines (thermal gradients).

A surface low riding along the warm side of a thickness gradient is an indication it is either developing or mature. When it crosses to the poleward side of the thickness gradient, it is likely to be occluding.

real-time Internet sources

wx.erau.edu/teaching/milsyn/
weather.cod.edu/forecast
www.weather.gc.ca/model_forecast

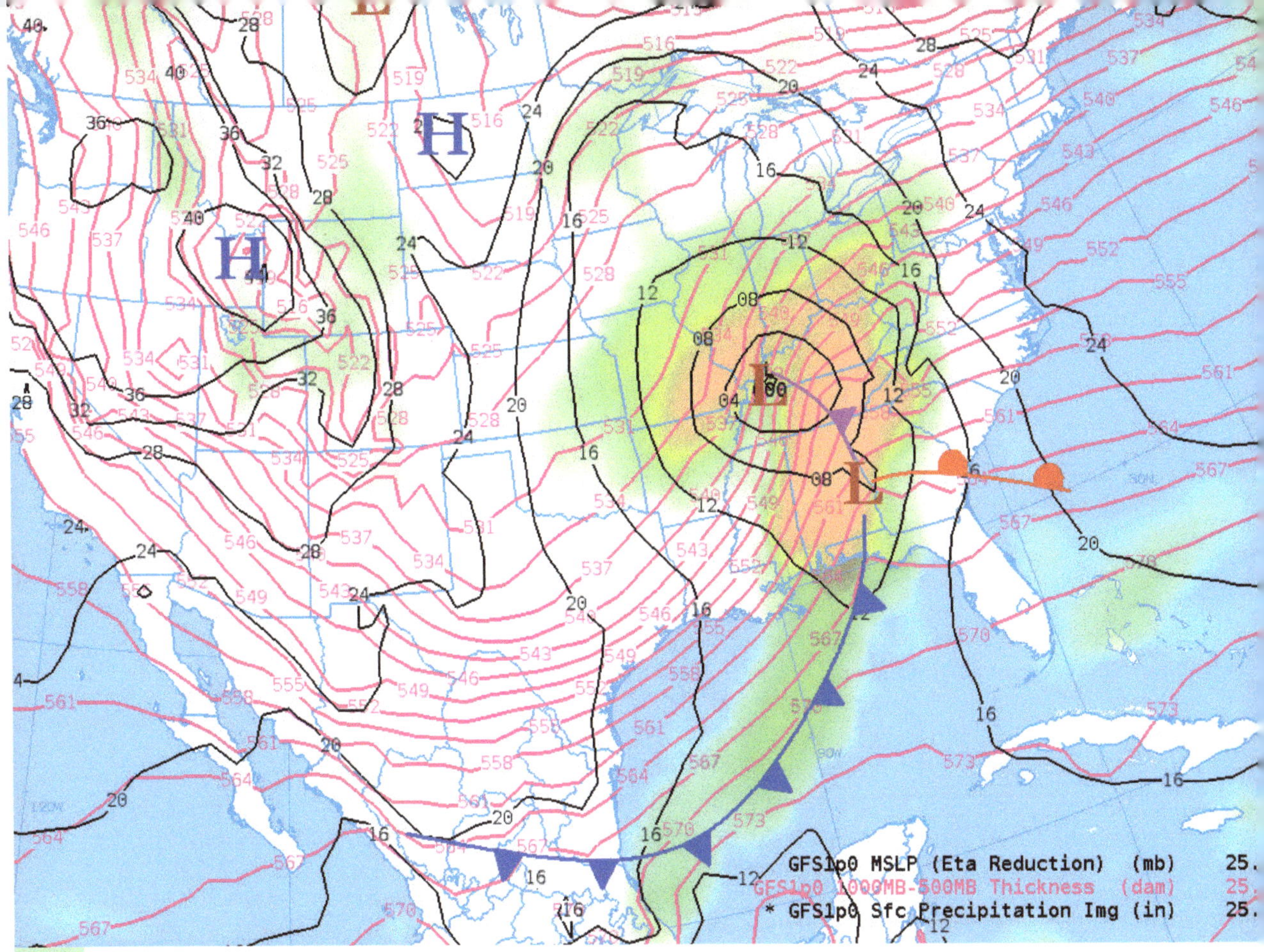

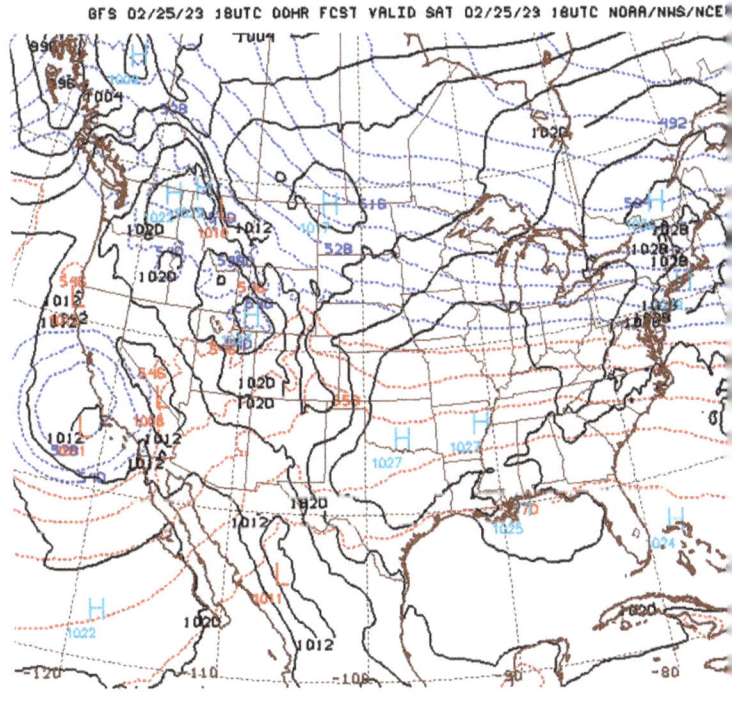

Above: GFS sea-level pressure and 1000-500 hPa thickness valid for 25 Jan 2023 at 1200 UTC. The thickness contours (red lines) represent a sort of averaged temperature in the lower and middle tropopause. The size of the boxes formed by the juxtaposition of thickness and sea-level pressure lines is, approximately, inversely proportional to the magnitude of thermal advection. *(AWIPS-II)*

Right: Official NCEP GFS output of sea-level pressure (black solid lines) and thickness (red and blue dotted lines). The thickness contours are closely related to 500 mb and 300 mb height fields, so the rounded "thickness bullseye" off the California coast very likely corrsponds to an upper-level low, and the large thermal troughing in the eastern regions are indicative of long wave troughing east of the Rockies. This makes the chart useful for a first guess at where upper-level jets, troughs, and ridges are.

Isentropic analysis

Isentropic analysis works on the understanding that a parcel of air does not move in a pure horizontal fashion but will cling to its own isentropic (potential temperature, i.e. theta) surface. This is true if there are no special thermal processes taking place involving addition or removal of heat, i.e. heating, evaporation, and condensation. As a result, vertical motion can be assessed by looking at the slope of the isentropic surfaces and the direction of the winds along the surface.

Surprisingly, isentropic analysis was widely used after WWII, but was largely abandoned with the first surge of model data in the early 1960s. It wasn't until the 1990s that the availability of instantaneous isentropic diagnosis on AFOS and personal computers helped revive the technique. It is now in use at most forecasting offices.

The basis of isentropic analysis is the rule that theta surfaces bend upward above a cold air mass (forming a sort of "cold dome" of theta surfaces) and dip downward over a warm air mass. This implies that when a mid-level parcel moves from a region of low-level warm air to another region with low-level cold air, it must rise to follow the upward bend in the isentropic surface. This is equivalent to the concept of "overrunning" ascent that occurs along warm fronts. Likewise, a movement of air from a cold to a warm region suggests isentropic descent, and this explains the rapid clearing that often takes place along cold fronts, particularly in their wake.

With the isentropic analysis chart, isobars will be displayed. These can be thought of as indicating the height of the theta field. For example, if we are looking at a 300 K isentropic level and a isopleth marked "700" crosses Vermont, we can conclude that the 300 K surface is at the 700 mb level in Vermont, which is about 10,000 ft above the ground. Therefore pressure contours with low values indicate high heights, and high values indicate low heights.

The second important component of the isentropic chart is wind barbs or wind vectors, which are valid for that isentropic level. So if a wind plot with southwest winds at 40 kt show over Vermont in the example above, it can be concluded that this particular wind sample exists at the 700 mb level.

A final valuable field is relative humidity, which indicates how close the air mass is to producing clouds (RH of about 70% or greater) and precipitation (RH of 90% or more). Other measures of moisture such as dewpoint or mixing ratio are not as useful since they give no indication how close to saturation the air mass is.

key details

Isentropic analysis shows the conditions only along an imaginary isentropic surface, which varies greatly with height depending on the density (controlled largely by temperature) of the air below it. Parcels have a tendency to follow this surface, so ascent or descent is determined by the wind direction.

Isentropic charts usually contain contours that show the height of the surface in millibars (lower values indicate larger heights) as well as winds at this level.

Relative humidity is often added to visualize how close the parcels are to being saturated. Parcels with relative humidities of 70% or greater are likely associated with cloud layers.

To see a higher elevation, choose a higher K isentropic level. To see a lower elevation, choose a lower one. Typically 300 K corresponds to the mid-troposphere in winter and the lower troposphere in summer.

Use warmer (higher K) charts in the summer, and colder (lower K) charts in the winter. Cold charts tend to intersect the ground during the summer, and warm charts are too high in the winter.

Isentropic analysis performs poorly in convective weather regimes such as with strong heating, strong air mass modification, and during the warm season. This technique should not be used in such patterns.

Isentropic analysis works best in the wintertime in stratified air masses with little or no convective precipitation, particularly with strong thermal gradients, strong wind flow, and weak air mass modification.

real-time Internet sources

weather.cod.edu/analysis
www.wxcaster.com/isentropic.htm
www.atmos.albany.edu/student/abentley

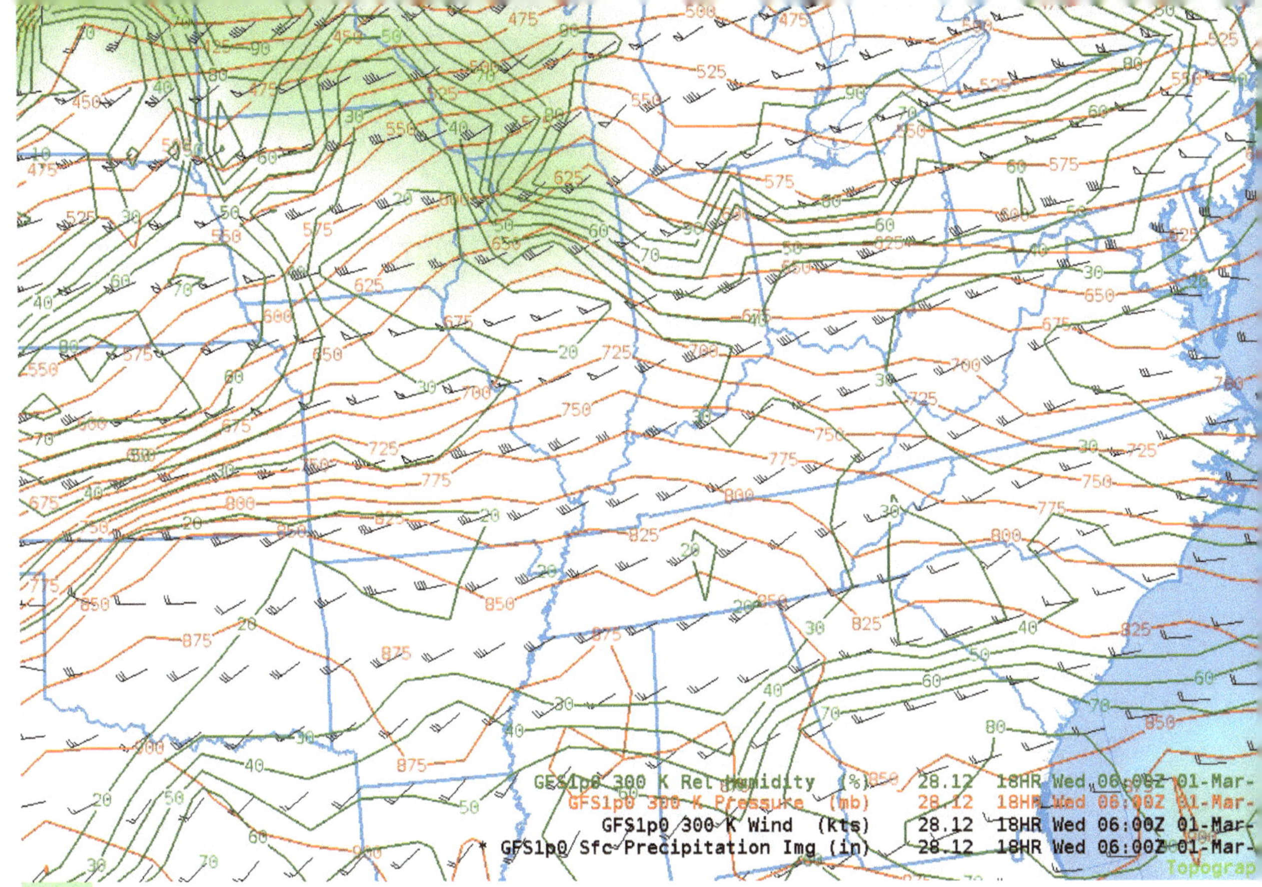

Above: Forecast isentropic analysis for the 300 K level on 1 March 2023. The pressure values are indicated by red lines, indicating that the 300 K level ranges from about 3000 ft MSL along the bottom of the chart to 20,000 ft MSL near the top. The wind flow across much of the Appalachian region suggests isentropic ascent, as the wind climbs to higher values. Note that the relative humidity (green) indicates low values here of 30 percent, so significant clouds are unlikely, but humidity is likely to increase as ascent continues. The tight pressure gradient in Kansas suggests the presence of a strong front in that area.

Below left: Cross section of isentropic surfaces across cold air mass (left) and warm air mass (right). Surfaces are higher over cold air and lower over warm air. Whether lift or subsidence is taking place depends on the wind flow with respect to these surfaces.

Below right: Data voids due to intersection with the ground. Data in and near these areas should be ignored.

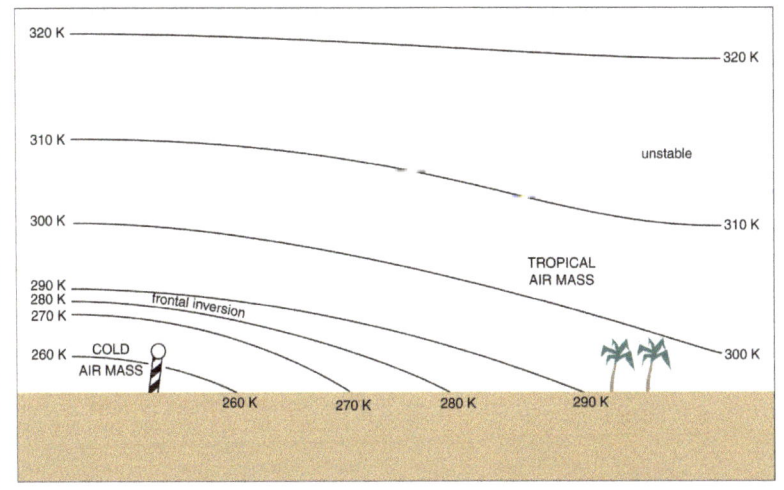

Vorticity diagnostics

The concept of vorticity is based on the technique of rearranging the quasigeostrophic equation to to solve for vertical motion based on rotation and thermal advection. Vorticity is a measure of the spin of a given parcel of air. Vorticity has two components: shear and curvature. Shear is caused by a difference in motion on opposite sides of the parcel. Curvature describes rotation of the parcel as a result of curved flow.

The most basic expression of vorticity is known as *relative vorticity*. This form is used mostly in mesoscale analysis. Rotation is measured with respect to the Earth's surface. Synoptic forecasting uses a version known as *absolute vorticity*. This is the sum of the spin of the parcel plus the spin of the earth. Both produce similar patterns.

Vorticity can be calculated for any level, but the most common use in forecasting is at the 500 mb (18,000 ft MSL) level. As this is near the level of non-divergence, it provides the best level for diagnosing vertical motion. Areas of high and low vorticity are referred to colloquially as "vort maxes" and "vort minima", and their position marked with "X" and "N" respectively.

Vorticity advection occurs where wind speed is strong (height contours close together) and vorticity gradient is strong (vorticity contours close together), with significant crossing of the two contours. The crossing of these lines visually form a series of boxes. The size of the box is inversely related to vorticity advection strength. New forecasters should familiarize themselves with the box method before continuing to charts where vorticity is displayed as a gradient (as shown on the page at right).

Cyclonic vorticity advection (CVA) is tied to ascent of air, which results in clouds and precipitation. On the other hand, anticyclonic vorticity advection (AVA) is associated with subsidence, which favors clear skies. In the Northern Hemisphere, positive vorticity advection is equivalent to CVA, while negative vorticity advection is the same as AVA. The meaning of PVA and NVA is hemisphere-dependent, so forecasters should use CVA/AVA in more formal settings and when engaging international readers.

The principle of tying vorticity advection to vertical motion is based on a number of assumptions. The first is that the atmosphere is responsive to restoration of geostrophic balance. The second is that the vorticity advection increases with height, which is not always the case. The third is that CVA is not negated by cold advection, or that AVA is not negated by warm advection, which often does occur.

key details

Vorticity diagnostics use 500 mb geopotential height every 30 or 60 m, and 500 mb absolute vorticity either isoplethed every 2×10^{-5} s^{-1}. or depicted as colored shading or bands.

The use of vorticity analysis for finding vertical motion involves several assumptions and should be used with caution.

Relative vorticity is pure wind field rotation relative to the ground. It is used for finding circulations on mesoscale charts, especially at low levels and at the surface.

Absolute vorticity includes the rotation of the earth, whose contribution increases with latitude.

In the northern hemisphere, PVA (positive vorticity advection) is referred to as CVA (cyclonic vorticity advection. Likewise, NVA (negative vorticity advection) is known as AVA (anticyclonic vorticity advection). The opposite is true in the southern hemisphere.

Areas of CVA are often associated with ascent. The ascent will be enhanced if warm advection is occurring in the lower troposphere. It will be negated (or even result in subsidence) if cold advection is occurring.

Areas of strong PVA/CVA should be shaded red. Areas of strong NVA/AVA should be shaded blue.

If you do not feel adept at finding areas of CVA and AVA, look for the boxes formed by the cross-overlap of height and vorticity isopleths.

real-time Internet sources

weather.cod.edu/forecast
wx.erau.edu/teaching/milsyn/
weather.ral.ucar.edu/model
www.pivotalweather.com/model.php
www.wxcaster.com/models_main.htm

ANALYSIS CHARTS

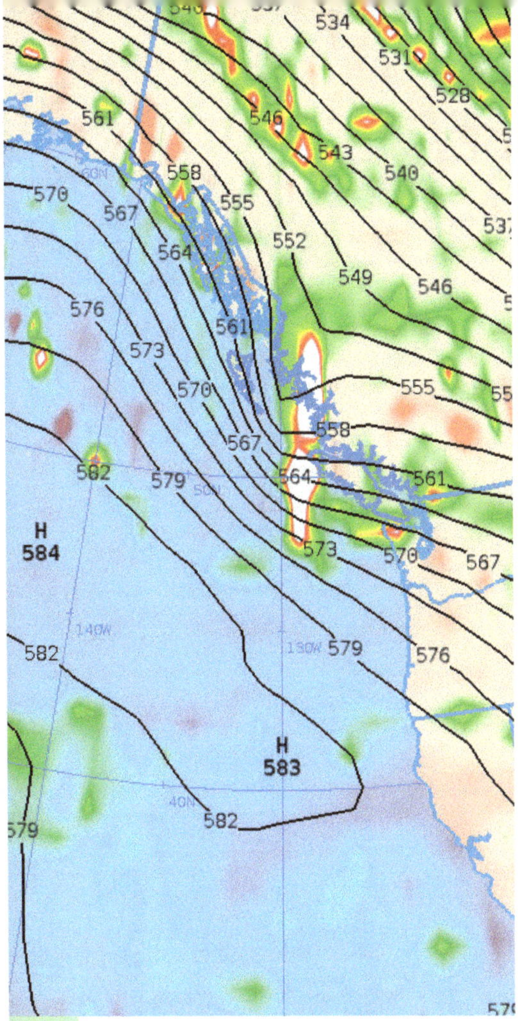

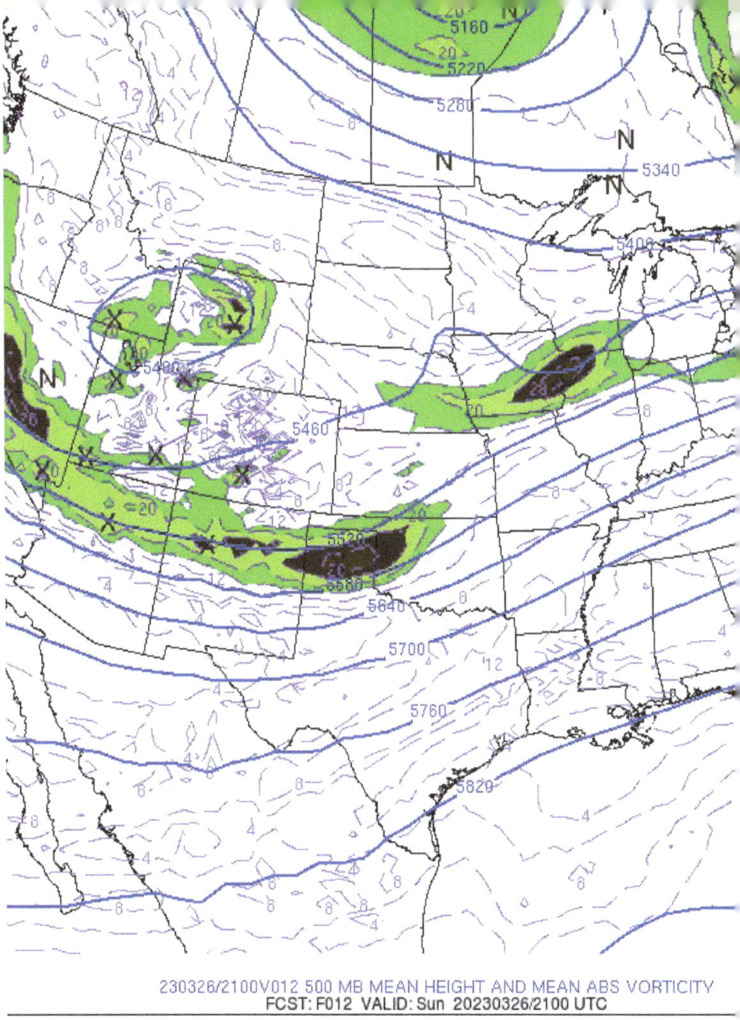

Above: 500 mb height and vorticity forecast fields for 27 January 2023 at 06Z. Shaded green, red, and white areas correspond to vorticity maxima, while red smudges indicate vorticity minima. *(AWIPS)*

Above: SPC SREF chart (www.spc.noaa.gov/exper/sref) showing mean ensemble 500 mb heights and vorticity. The polar jet is found on the equatorward side of the elongated cyclonic vorticity channel (green). *(NOAA/SPC)*

Right: Interpretation of key features on a 36 hour GFS forecast valid for 1800 UTC 26 February 2023. A large compact area of high positive vorticity (blue) in southern Arizona and a pinched appearance to the height contours indicates the presence of a jet max. The left-front quadrant (LFQ) and right-rear quadrant (RRQ) are associated with upper-level divergence and rising motion, while the left-rear quadrant (LRQ) and right-front quadrant (RFQ) are associated with upper-level convergence and sinking motion. Due to the strong curvature, only the LFQ and LRQ regions are dominant. A small-scale area of enhanced vorticity is coming onshore in Oregon. The axis of high vorticity crosses the contours, defining what is often referred to as an advection lobe, and representing a classic short wave trough. *(AWIPS-II)*

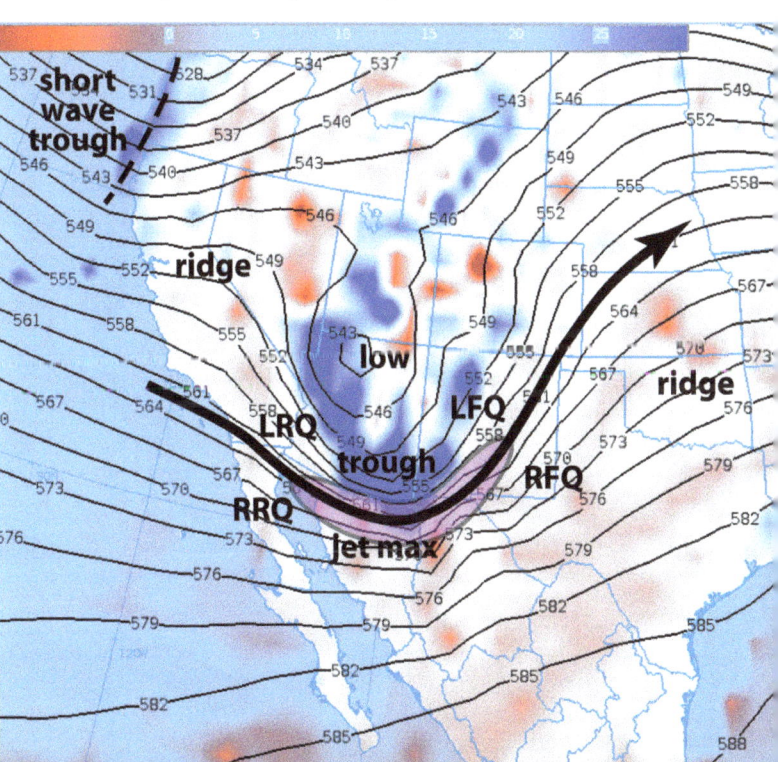

Vertical velocity

The vertical velocity product, often abbreviated VV (vertical velocity) or UVV (upward vertical velocity), is an instantaneous value derived from the forecast model. It is typically given at 700 mb, which is close to the level of non-divergence.

Vertical velocity cannot be measured directly in part because the magnitude is so small compared to the horizontal motion of air and large variations in the form of gusts. Therefore dynamical models are required for resolving these motions. Vertical motion which produces widespread areas of light rain and snow, for example, may be on the order of millimeters or centimeters per second. Over many hours, however, this motion adds up to kilometers of ascent.

Vertical velocity is most commonly expressed in microbars per second (μb/s), exactly equivalent to decipascals per second (dPa/s). A microbar is one thousandth of a millibar. One microbar per second in the low levels is roughly one centimeter per second of ascent or subsidence. A *positive value of μb/s indicates subsidence*, while a *negative value indicates ascent*. Sometimes vertical velocity is given in geometric units as m s^{-1}.

Most "vertical velocity" charts found on the Internet indicate a direct output of vertical motion as given by the dynamical model. These fields are often somewhat noisy and are contaminated by small-scale processes such as convection and gravity waves.

A few websites offer "omega diagnostics" in which the components of the quasi-geostrophic (QG) omega equation are divided into the vorticity and thermal advection constituents. This equation specifies that when cyclonic vorticity increases with height and/or warm advection is present, upward motion should occur to restore the atmosphere to geostrophic and hydrostatic balance. The amount of each quantity is proportional to the vertical motion.

These two components often cancel each other out, so other diagnostic tools such as Q vectors are recommended for extended analysis of the vertical motion field. However where model-derived vertical motion fields indicate strong ascent or descent, omega diagnostics can be used to inspect the components producing these vertical motions, establish their location and which vertical layers these processes are strongest in, and to evaluate their change over time.

key details

Vertical velocity is used for assessing ascent or subsidence in the air mass. Omega diagnostics provide the forecaster with additional information.

Vertical velocity units may be given in one of two forms:
- pressure units (dp/dt), microbars per second, where positive values indicate subsidence.
- geometric units (dz/dt), meters per second (m s^{-1}), where positive values indicate ascent.

Vertical velocity is usually paired with the 700 mb geopotential height, whose lines are plotted every 30 or 60 m. The recommended vertical velocity contour interval is 2 microbars per second or 0.02 m s^{-1}. Relative humidity at 700 mb may be added in green, every 10 or 20%.

Each microbar per second is roughly equal to 0.01 meter (1 centimeter) per second, or 1.97 fpm, or 0.02 kt. At 700 mb, each millibar of height is about 12 meters.

Some typical values of lift in microbars per second. Convective speeds are provided only for comparision; model omega only accounts for large-scale ascent and does not factor in convection.

Velocity	Typical weather
0.5	Stratus
5	Light rain
50	Heavy rain, cumulus
500	Thunderstorm
5000	Tornadic thunderstorm

Relative humidity may be paired with this chart to assess whether clouds and precipitation are imminent.

real-time Internet sources

weather.cod.edu/forecast (VV)
weather.ral.ucar.edu/model (VV)
www.pivotalweather.com/model.php (VV)
wx.erau.edu/teaching/milsyn/ (diagnostics)
inside.nssl.noaa.gov/tgalarneau/real-time-qg-diagnostics/ (diagnostics)

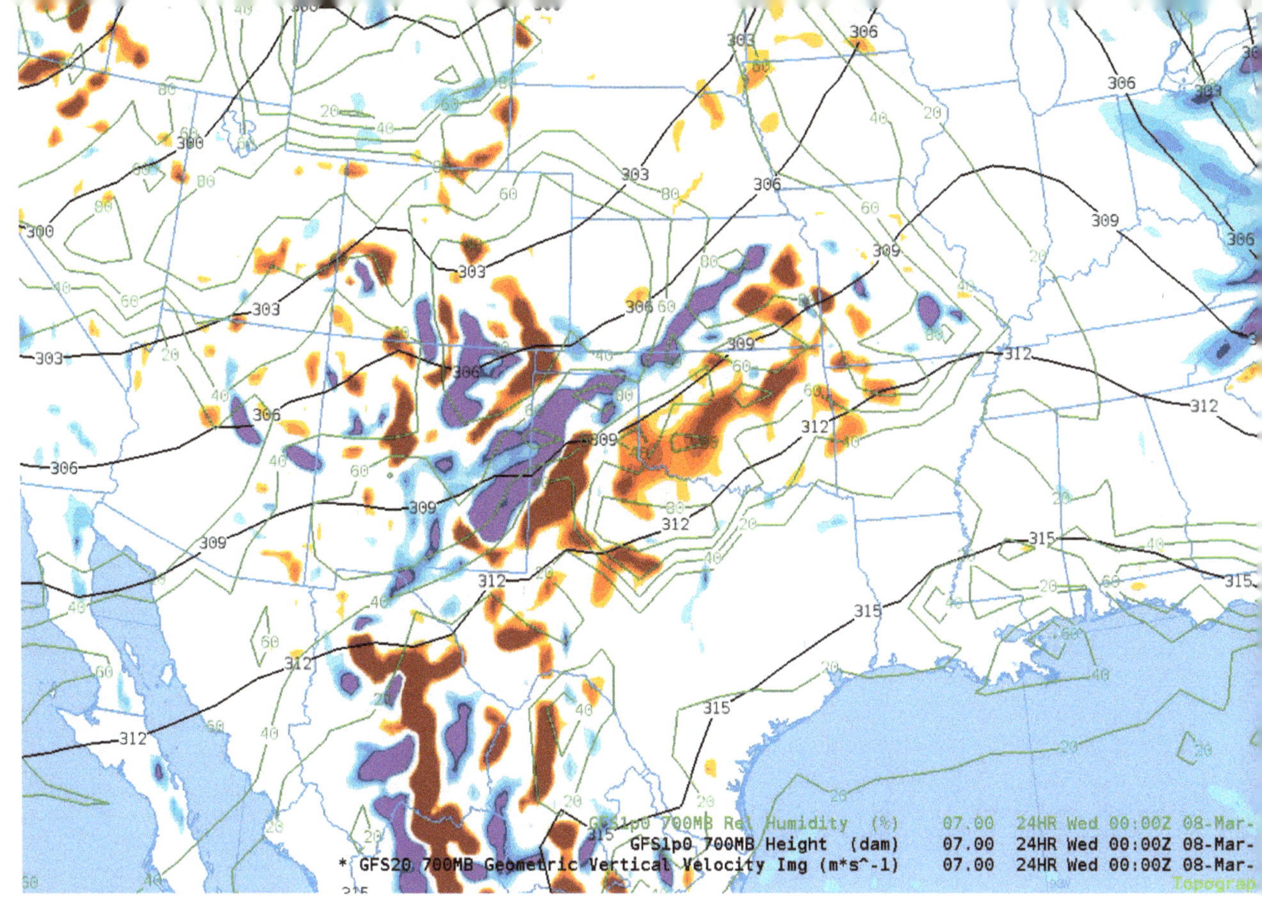

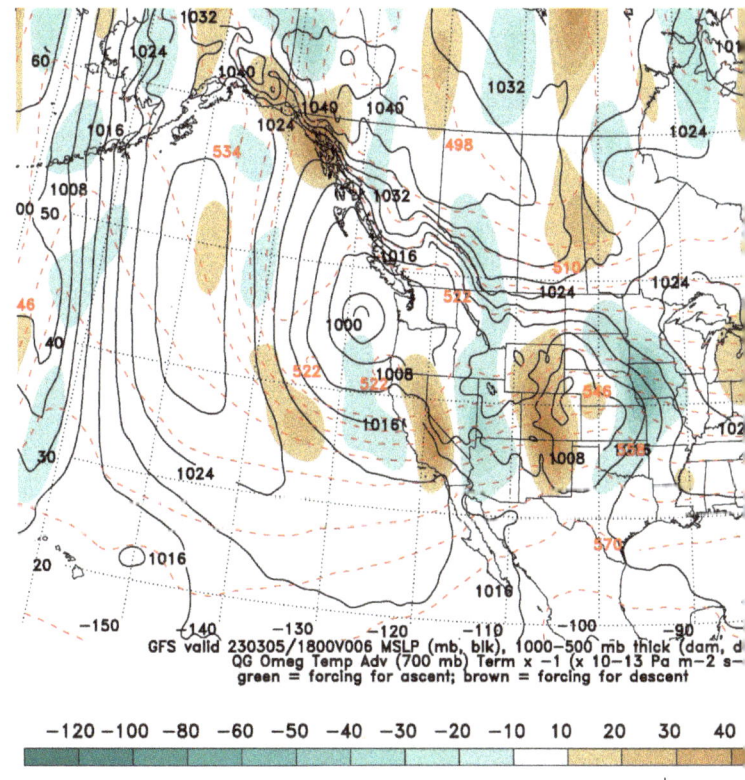

Above: GFS forecast for 700 mb height (black lines), relative humidity (green lines), and vertical velocity (red shading for ascent, blue shading for subsidence). The vertical motion fields are unsmoothed and noisy due to the high resolution of the model and the influence of gravity waves near the Rocky Mountains. Nevertheless a distinct zone of ascent can be identified over Oklahoma and Texas. *(AWIPS-II)*

Right: Temperature advection chart for 700 mb, at the midpoint of the 900-500 mb layer. This is one component of omega. Green indicates warm advection and brown cold advection. In southern California, differential negative vorticity advection (above) has coupled with cold advection to produce downward motion. However in Colorado, differential positive vorticity advection is cancelled by significant cold advection. *(ERAU)*

Q vector diagnostics

The Q vector technique is a rather new addition to the meteorological toolbox, dating to 1978 in a paper by British meteorologists Hoskins, Draghici, and Davies. In the 1980s the technique was advanced in the United States by Stanley Barnes. It estimates *large-scale baroclinic vertical motion*. Q vectors have a significant advantage of bypassing the traditional quasi-geostrophic (QG) formulation problem based on vorticity and thermal fields, which frequently cancel each other out.

A Q vector describes the rate of change of the horizontal potential temperature gradient. Looking at a single Q vector by itself, whenever its magnitude is high, a strong horizontal ageostrophic wind is implied. This is interpreted as a strong response by the atmosphere to restore the thermal wind balance. Vertical motion is likely to develop where the Q vectors converge or diverge.

In a convergent Q vector pattern, the vectors tend to point at one another. Q vector convergence implies ascent at that level. In a divergent pattern, the Q vectors point away from one another. Q vector divergence implies subsidence at that level. Some panels and software displays measure the magnitude of convergence or divergence, eliminating much of the guesswork. The convergence or divergence magnitude corresponds to the strength of ascent or descent, respectively.

Q vectors are often overlaid on top of a product showing isotherms for that level, or thickness for a layer centered on that level. When Q vectors point across isotherms from cold to warm air, the thermal gradient is strengthening and frontogenesis is implied. When Q vectors point from warm to cold air, the thermal gradient is weakening and frontolysis (weakening of a front) is implied.

There are some caveats to using Q vectors. It assumes that the winds are in near geostrophic balance. The presence of mesoscale weather systems, and even centrifugal force from strongly curved flow can disrupt the balance of forces and cause the upper lift to differ from that suggested by the Q vectors.

In daily operations, Q vectors are seldom used with today's high-resolution models, since they capture significant amounts of mesoscale detail and require extensive filtering. Without filtering, the Q vector fields appear noisy and are difficult to use. Properly processed Q vector output gives one of the best measures of the *synoptic-scale forcing* that leads to vertical motion.

key details

Q vectors are used to assess the potential for vertical motion that arises from the atmosphere trying to restore geostrophic balance. This provides a measure of synoptic-scale forcing. This differs from model vertical velocity, an instantaneous sample where vertical motion comes from all types of lift.

Modern high-resolution models produce very noisy Q vector fields, which is part of the reason this tool is not common. These artifacts become noticeable as model resolution drops below 100 km. Careful postprocessing is required to get usable data. Nearly all of today's models are run at about 3 to 15 km resolution, so direct viewing of Q vector fields will not give satisfactory results.

Q vector convergence is associated with ascent. Q vector divergence is associated with subsidence.

The Qs vector is the Q vector component parallel to the thermal (or thickness) contours, useful for measuring advection. A large Q vector parallel to the thermal lines means significant thermal advection is taking place.

The Qn vector is the Q vector component perpendicular to the thermal (or thickness) contours, which measures frontogenesis or frontolysis. A large Q vector pointing from cold to warm air indicates frontogenesis. A large Q vector pointing from warm to cool air indicates frontolysis.

Layer Q vectors are preferred over single-level Q vectors. This is because they smooth out noise, which can seriously degrade the value of the product. However single-layer Q-vectors tend to be common.

real-time Internet sources

wx.erau.edu/teaching/milsyn/
inside.nssl.noaa.gov/tgalarneau/real-time-qg-diagnostics/ (diagnostics)

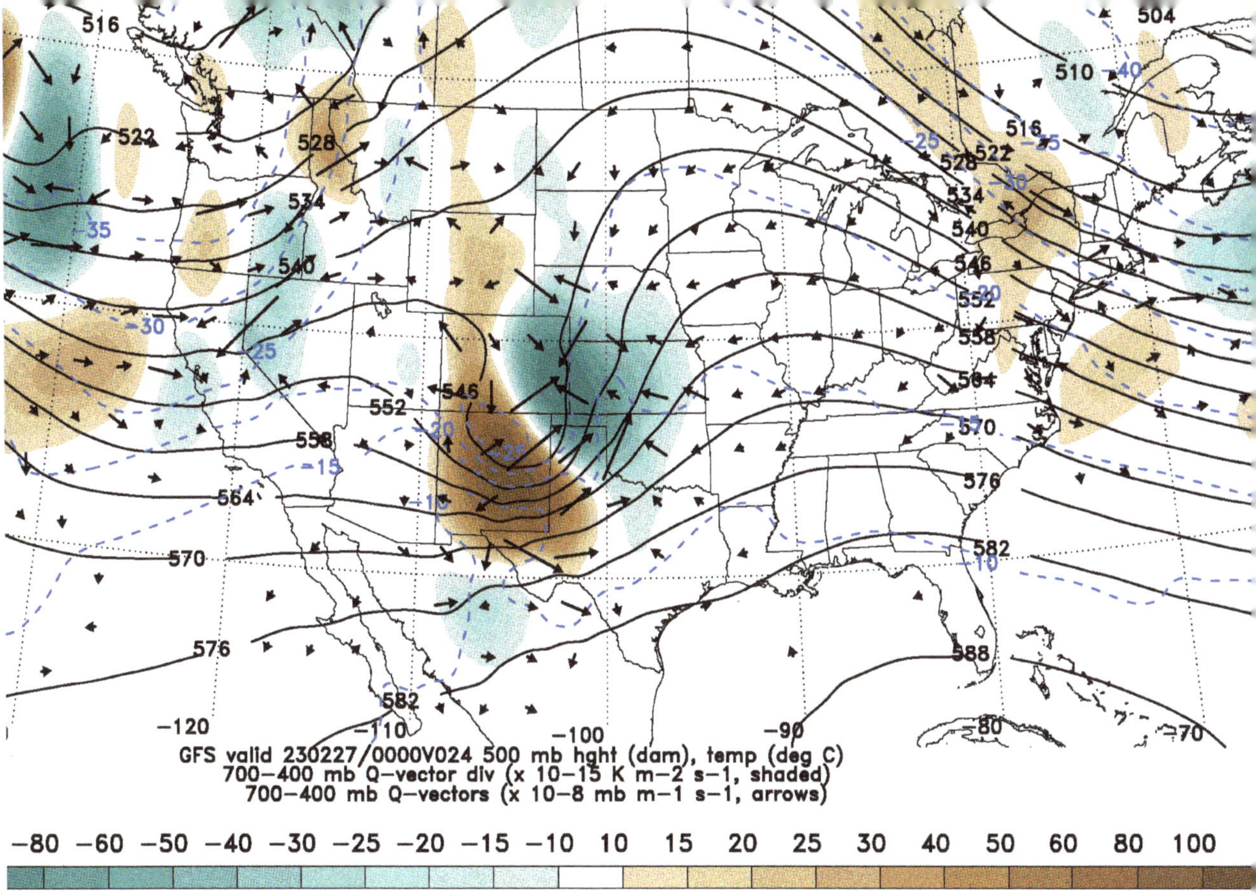

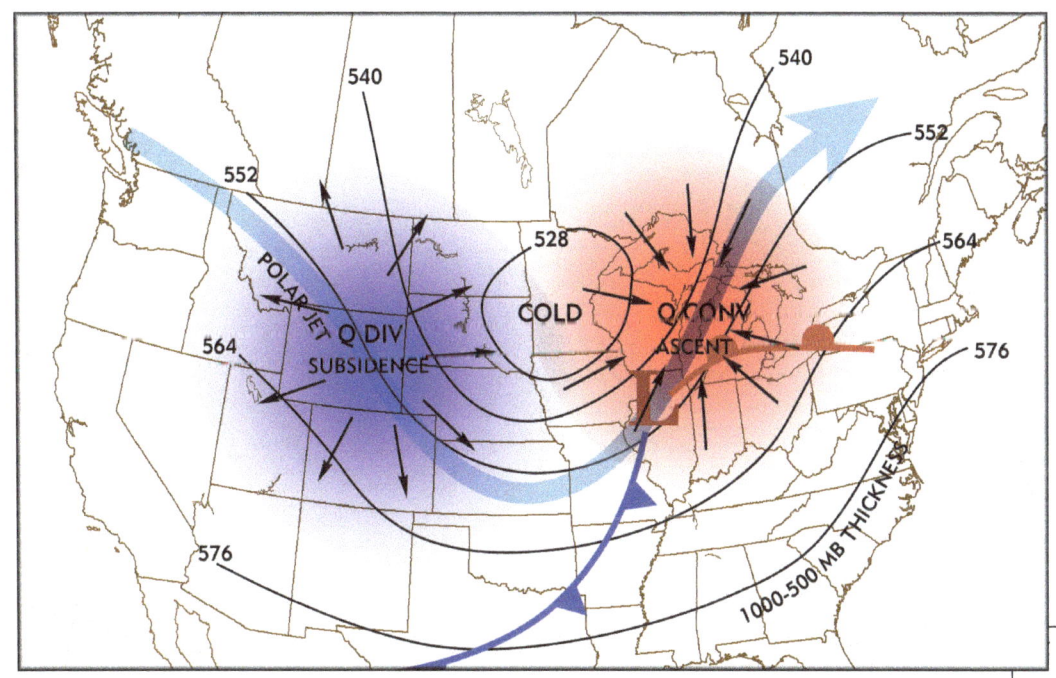

Top: Embry-Riddle University's meteorology website offered some of the best diagnostics on the Internet at press time. This example shows Q vectors, highlighting a major weather system over New Mexico and adjacent areas of the Great Plains. Note the Q vector convergence ahead of the system indicating ascent.

Below: Simplified conceptual diagram showing the typical relationship of Q vectors to the upper level jet and to a surface system and the 1000-5000 mb thickness field.

Precipitable water

Precipitable water (typically abbreviated PW) is one of the key indicators commonly used in precipitation forecasting. The AMS Glossary definition describes precipitable water as the total atmospheric water vapor contained in a vertical column between two specified levels. In operational meteorology this is typically from the surface to the tropopause or to the top of the model volume, usually some distance into the stratosphere. At jet stream level and higher, the atmosphere is dry enough to contribute only a negligible amount to the total value.

Precipitable water provides an indication of the *potential* for heavy rain in a given weather situation. Values of 0.75 inch or greater are a threshold value for heavy precipitation and flash flood risk. Such threshold values also depend on the region. Whether heavy rain is actually realized depends on the strength of forcing mechanisms, the amount of instability, the orientation of low-level boundaries to the prevailing flow (parallel orientations favor "training" of storm cells), and the speed of the tropospheric flow (fast movement favors faster clearance of storm cells but can also augment organization of convective weather systems as well as larger scale development of weather systems).

Some software packages and websites (such as Pivotal Weather) provide *precipitable water anomaly*. This can be provided as *difference* between the current and climatological normal, as a *percentage* relative to normal, or as a *percentile* value that evaluates the percent of observations that were drier than the current value. Normal values may be based on the calendar date or on the month. Anomaly products are useful for not only detecting the degree of anomaly for a given location, but for rapidly identifying locations across a broad area where heavy precipitation is favored. The SPC Sounding Climatology Page (see margin) provides detailed percentiles and maximum values for all stations in the United States. These values are often referred to in NWS prognostic discussions.

Precipitable water is sometimes confused with vertically-integrated liquid water (VIL) as detected by radars such as the WSR-88D. However these quantities are not the same. VIL is unable to evaluate water vapor in the atmosphere, but rather estimates the amount of liquid precipitation in a vertical column in the cloud. This estimate can also be distorted by the presence of hail. As liquid water can be concentrated in a small volume by storm-scale convergence, it has only a loose relation to the atmospheric precipitable water.

key details

Precipitable water (PW, PWAT, PWV), total precipitable water (TPW), or column precipitable water (CPW) is the vertical integral of the absolute humidity. In simple terms, it provides a measure of the total available water vapor through a vertical column.

To convert inches to kg m², multiply by 25.4. For the reverse, multiply by 0.039370.

Suggested settings are as follows:
* Precipitable water contours every 0.25 inch, or a banded or gradient color scheme.

Common color gradients for precipitable water are as follows:

0 to 0.5 in	Browns
0.5 to 1.0 in	Greens
1.0 to 1.5 in	Blues
1.5 to 2.0 in	Purples
2.0 to 2.5 in	Lavenders
2.5 to 3.0 in	Pinks

Values of 0.75 inch or more have the potential to support heavy rain and localized flooding, while values of over 1.5 inches often accompany significant rain events. The actual rainfall depends on the available lift mechanisms, the orientation of boundaries, the duration of precipitation, and the strength of the tropospheric flow.

The depth of the layer for calculating precipitable water is from the surface to the top of the model (10 mb for the GFS, 15 mb for the HRRR), or from the ground to 500 mb (the discontinued NGM). The contribution to precipitable water from the upper troposphere and the stratosphere is low to negligible.

real-time Internet sources

wx.erau.edu/teaching/milsyn/
weather.cod.edu/forecast
www.weather.gc.ca/model_forecast
www.spc.noaa.gov/exper/soundingclimo/
www.weather.gov/unr/uac
www.pivotalweather.com/model.php

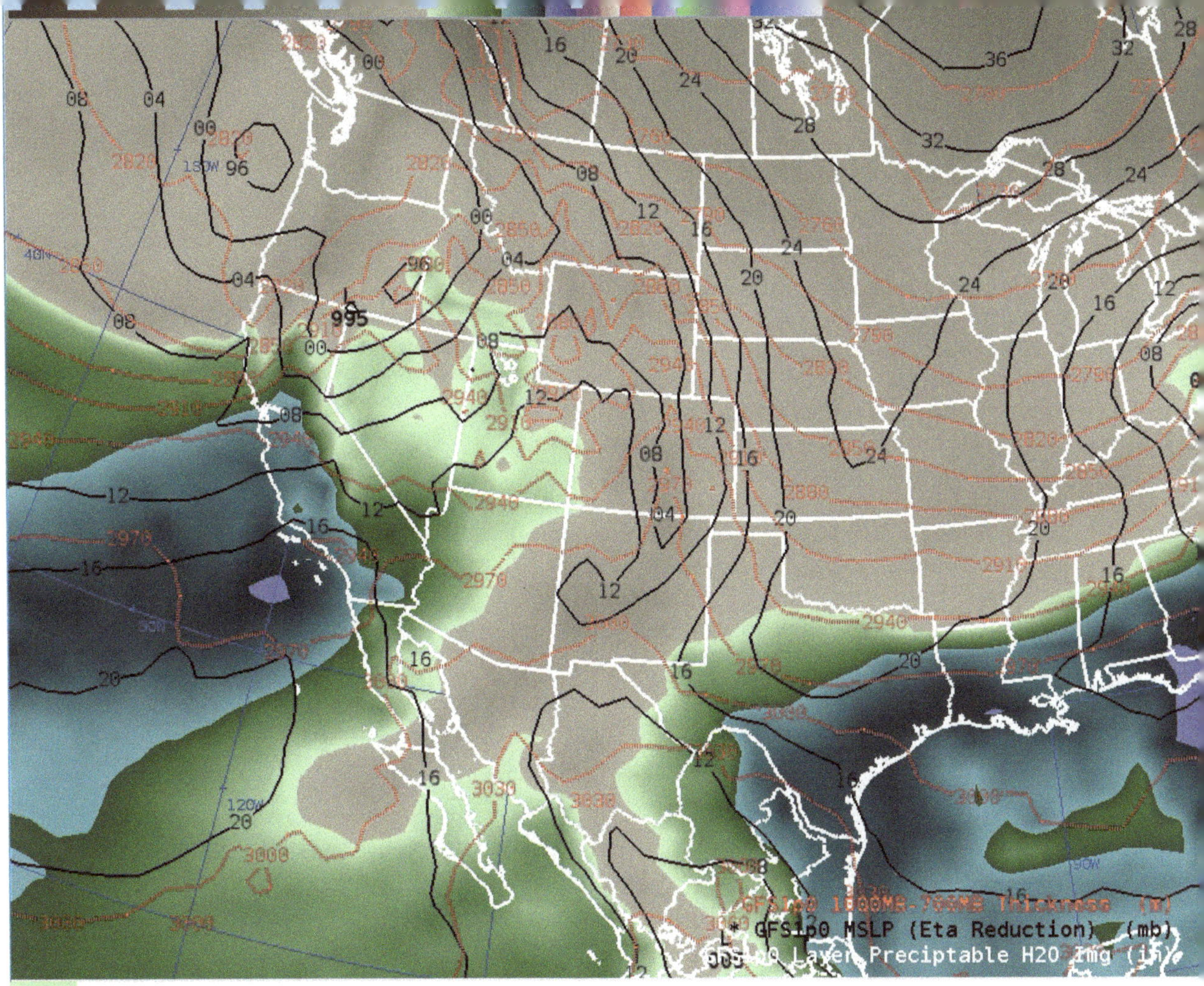

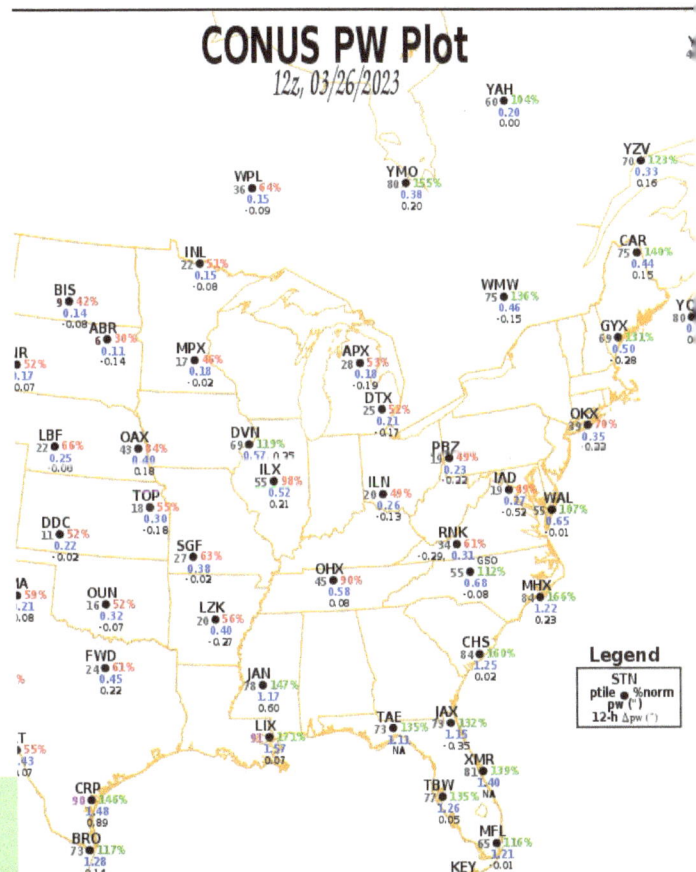

Above: Precipitable water GFS 54-hour forecast for 1800 UTC 10 March 2023, when strong warm air advection combined with high precipitable water and a record snowpack in the Sierra Nevadas, producing rapid snowmelt and flash floods in canyons and foothills across California. The green bands are 0.5 to 1 inches, the blue bands 1 to 1.5 inches, and purple 1.5 to 2 inches. *(AWIPS)*

Right: A portion of the precipitable water analysis from www.weather.gov/unr/uac. The actual precipitable water is found at the bottom of each plot along with the 12-hour change. The number at right shows the anomaly relative to normal as a percentage, and the number at left shows the anomaly as a percentile (i.e. percentage of observations at that station for the month which were drier than the current value; below 10 and above 90 are highly significant). *(Pivotal Weather)*

Potential vorticity

Potential vorticity (PV) is a function of the absolute vorticity divided by a stability parameter. It is widely expressed in potential vorticity units (PVU, see margin), which are shorthand for more complex units. Low PVU values are correlated with unstable air, while high values are associated with stable air. This is especially significant for stratospheric meteorology, where the atmosphere is heated by a combination of ozone and solar radiation, providing warming which reduces the lapse rate significantly.

The entire stratosphere acts as a stable layer, so it has very high PV. Most depictions of PV focus on this characteristic, where PV rises from about around 1 potential vorticity unit to 1.5 to 2 PVU near the tropopause, and then increases rapidly to about 10 PVU near 60,000 ft altitude (18 km). Due to the thermal gradient above the tropopause, there is very little mixing between the troposphere and stratosphere.

Since the potential vorticity value is near 1.5 to 2 PVU at the tropopause, it can be used as an indicator of the *dynamic tropopause* height. Cross section diagrams from polar to equatorial latitudes, enhanced to show the 1.5 or 2 PVU altitude, will show the differing tropopause heights between polar and tropical air masses. This can also be depicted on horizontal plots showing the altitude or pressure of the 1.5 or 2 PVU contour.

Potential vorticity is conserved in adiabatic, frictionless conditions. The work of Reed and Sanders in the 1950s showed that this makes PV an excellent tracer, in other words, when the atmosphere is undergoing subsidence we will see the higher values in the stratosphere descending to lower levels.

Because of this, PVU is useful in showing tropopause folding, which occurs on the poleward side of a polar front jet, especially during baroclinic development and cyclogenesis. Higher PVU values fold in a downward direction and toward the warm air mass, forming a *positive PV anomaly*. Water vapor imagery will show dry conditions within this anomaly. Measurements within these troposphere folds have revealed the presence of ozone, demonstrating the intrusion of stratospheric air to low levels. In especially strong weather systems, the 2 PVU surface will descend below 20,000 ft (3 km) and reach higher terrain.

Ascending motion will be found ahead of upper-tropospheric positive PV anomalies, with subsidence behind. Gradients of PV have also been correlated to regions where significant weather occurs, such as high winds at the surface.

key details

Potential vorticity measures the relation of absolute vorticity to stability. It tends to be a conserved property, so PVU surfaces that show deformation in the vertical or a change over time at a given level indicate vertical motion.

Potential vorticity is expressed in potential vorticity units (PVU), where 1 PVU equals 1×10^{-6} m^2 s^{-1} K kg^{-1}.

Potential vorticity is useful for identifying the dynamic tropopause level. This normally correlates to 1.5 or 2 PVU. It is recommended to display either the pressure or the altitude of the PVU surface in horizontal charts.

Troposphere folds are associated with transport of ozone from the stratosphere into the troposphere. Likewise, tropospheric pollutants, CFCs, and associated gases can be transported into the stratosphere. This can include ozone precursors such as methane. Troposphere folds are responsible for a significant amount of the ozone present in the stratosphere.

Deep convection, such as from thunderstorms, will interfere with potential vorticity, producing localized areas of anomalous data.

real-time Internet sources

wx.erau.edu/teaching/milsyn/
weather.cod.edu/forecast
www.weather.gc.ca/model_forecast

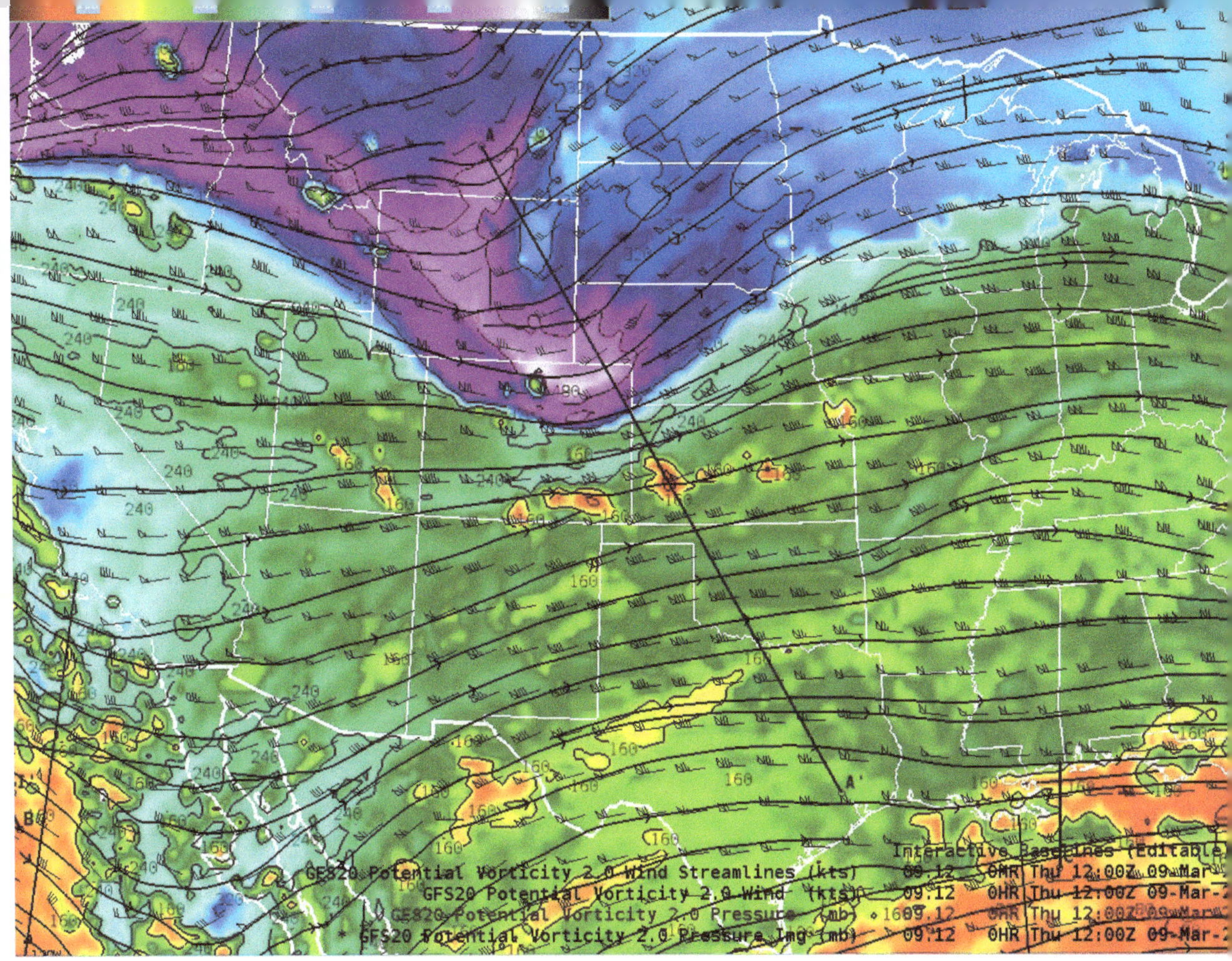

Above: GFS depiction of the 2 PVU field expressed as the altitude in pressure units (shading and gray contours), along with the wind field at that level. *(AWIPS-II)*

Right: Cross section along the black line above from Montana to Texas. This shows PVU as color bands, with 2 PVU near the transition from purple to blue. Note how this depicts the lower tropopause north (left) of the polar jet, with a lower altitude for the stratosphere indicated by dense potential temperature packing, solid red lines. *(AWIPS-II)*

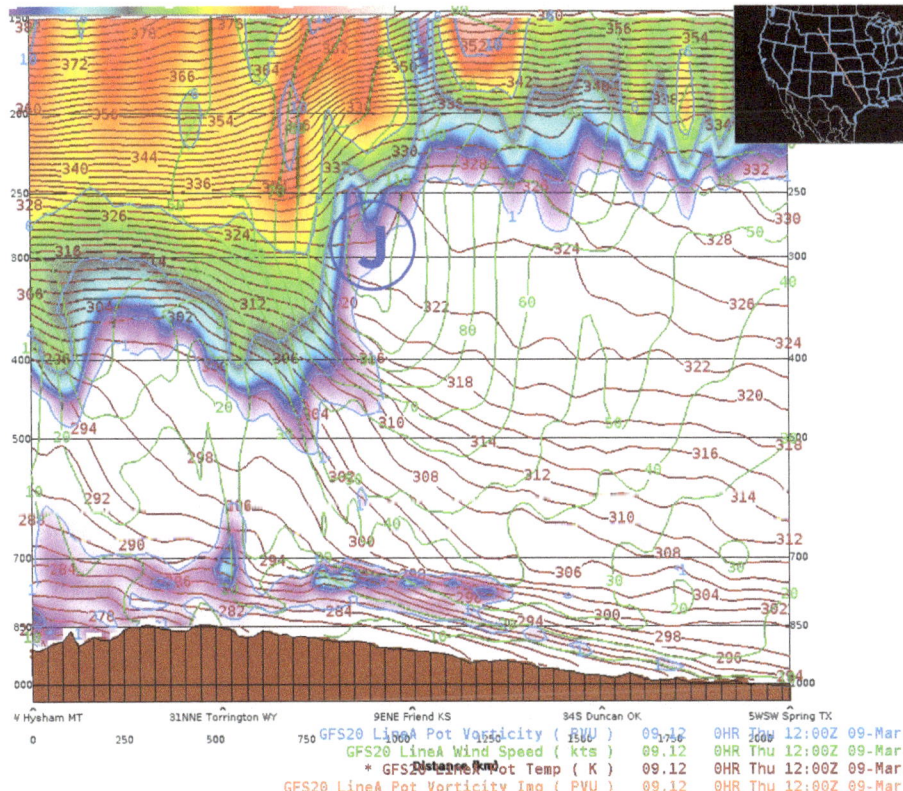

Cross section analysis

One of the benefits of analysis in the modern computer age is the ability to easily construct vertical cross sections. Instead of being confined to horizontal charts or a one-dimensional sample as with Skew-T's, cross sections provide a two dimensional picture in a vertical plane. This greatly supplements the analysis and forecast process.

The cross section uses a horizontal baseline on the x-axis, with a vertical coordinate on the y-axis. This makes the diagram very much like transecting a weather system and viewing it from the side. During the middle and late 20th century, cross sections were often constructed by joining up lines of radiosonde stations and plotting the observed data. While this is still a feasible method for analyzing a forecast region, numerical models have advanced significantly. Data assimilation schemes of the 2010s and 2020s are highly robust and provide very accurate initial conditions for dynamical models. It is more practical to start with a high-resolution model, then note any inconsistencies with sounding, satellite, and radar data.

An excellent use of the cross section is identifying air mass transitions in the vertical. Frontal transition zones are indicated by tight spacing of isentropes (potential temperature), while widely-spaced isentropes indicate regions of steep lapse rates and instability.

Additional parameters can also be plotted. Potential vorticity can be used to identify the tropopause and locate tropopause folds. Specific humidity helps locate interfaces between continental and maritime air masses, both in the horizontal and the vertical. Isotachs help locate jets and allows the forecaster to relate them to processes taking place across the region of interest. A cross section may also provide some information for isentropic analysis if it is oriented parallel with the wind field that is of interest.

The forecaster must be especially careful when interpreting wind on this chart, as the compass and cross section coordinate systems differ. In a standard cross section wind plot, left is a west wind and up is a north wind, exactly as if the diagram was a standard surface chart. In a section-normal plot, left is the left-to-right component along the plane and up is the component out of the plane.

Scalar values for wind speed, such as isotachs, may be plotted as absolute values (exactly as observed) or may be plotted as cross-section relative components: parallel with the cross section ("comp along", positive is left-to-right) or in/out of the cross section (in a "comp into" chart, positive is into the plane). A circle with an x, by convention represents maximum wind into the page, while a circle with a dot represents flow out of it.

key details

The cross section analysis depicts weather in a vertical plane where pressure, height, or altitude comprises the y-coordinate, and distance is used for the x-axis.

The chart is drawn as follows. All fields except isentropes are optional and may be selected according to the prevailing weather pattern):
- Isentropes (potential temperature, or theta) every 2 or 5 K.
- Relative humidity every 10%.
- Specific humidity every 1 or 2 g kg^{-1} instead of relative humidity when forecasting deep convection situations.
- Isotachs as purple dashed or solid lines every 10 or 20 kt, or as color banding or color shading. The forecaster may select the magnitude (spot velocity) or may choose the velocity component into the plane or parallel (along) to it.
- Potential vorticity every 0.5 PVU. The 2.0 PVU line may be outlined with a bold mark to delineate the dynamic tropopause.
- Terrain outline if available to depict surface elevation along the x-axis.

Stable layers. Stability is indicated by spacing of isentropes in the vertical. Where they are close together, lapse rates are weak or an inversion is present.

Tropical cyclones. Cross sections of dynamical model output are useful for assessing the state of tropical cyclones. An anticyclone near the top of the tropopause is indicative of a tropical cyclone. If a cyclone exists at the top of the tropopause, this indicates an extratropical system. Subtropical cyclones may have neither, and may show a shallow potential vorticity tower.

real-time Internet sources

www.weathernerds.org/models
www.tropicaltidbits.com/analysis/models/

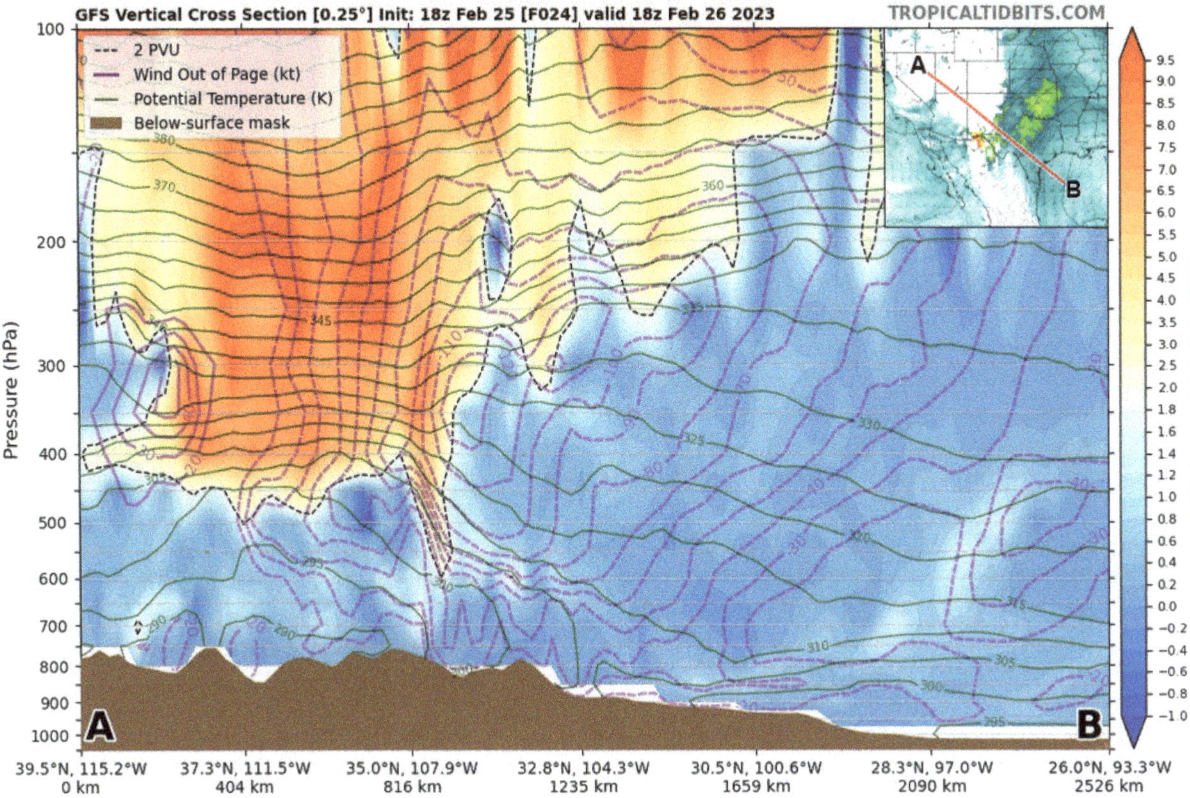

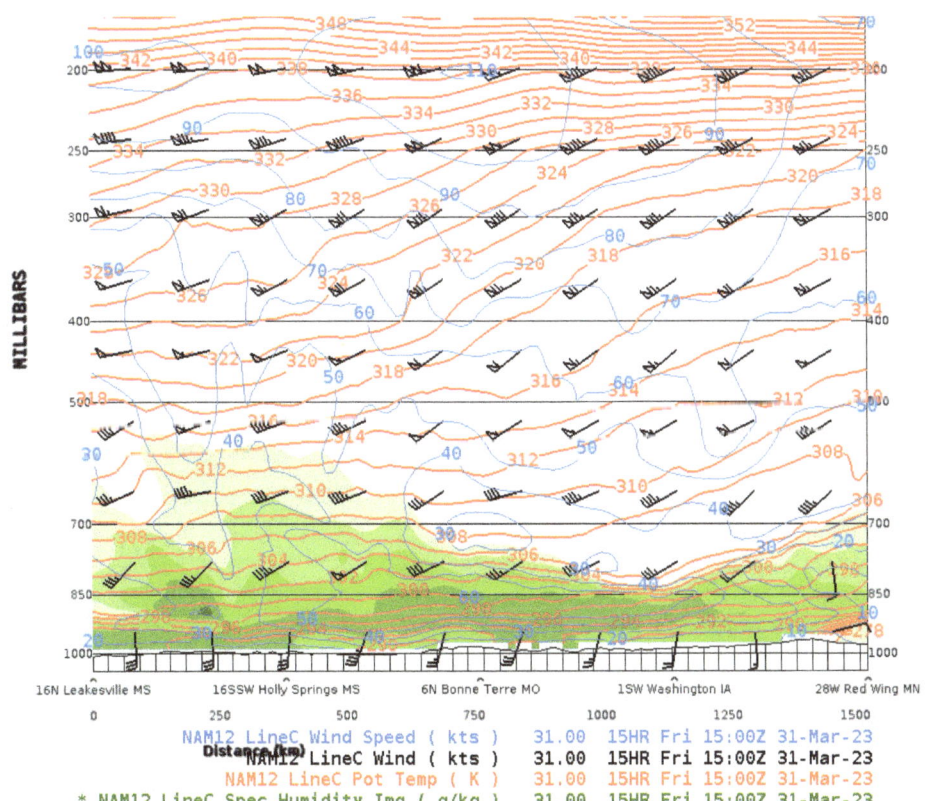

Top: Cross section of the GFS model for 26 February 2023 showing a cross section through an active baroclinic weather system. This plot includes a dynamic tropopause (dotted line separating reds from blues), isentropes (green lines), and the magnitude of the wind vector flowing into the page (purple). *(TropicalTidbits)*

Below: AWIPS-II cross section roughly parallel to the low-level jet just before the severe weather outbreak of 31 March 2023. Wind plots are standard, in which up is north and left is west. Green shading is specific humidity, which is comparable to dewpoint. This clearly shows the depth of the moist layer. *(AWIPS)*

Thermodynamic diagram

The thermodynamic diagram shows the profile of temperature and dewpoint above a given weather station with height. The X-coordinate is always temperature, and the Y-coordinate is height. Using these coordinates, observations of temperature and dewpoint at various heights are plotted. Since dewpoint is always equal to or lower than temperature, the dewpoint trace is usually to the left of the temperature trace. It is typically shown as a dashed rather than a solid line.

In the *skew T diagram*, widely used by North American meteorologists, the temperature coordinates are skewed 45 degrees to the right to help make changes in the slope of the temperature profile easier to see. In other words, the background coordinates of height and temperature do not form a grid like classic graph paper; rather the vertical lines are turned diagonally by about 45 degrees to the right.

The most important use of the sounding is to assess the degree of instability present in the atmosphere. If any part of the sounding leans sharply to the left with height, it is assumed to have a large lapse rate (a great temperature decrease with height). If the dewpoint trace indicates significant moisture in the lower levels, the moisture and its potential for latent heat release will combine with this large lapse rate to produce an unstable atmosphere.

The simplest measure of instability is the Showalter Stability Index (SSI) and the Lifted Index (LI). However, both of these rely on a simple comparision of the parcel with the environment at one level. A far more accurate measure is Convective Availability of Potential Energy (CAPE), which assesses the parcel-environment temperature differential throughout the entire vertical column. A CAPE value should always be used except when it is not available. All instability calculations depend on an accurate representation of parcel temperature and moisture, which in turn requires a representative integration of low-level moisture and an accurate temperature forecast. Human manipulation of the parcel attributes are always worthwhile, and are easy to do on paper SKEW-T diagrams and in certain weather software applications.

The basic concept for determining the type of precipitation that reaches the surface is to begin in the mid-levels of the troposphere, determine what type of precipitation it begins as, and observe the temperature regimes that affect the particle as it falls downward. As a rule of thumb, 1200 ft of warm air is considered ample to completely melt snow, with 400 ft of cold air considered enough to freeze liquid precipitation.

key details

The thermodynamic diagram is used for establishing the character of the entire air mass above a given station. It consists of a plot of temperature, moisture, and wind.

Height lines are horizontal and are usually calibrated in millibars. The top of the chart is usually 100 mb (about 53,000 ft) and the bottom is usually 1050 mb (about minus 300 ft MSL).

Temperature lines slope up and to the right. They are calibrated in degrees Celsius with an interval of every 10 C°.

Dry adiabats slope up and to the left. These indicate how a dry parcel will cool as it rises.

Moist (wet) adiabats slope upward and then curve sharply to the left. These indicate how a saturated parcel will cool as it rises.

Mixing ratio lines slope upward and to the right. They are more vertical than the temperature lines, and are typically omitted in the upper portion of the sounding. A rising parcel's dewpoint will follow these lines upward until it saturates.

A software package dedicated to working with soundings is highly recommended and makes quick, accurate work of parcel estimates and daytime heating influences.

The "bible" for the SKEW-T is "The Use of the Skew-T Log-P Diagram in Analysis and Forecasting" by Robert C. Miller, AWS TR-200, published in 1972 and changed very little since then.

real-time Internet sources

www.rap.ucar.edu/weather/upper
weather.unisys.com/upper_air/skew
weather.cod.edu/analysis
weather.uwyo.edu/upperair/sounding.html

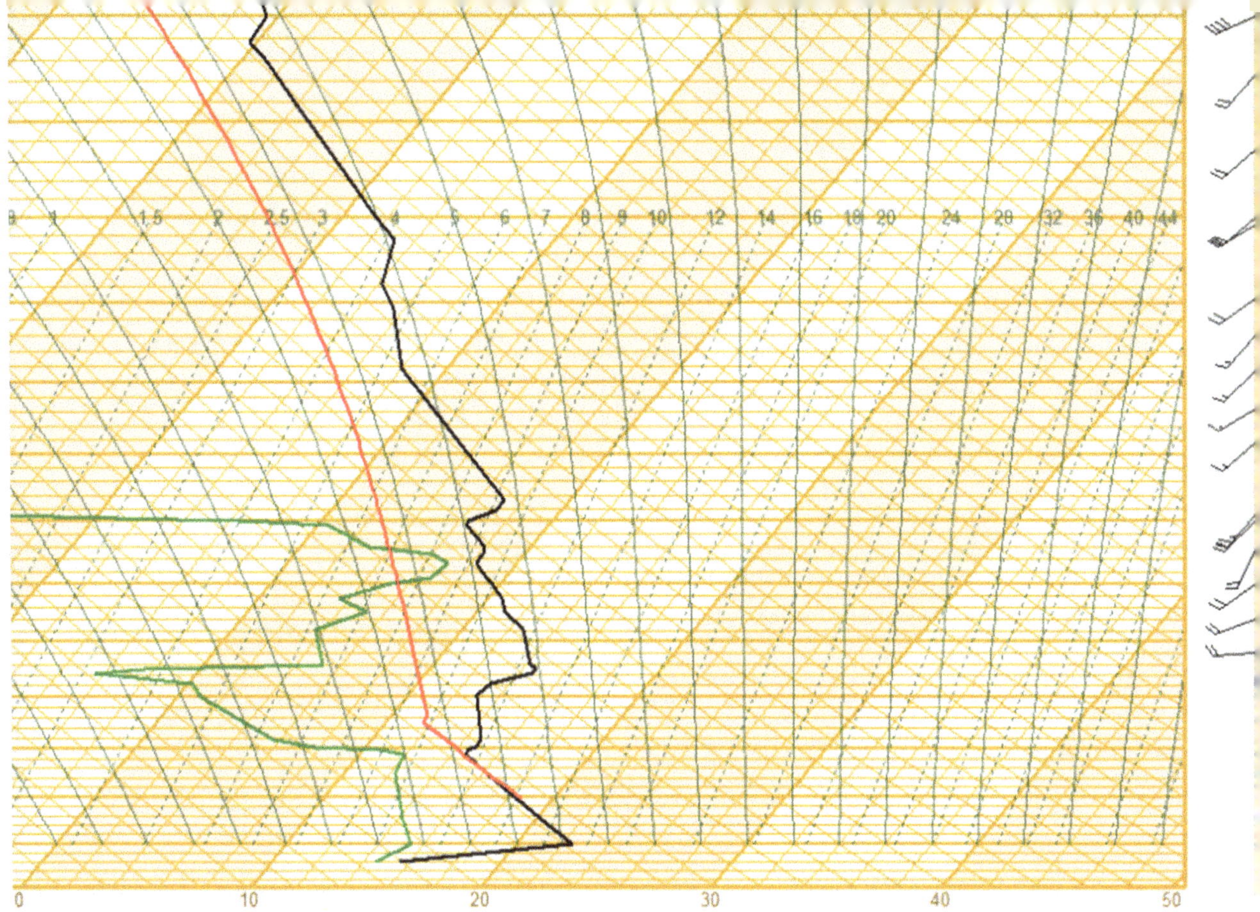

Above: Detailed sounding generated by the Windows analysis software Digital Atmosphere. Detailed diagrams like this are useful in meticulous severe weather forecasting work. Note the hypothetical parcel lift outlined in red, which is a parcel constructed from the mean mixing ratio and mean potential temperature in the lowest 150 mb of the atmosphere.

Right: Sounding from the UCAR weather page for Barrow, Alaska in late summer. It shows a classic frontal inversion at 870 mb (about 4500 ft MSL). Thick cloud layers are also indicated above 730 mb (about 9500 ft MSL). These indicators make the chart very useful for aviation forecasting.

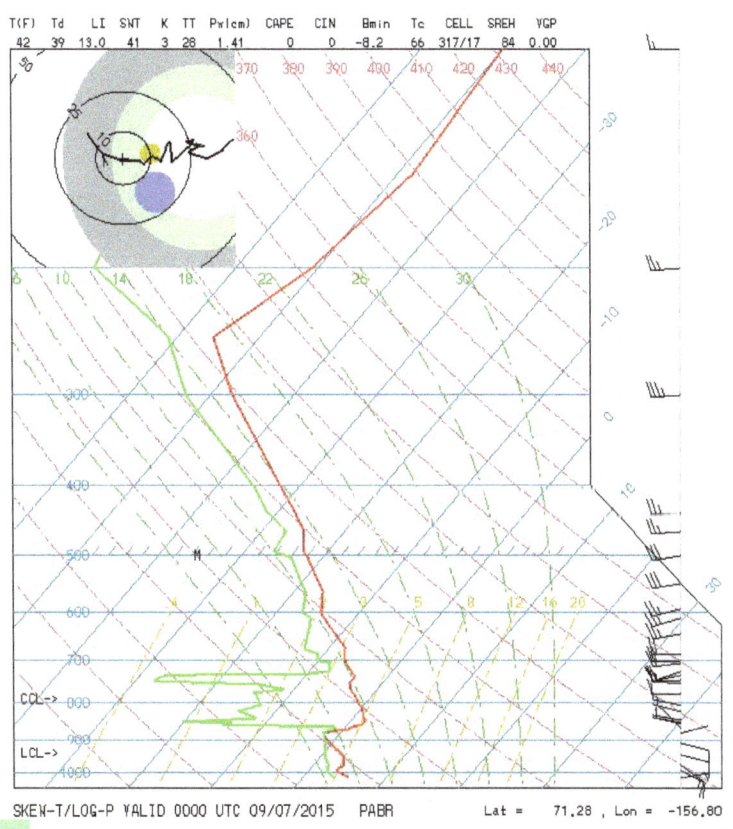

Wind profiler

A wind profiler is a special type of Doppler radar that measures wind speed with a fixed antenna. The radar is able to project in two perpendicular locations off the vertical axis in order to measure the horizontal wind components. The advantage of a wind profiler is that it can sample winds continuously without the quantity of scatterers needed by a WSR-88D producing a VAD/VWP product. They have fewer moving parts, making them more reliable.

Profilers operate in a high mode and a low mode. The low mode usually covers the lower and middle troposphere (up to about 9 km), while the high mode covers the upper troposphere (from about 7 to 16 km) in a much more sensitive detection mode.

A piggyback technology incorporated in most soundings is called *Radio Acoustic Sounding System (RASS)*. Its main function is to estimate temperature at various layers. Since the speed of sound is dependent on temperature, RASS allows temperature to be measured. A 900 Hz burst, one octave above the musical middle A, is transmitted upward into the atmosphere and its echo is measured. This data is not as accurate as radiosonde measurements so it should be used with caution in convective forecasting.

Discontinuation in the United States

The United States National Profiler Network was shut down in August 2014. According to NOAA, "A recent study concluded that wind, temperature, and moisture observations measured by sensors on commercial aircraft [ACARS] and those taken by equipment on weather balloons provide data that are superior to the NOAA 22-year-old Wind Profiler Network (NPN) for short-range weather forecasts. The combination of tight economic conditions and wind profiler system obsolescence, in addition to the limited geographic scope of the NPN, led NOAA to recommend discontinuing the use of the Wind Profiler Network in a report to Congress in May 2013."

Elsewhere in the world

Europe has continued to invest in wind profiler technology and currently maintains a network of about 26 wind profilers stretching from northern Italy and Spain to Germany, the United Kingdom, and Norway. The data is available on the Internet without restriction at the UK Met Office website <www.metoffice.gov.uk/research/interproj/cwinde/profiler>.

key details

Wind profilers are a special type of clear-air Doppler radar with no moving parts. They often come with RASS units which measure temperature by projecting acoustic beeps toward the zenith.

Wind profilers usually require six minutes for each cycle. This is comprised of three 2-minute samples: one at the zenith, one tilted in the X direction, and one tilted in the Y direction. Within each of these modes, there is a 1-minute sample in low mode and another in high mode (see text).

The NOAA 404 MHz wind profilers were shut down in 2014 due to budget cuts. The dense network of WSR-88D radars provides a substitute for wind profilers in the form of Velocity Azimuth Display Velocity Wind Profiles (VAD/VWP). These are available for viewing on most single-site radar pages.

Strips of bad data may be caused by aircraft flying over the profiler.

Scatterers are required to provide measurements. As a result, the displays will typically show cloud layers along with layers of biological scatterers (birds, insects, and so forth). The displays can also be used to get an idea of the height of these layers along with their depth and persistence.

One overlooked method of viewing profiler data is using horizontal plots of multiple stations at a given level. This can fill in data between radiosonde stations. The UCAR Weather website was one example of a site that offered such plots. Tools like GEMPAK and McIDAS are capable of creating such maps.

real-time Internet sources

www.metoffice.gov.uk/science/specialist/cwinde/profiler/

ANALYSIS CHARTS

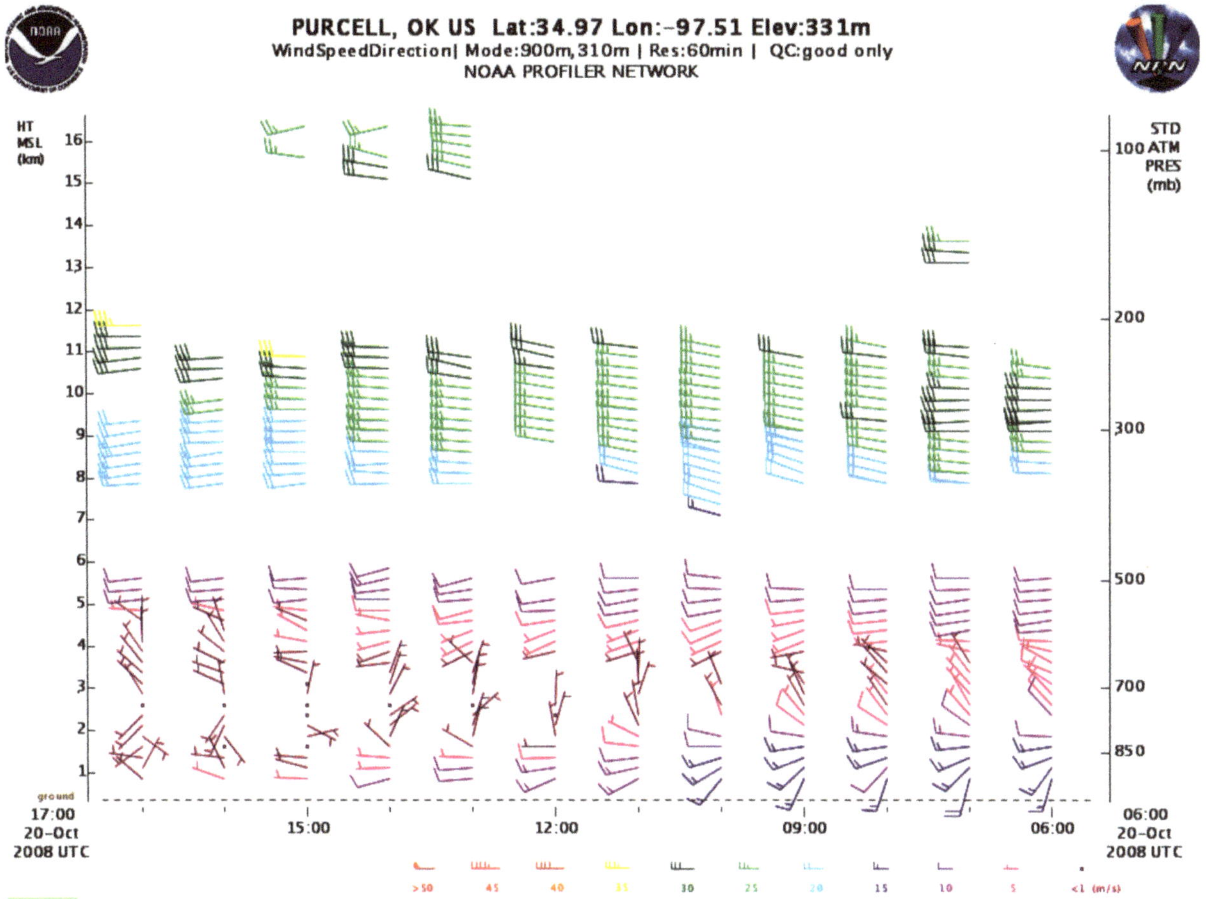

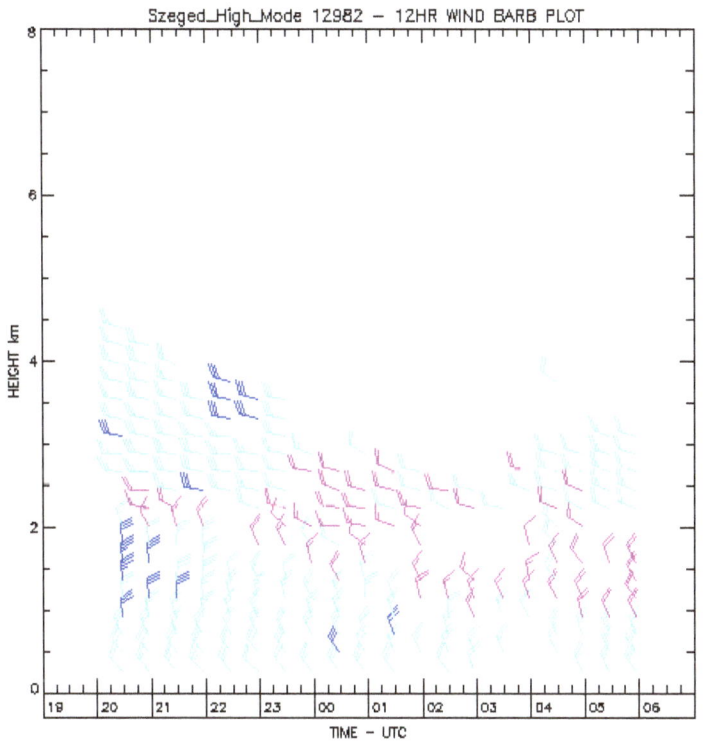

Above: NOAA profiler chart for 2008, when the network was in operation. The coordinates are reversed time (X-axis) and height (Y-axis). The newest plot is on the left side. Most of this functionality can be obtained from WSR-88D VAD wind profiles. *(NOAA)*

Right: An example of a profiler plot from Szeged, Hungary for 7 September 2015. The graphics are small and can be difficult to read, but the MetOffice website also offers a profiler diagnostic plot which provides a graph of direction and speed versus height. *(COD)*

Lightning detection

Lightning detection has been part of operational forecasting in the United States since the mid-1980s, when the National Lightning Detection Network (NLDN) was developed and data was fed into early display systems at NWS offices. This system was based on radio direction-finding techniques using a network of ground stations. By 1989 the nation was fully covered. Since this data was maintained by private firms, it was not extensively used outside government, research, and broadcast activities.

Unrestricted lightning data was made possible with the launch of the GOES-R series of satellites in 2016. These satellites carry the Geostationary Lightning Mapper (GLM) sensor package which allows realtime observations of lightning from space. Rather than using radio techniques, a near-infrared (777 nm) optical detector senses lightning flashes.

This instrument has an effective resolution of about 10 km, with degradation near the edge of the Earth's disc due to the effects of low slant angle. Note that because this is similar to visible imagery, the sensor cannot see individual lightning strikes beneath a cloud; it can only detect the burst of light within the cloud at that location. It also cannot detect flash polarity. However a significant advantage offered by the GLM is consistent coverage of most of the viewable Earth between 54°N and 54°S, even in remote areas.

Although plots of lightning "strikes" are the most familiar type of lightning diagnostic tool, the GLM system also offers a number of other measurements described in the inset at right. The GLM will occasionally present flashes from false sources, including sun glint, stray solar light, cloud reflections, and meteors.

Lightning detection provides an excellent measure of the strengthening and weakening of thunderstorms. The trends TOE, MFA, and FED provide further data that may characterize storm behavior.

Land-based radio triangulation networks like NLDN and LightningMaps provide much greater positional accuracy of lightning strikes, providing there are multiple sensors within a few hundred miles. The NLDN network can determine cloud-to-ground (CG) lightning polarity (charge of lightning relative to the ground). The vast majority of strikes are negative (–CG). Positive CG strikes (+CG) in large numbers near a strong updraft are usually associated with supercellular structures, but are also common in the trailing stratiform region of large, established MCSs even when no severe weather is occurring.

key details

The Geostationary Lightning Mapper (GLM) is currently the most common and easily available source of lightning data for the US.

GLM lightning data is processed by grouping all connected pixels that illuminate during each 2-millisecond cycle of the sensor. This forms a single "group". All groups within 16.5 km within a 330 millisecond window are then assembled into a single "flash".

Flash point
The centroid of the processed flash.

Total optical energy (TOE) (fJ)
The sum of all optical energy during a given time interval at each grid box. A high value indicates numerous bright flashes. This product has been compared to a light bulb brightening and dimming as a storm grows and decays.

Minimum flash area (MFA) (km^2)
The areal size of the smallest flash at each grid box during a given time interval. Small MFA values may indicate strengthening convection, if accompanied by high flash rates. High MFA values may signify long complex flashes that commonly occur in the trailing stratiform region of an MCS.

Average flash area (AFA) (km^2)
The average area of all connected grid boxes reporting a flash. This product has been replaced by MFA.

Flash extent density (FED) (flashes/time)
The number of detected flashes within a given time interval at each grid box. Rapidly increasing values indicate an increase in the number of lightning strikes.

Radio-based lightning detection networks are available free of charge on the Internet at <lightningmaps.org>. This can complement other sources of information.

real-time Internet sources

weather.cod.edu (GLM as satellite/radar overlays)
www.star.nesdis.noaa.gov/goes
rammb2.cira.colostate.edu/ramsdis/online
lightningmaps.org

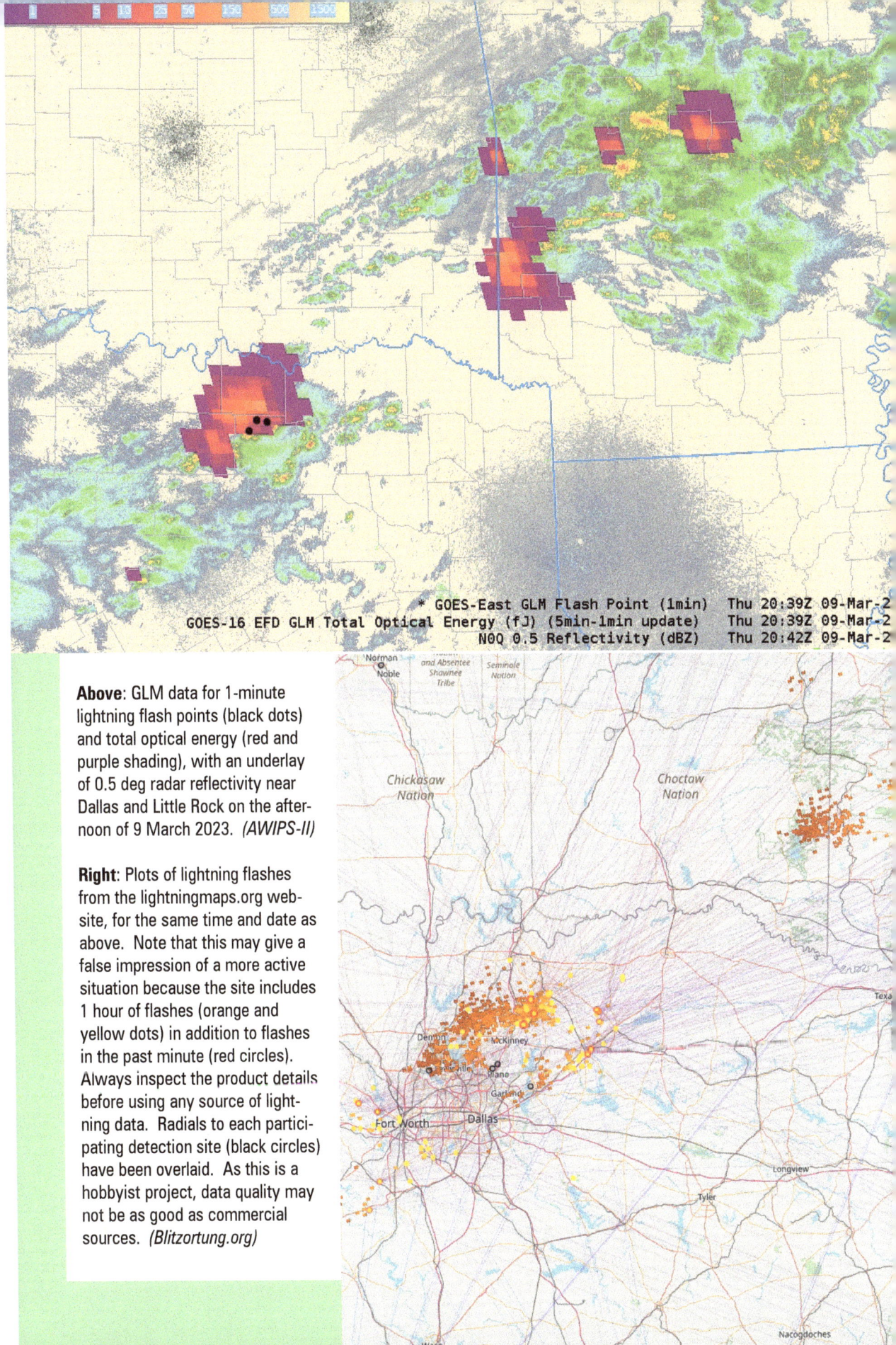

Above: GLM data for 1-minute lightning flash points (black dots) and total optical energy (red and purple shading), with an underlay of 0.5 deg radar reflectivity near Dallas and Little Rock on the afternoon of 9 March 2023. *(AWIPS-II)*

Right: Plots of lightning flashes from the lightningmaps.org website, for the same time and date as above. Note that this may give a false impression of a more active situation because the site includes 1 hour of flashes (orange and yellow dots) in addition to flashes in the past minute (red circles). Always inspect the product details before using any source of lightning data. Radials to each participating detection site (black circles) have been overlaid. As this is a hobbyist project, data quality may not be as good as commercial sources. *(Blitzortung.org)*

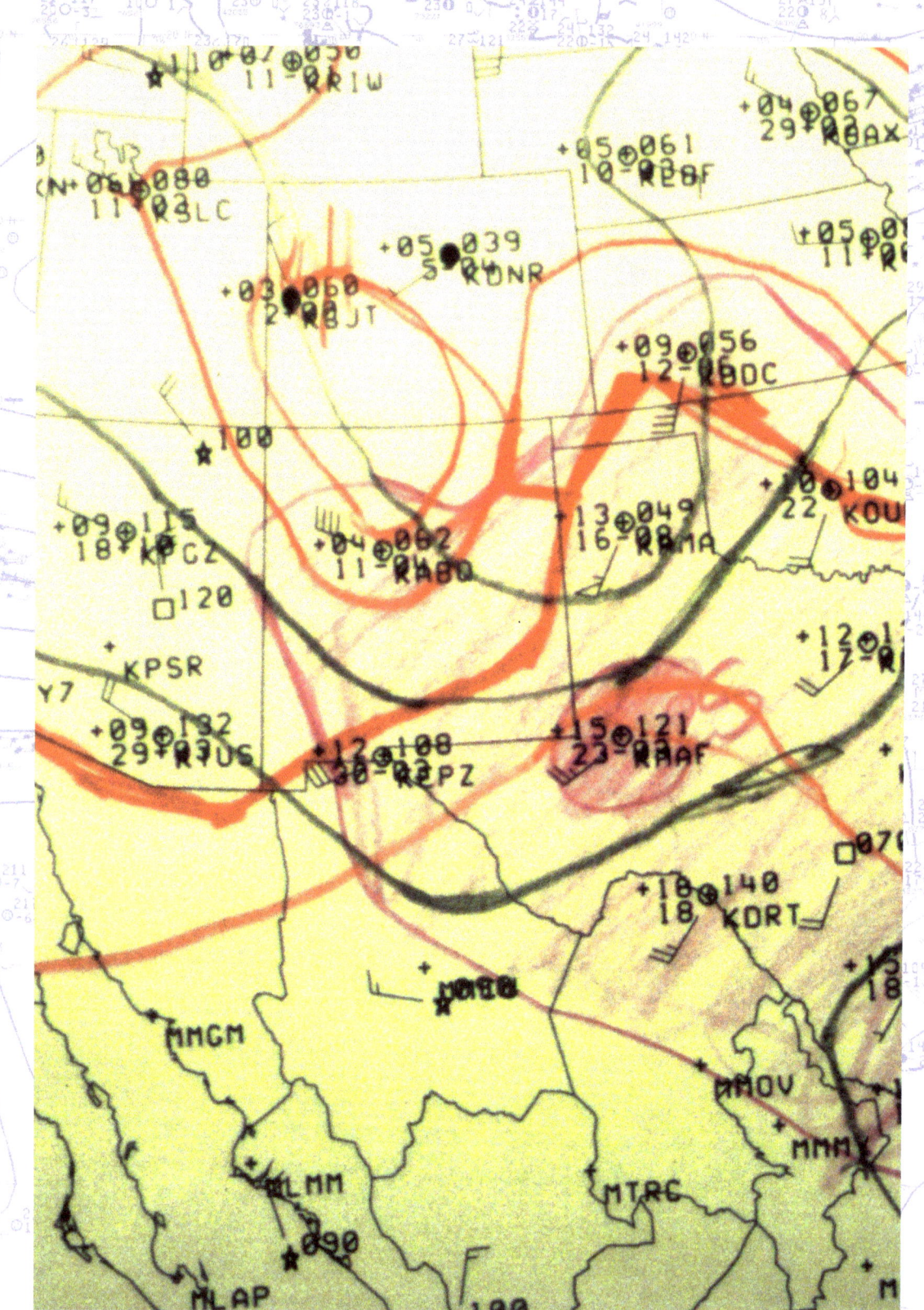

Hand analysis of 700 mb conditions photographed on the operations desk at WSFO Fort Worth during a severe weather outbreak in 2003. *(Tim Vasquez)*

CHAPTER 2
SATELLITE IMAGERY

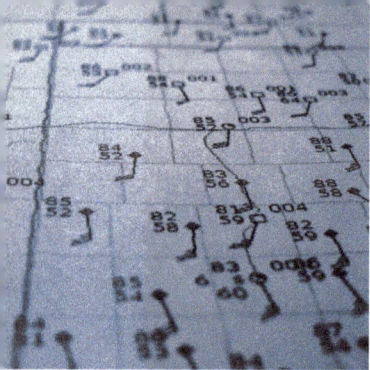

Satellite

In April 1960, the world's first weather satellite was launched, named TIROS (Television InfraRed Observation Satellite). This system was little more than a television camera mounted on a small orbiting platform. It was only during the 1970s when more sophisticated imaging techniques were adopted. Another milestone came in the mid-1990s, when detailed satellite imagery became available freely to anyone who had access to the Internet.

Weather satellite imagery consists primarily of visible, infrared, and water vapor imagery. There are also complex multispectral products available. Each will be covered in the sections ahead. First, we will take a look at the major classes of weather satellites: *geostationary* and *low-orbiting*.

Geostationary earth orbiting (GEO)

Geostationary satellites are placed in orbit 19,323 nm (22,236 sm or 35,786 km) above sea level. At this altitude it is possible to put an object in a semipermanent "free fall" orbit around the earth with an orbital speed that coincidentally matches the earth's rotation. By doing this, the weather satellite can be parked above a given spot on earth.

The weather satellite imager scans the Earth in a raster format, using the spacecraft's spin for imaging along one axis and a stepping mirror for incrementing the imager along the perpendicular axis. This allows the spacecraft to build an image with a minimum of moving parts.

The first geostationary satellite was ATS-1, launched in December 1966. This was not intended for operational use, and was meant to test the concept. The first official GOES (Geostationary Operational Environment Satellites) satellite was SMS-1, launched on May 5, 1974. It was operational from 1974 to 1981. SMS-1 was initially a NASA project but was turned over to NOAA. Since then, 18 satellites have been entered service, and are referred to as GOES. Nearly all of these satellites have enjoyed long service lives, and arguably represent one of the most successful technological programs ever created by the U.S. government.

A major upgrade came to GOES imagery in 2017 with the commissioning of the GOES-R series of satellites. Infrared and visible imagery increased significantly in areal resolution. Images were made available every 5 minutes instead of every 15 minutes, and mesoscale floater sectors offered 1-minute imagery. For forecasters, this was one of the most significant milestones in the program's history.

Japan and Europe both launched their first geostationary weather satellites in 1977, maintaining consistent reliable coverage ever since. The Japanese satellite is known as GMS, while the European satellite is called METEOSAT. India,

---

Below: One of the first TIROS images: May 1960, showing a line of severe thunderstorms moving through northern Missouri. Satellite imagery would revolutionize meteorology during the coming decades. *(NOAA)*

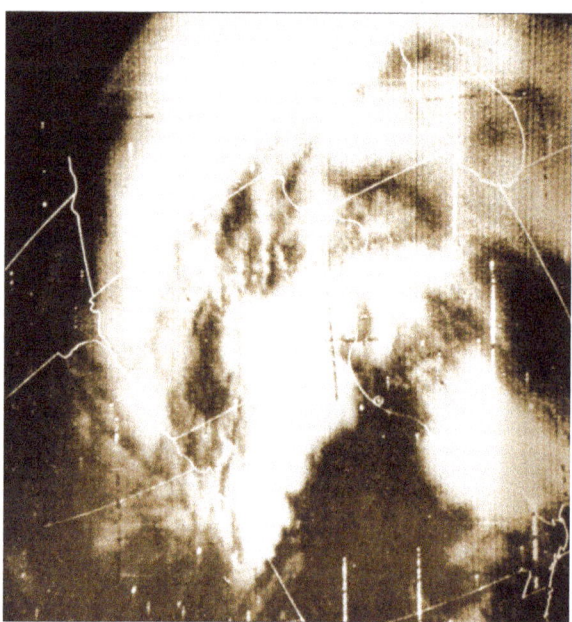

Russia, and China have launched geostationary weather satellites for their parts of the world with varying degrees of success.

Low Earth orbiting (LEO)

The term "low-Earth orbiting" is a name given to weather satellites which orbit only about 500 miles above the ground. They are often called *polar orbiters*. A rotation is completed about once every 1.5 hours, and the orbit is typically *sun-synchronous*, which means that if the Earth's rotation was completely stopped with respect to the sun and the satellite's path was marked on a world map, the spacecraft would never deviate from that path. The Earth's rotation underneath the sun-synchronous orbit allows different areas of the Earth to sweep by underneath, each part passing underneath every 12 hours: once on the day side and once on the night side. Polar orbiters have an inclination that takes them over polar regions, giving them unprecedented ability to image polar regions such as Alaska, northern Canada, and Antarctica, all of which are seen at too much of a slant on GOES satellites to give useful images.

Nearly two dozen civilian LEO satellites have been launched by the U.S. government. They are all referred to as the POES (Polar Orbiter Environmental Satellite) series, and are comprised of satellites designated "NOAA" (older ones were known as TIROS). The U.S. military also operates a network of LEO weather satellites known as DMSP (Defense Meteorological Satellite Program), currently being phased out and replaced with the Weather System Follow-on Microwave (WSF-M) program. Though these satellites originally had classified capabilities, they have become increasingly tailored for the military's space weather and oceanographic missions.

NASA launched the MODIS (Moderate-Resolution Imaging Spectroradiometer) system on board the Terra and Aqua satellites in 1999 and 2002, respectively. Terra and Aqua are LEO birds orbiting at an altitude of about 700 km. The result is LEO weather imagery with extremely high resolution (0.25 km). They have been supplemented with Suomi NPP in 2011 and NOAA-20 in 2017. The images are easily viewed at <*worldview.earthdata.nasa.gov*> and <*lance.modaps.eosdis.nasa.gov/imagery-apps/*>.

Above: Highest possible infrared resolution for the GOES-10 geostationary satellite (top) compared to the equivalent Chinese Feng-Yun 10 polar orbiter (bottom). The poor quality GOES imagery is caused by a low slant angle and a coarse 4 km infrared imager. The polar orbiter offers a high overhead view angle and 1 km resolution, making polar orbiters a serious tool in polar regions. *(NOAA/ARH)*

Though LEO imagery has the advantage of mapping anywhere on earth without slant angle degradation in the polar regions, it does have some drawbacks. It can only image the same spot on earth twice a day, and the view always requires complicated georeferencing (adding borders and geography). This adds processing time, with LEO images already commonly delayed by 1 to 3 hours. The images also tend to have some distortion since the satellite is so low. It should be noted that the NOAA LEO series has much better infrared resolution than that onboard the GOES satellite.

Imagery on the Web

Perhaps the most important consideration besides convenience and timeliness is the utility of the images themselves. Make sure you are getting the best possible image size and resolution for the job. Many weather web sites still are optimized for the "old Internet" or mobile browsing, offering coarse resolution or small images. Such photographs are difficult to use in operational forecasting. For full customization, a client-side program like McIDAS (free download) offers the highest resolution available.

Official web sites for government weather satellite programs include <www.goes.noaa.gov>, <www.ospo.noaa.gov>, , <www.eumetsat.de>, <www.data.jma.go.jp/mscweb/en/>, <www.cma.gov.cn/en2014/satellites>, and <satellite.imd.gov.in>, and <smiswww.iki.rssi.ru>.

Right: Tornadic storms developing over Iowa and Missouri during the 31 March 2023 "high risk" event in the Mississippi River region. Little Rock had been hit by a large tornado about 30 minutes earlier. *(NOAA/NESDIS)*

Below: The world's first geostationary satellite image. Dating back to 11 December 1966 (when Beach Boys' "Good Vibrations" was at the top of the music charts), this image was taken by ATS-1, a satellite operated by NASA. This satellite was used to relay color television and White House communications across the continent. Early weather satellite technology was largely hindered by a lack of suitable display technology at each field office. Television screens were far too coarse, high-definition computer monitors were still 15 years away, and laser printers would not be invented until 1975. Therefore film recorders were a mainstay in the early years. These systems offered marginal quality and the consumables were expensive. *(NOAA)*

31 Mar 2023 20:06Z - NOAA/NESDIS/STAR - GOES-East - GEOCOLOR Com

Visible imagery

Visible satellite imagery is the most intuitive type of satellite imagery, as it detects exactly what an astronaut would see from space. But to better understand the scientific principles underlying the imagery, we have to understand what is being measured. Visible imagery measures brightness, which is a function of illumination and albedo. Illumination is the visible radiation reaching the scene (which is nearly zero at nighttime!), while albedo is the percentage of this illumination which is reflected into space and detected by the satellite.

Clouds composed of water droplets have some of the highest albedo of any meteorological object; definitely much higher than that of ice crystal clouds. They are capable of reflecting over 90% of the visible light back to space! This means that newly emerging cumulonimbus clouds will be extremely bright, while cirrus clouds will tend to be grayish to light gray. This has interesting consequences: for example the thin veil of outflow cirrus around hurricanes is strongly detected by to infrared imagery, giving the storm a formidable appearance, but is rather translucent on visible imagery, making the storm look smaller and less potent.

Visible imagery as a rule is not available at night. For the past few decades, the military DMSP satellites have offered nighttime moonlight imagery using sensitive night vision optics. It would seem this would make the technology useful for fog detection, however DMSP imagery is a polar orbiter technology, which introduces delays and significantly increases the complexity on weather websites. As of 2015 only archive products were available.

Due to its high resolution, visible imagery can be used for assessing wind direction. If an air mass is unstable (such as in tropical air masses equatorward of frontal systems), these bands will form cloud streets that are aligned *parallel* to the mean low level wind flow. If the air mass is stable, transverse bands will form instead, in which case the bands will be *perpendicular* to the wind flow through the layer. In many cases it can be difficult to tell whether an unstable or stable regime is present, so forecasters often need to fall back on experience or a detailed surface analysis to resolve what the alignment of cloud bands means.

Benign or subtle low-level boundaries, such as cold fronts and thunderstorm outflow boundaries, are extremely important features seen on visible imagery. Assuming there are no high cloud layers obscuring the region, these boundary cloud formations can be extremely important indicators in the hours just before storm development.

key details

Visible imagery is extremely valuable for finding small-scale cumuliform boundaries, for assessing the character of broken cloud layers for stability, for identifying areas of developing convection, and for locating features nearly invisible on infrared imagery such as fog and stratus. It has limited use for convective forecasting at resolutions coarser than 1 to 2 km due to the need to resolve the fine scale structure of cumulus fields.

Forecasters should strive to get the highest resolution possible, except when assessing synoptic-scale features. High-resolution detail is essential for accurate mesoscale forecasting, and improves the overall forecast process.

Geostationary visible imagery has improved to 0.5 km in North America, thanks to the GOES-R series upgrade. From the 1970s to 2017, 1-km imagery was the standard.

Most polar orbiter visible imagery has a resolution of 0.5 to 1 km. Since the 1970s hobbyists have used American polar orbiters as one way of circumventing lack of access to METEOSAT and GMS. Images are available immediately but processing may add delays, particularly on web sites that need to georeference the images.

NASA MODIS imagery from the Terra and Aqua satellites has a resolution of 0.25 km, but due to processing delays it may take 1 to 3 hours to get an image.

Visible imagery is sensed in the 0.52 to 0.72 micron range, which is all of the visible spectrum except blue. The eye can see everything between 0.4 and 0.7 microns.

real-time Internet sources

star.nesdis.noaa.gov/goes
weather.cod.edu/analysis
weather.ndc.nasa.gov/GOES
weather.ral.ucar.edu/satellite
www.ssec.wisc.edu/data

Above: GOES-18 0.5-km visible image. Supercells develop in the Midwest region, some of which would go on to produce significant tornadoes. This is actually a multispectral "true color" image which is a combination of red, blue, and veggie channels, but serves as an effective visible image. *(GOES via College of DuPage)*

Below: GOES-18 visible image showing a duststorm and severe weather in Texas on 13 March 2021. *(GOES image via College of DuPage)*

Infrared imagery

Infrared imagery looks a lot like visible imagery, but it is actually temperature we are seeing. Also there's a significant difference: infrared imagery is available 24 hours a day, whereas visible imagery goes away at night.

Dark areas correspond to areas of high thermal radiation, while white areas indicate areas of low thermal radiation. Therefore anything dark is warm, and anything white is cold. In fact, since infrared imagery helps to measure emitted radiation, it is possible to use a scale to find the temperature of any pixel in an image! For decades, weather offices in citrus farming districts have even looked for specific color ranges matching freezing temperatures.

Infrared imagery is often enhanced, which refers to the technique of adding false color banding or artificial color gradients to allow certain temperature ranges and cloud patterns to stand out. The most basic enhancement was known by decades as the ZA curve, which simply increases contrast . A more sophisticated enhancement is the MB curve, used by the National Weather Service since the 1970s to highlight temperatures below minus 32°C in various shades of gray and extract detail from cold, amorphous stratiform tops. Various Internet sites often implement their own enhancement schemes.

Infrared satellite imagery from the new GOES-R satellite series is 2 km in resolution. Since the visible satellite channel uses 0.5 km resolution, this means that there is only one infrared pixel for every 16 visible pixels. Therefore it is better to use visible imagery when practical.

Clouds may be invisible on infrared imagery when they take on the same temperature as the ground. This is true of fog and stratus; the full extent of fog is often not known until the morning hours when visible imagery is available. This sort of masking also affects mid-cloud layers in strong polar air mass outbreaks, and cirrus in arctic air masses and polar air source regions.

Infrared radiation encompasses a very wide range of wavelengths. While most infrared imagery we are familiar with is sampled at about 10 to 11 microns, another type of imagery sampled at 3 to 4 microns is known as shortwave infrared. Fog can be distinguished more easily with shortwave infrared images. It is especially sensitive to hot spots caused by wildfires. In between the short and long wavelengths we find the water vapor channel at 6 to 8 microns. This corresponds to the wavelengths at which water vapor attenuates ingoing and outgoing radiation. This is covered in the next section.

key details

Infrared imagery measures infrared radiation emissions from clouds, landmasses, and oceans. This is strongly correlated with temperature.

The next generation of geostationary satellites (GOES-R) entered service in 2017 and has been providing infrared and water vapor imagery at 2-km resolution. Previous GOES satellite resolution from the 1970s to 2017 was 4 km.

Infrared imagery is unable to monitor low stratus and fog consistently, because of the similarity in temperature with the underlying surface. For this reason cloud layers in arctic air masses will also appear dull gray. Visible and nighttime multispectral products are more appropriate for both of these patterns.

In large arctic air masses and in polar latitudes, cirrus clouds, which are very cold, are often camouflaged against the cold landmass. masked by the similar temperatures. Different ABI bands or multispectral techniques may be needed to resolve these cloud forms.

Near infrared imagery (near IR) is a special type of infrared imagery that uses the 3.9 micron band. It shares a lot in common with visible imagery and is used to differentiate low-level features that are often masked, such as snow, fog, and stratus.

Infrared imagery should be used to find overshooting tops in both thunderstorms and tropical cyclones.

24-hour satellite imagery including the loops often shown on television of hurricanes are almost always constructed with an infrared channel. Visible channels are not available during the day.

real-time Internet sources

star.nesdis.noaa.gov/goes
weather.cod.edu/analysis
weather.ndc.nasa.gov/GOES
weather.ral.ucar.edu/satellite
www.ssec.wisc.edu/data

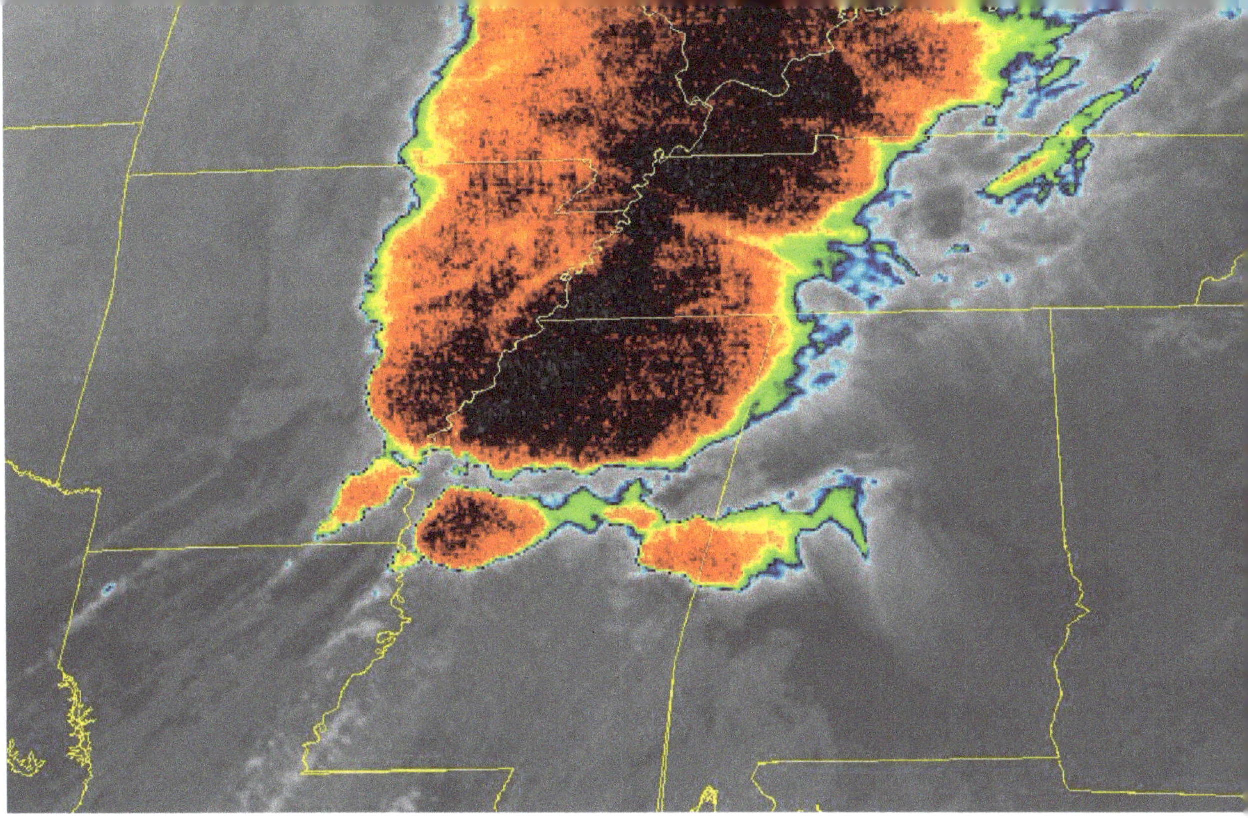

Above: Infrared imagery at the exact time of the 24 March 2023 Rolling Fork, MS tornado. *(NOAA/NESDIS)*

Below: Infrared imagery for Hurricane Irma as it passed Haiti on 8 September 2017. Hurricane forecasters often remap the infrared data to a Dvorak enhancement curve and this is used as a basis for estimating hurricane strength. Some of the key indicators normally found on the Dvorak curve for a strong hurricane can be seen here, such as the symmetric, almost circular eyewall and a distinct eye at the center. *(NOAA)*

Water vapor imagery

The radiosonde observation network only detects small pieces of the moisture field across a given continent. Because of this, there is tremendous interest in using satellites to map out the extent of water vapor. The assumption is that moisture is associated with synoptic-scale ascent, while dry areas are associated with synoptic-scale subsidence.

Water vapor, particularly in the mid troposphere, tends to absorb radiation in the 6 to 7.5 micron range. The new GOES-R series ABI provides three channels in this range. The satellite sensor is unable to find water vapor directly, instead it measures attenuation of infrared radiation in different parts of the scene. This allows "water vapor imagery" to be produced. The scale is set up so that dry air is indicated by high (warm) values of IR radiation and cold air is suggested by low (cold) values.

Where a bright area (usually white or blue) is shown, it suggests the presence of moisture. This is actually a weak radiation signature, suggesting that enough moisture was present to absorb radiation at this wavelength from the surface. Thus we can conclude that moisture exists in the mid-levels of the atmosphere. However this weak radiation signature can also come from a cold region, which does not emit much radiation at all to begin with.

A dark area (dark gray, orange, or red) indicates that large amounts of radiation arrived at the satellite without being attenuated. Thus we can conclude two things: these areas are warm, and very little moisture absorbed the outgoing radiation. Thus the atmosphere is likely to be dry, at least in part of the column.

Here are some limitations with water vapor imagery:

WV imagery works best in the middle troposphere, mostly between 350 and 650 mb (12,000 and 25,000 ft). Therefore, dark areas (associated with dryness) may be indicated even if low-level moisture is present.

WV imagery works best in a warm atmosphere. Brightness, the result of a lack of radiation, will also be produced in areas where it is so cold that no radiation can be emitted into space. This makes the imagery least useful in northern latitudes during the winter, where everything appears white in the 6.2 to 7.3-micron band.

WV imagery is degraded by the presence of clouds. Brightness can indicate a thin layer of water droplets or ice (clouds), rather than a rich layer of water vapor.

Cold regions are biased towards bright values, due to the abundance of cold air with weak infrared signatures.

key details

Water vapor imagery is used for assessing mid-level and upper-level moisture in warm tropospheric environments. It was introduced starting with Meteosat-1 in 1977 and GOES-4 in 1980.

GOES water vapor imagery is sensed at a wavelength of 6 to 7.5 microns and has a resolution of 2 km. Some polar orbiter systems do not produce water vapor imagery.

Clouds do not need to be present to show up on water vapor imagery. All that is needed is water vapor (a gas) in the middle troposphere to affect the infrared signature. Therefore moist bands can show up well before clouds begin forming on infrared or visible imagery.

A dark pixel indicates that a strong radiation emission successfully propagated from the ground to the satellite without being interfered with by mid- or upper-level moisture.

A bright pixel indicates low levels of radiation were received. The cause of this can either be poor ground radiation (cold temperatures) or absorption (strong radiation being absorbed by moisture in the mid- or upper troposphere).

Water vapor imagery is excellent for picking out the position of the subtropical jet (STJ). The STJ usually lies along the poleward periphery of a broad bulge of tropical moisture in the subtropical latitudes.

Very dark bands on water vapor imagery in the wake of a baroclinic storm system may lie in proximity to a very strong jet max. This can help refine the jet max position. The location is usually on the periphery of the dark band (poleward) and the brighter area (equatorward).

real-time Internet sources

star.nesdis.noaa.gov/goes
weather.cod.edu/analysis
weather.ndc.nasa.gov/GOES
weather.ral.ucar.edu/satellite
www.ssec.wisc.edu/data

Above: Water vapor imagery on 2 April 2023 at 0341 UTC, showing clear evidence of the polar front jet: roughly between the yellow "dry band" over the Carolinas and the moist blue, white, and green bands in the Atlantic, and stretching from west-southwest to east-northeast. Water vapor loops are essential tools for tracking quasi-geostrophic disturbances and monitoring the influx of atmospheric rivers into a region. *(NOAA/ NESDIS)*

Below: ABI Band 10 (low level water vapor) on 5 April 2023 at 1621 UTC, showing a large MCS from the Great Lakes to Texas. (NOAA/NESDIS)

GOES ABI bands

Although the traditional "visible", "infrared", and "water vapor" products suffice for most general forecasting, forecasters should be familar with all the channels offered by the GOES-R Advanced Baseline Imager (ABI). Titles given are not official channel names, but they are descriptive titles that have entered common use.

- **Band 1** (Blue) (0.47 nm). One of the visible channels but is more sensitive to scattering from smoke, dust, and aerosols, providing better contrast if a suitable multispectral product is not available.
- **Band 2** (Red) (0.64 nm). The traditional visible channel and is provided at the highest resolution (0.5 km).
- **Band 3** (Veggie) (0.86 nm). Vegetation reflects strongly in near-infrared (NIR) and appears bright. Useful for detecting flooded regions.
- **Band 4** (Cirrus) (1.37 nm). NIR channel sensitive to cirrus, including very thin layers.
- **Band 5** (Snow/Ice) (1.61 nm). Ice reflects very poorly at this wavelength, so it appears darker. Fires emit strongly on this channel and will be seen at night.
- **Band 6** (Cloud Particle Size) (2.24 nm). Large cloud particles appear darker on this channel.
- **Band 7** (Shortwave IR) (3.9 nm). Useful for locating fog and low clouds at night, as well as fires.
- **Band 8** (Upper WV) (6.2 nm). Sensitive to upper-level water vapor, primarily at 20-30,000 ft MSL.
- **Band 9** (Mid WV) (6.9 nm). Sensitive to mid-tropospheric water vapor, especially near 15-25,000 ft MSL.
- **Band 10** (Lower WV) (7.3 nm). Sensitive to lower mid-tropospheric moisture especially at 8-18,000 ft MSL.
- **Band 11** (IR Cloud Phase) (8.5 nm). Used primarily for monitoring volcano activity.
- **Band 12** (Ozone Band) (9.6 nm). Sensitive to ozone absorption and is used in atmospheric chemistry.
- **Band 13** (Clean Longwave IR) (10.3 nm). Classic infrared band (old GOES technology often used 10.7 nm) that is least sensitive to moisture absorption.
- **Band 14** (Longwave IR) (11.2 nm). Longwave IR band more influenced by moisture absorption. It is also a somewhat "dirty" channel similar to Band 15.
- **Band 15** (Dirty IR) (12.3 nm). Extensive moisture produces "colder" images. Band 13 is usually preferred for general analysis.
- **Band 16** (CO_2) (13.3 nm). The longest wavelength IR band is used mostly for automated analysis of cloud top height, temperature profiles, and ash detection.

key details

The GOES-R series of satellites starting with GOES-16 entered service in 2017, providing 16 bands of imagery. Previous satellites were limited to 5 bands or less. The new GOES-R imagery is referred to as the Advanced Baseline Imager (ABI).

Resolution is 1 km for Band 2 (Red), 1 km for bands 1, 3, and 4, and 2 km for all other bands. This is "ideal resolution" measured on the equator under the satellite; actual resolution will be coarser elsewhere due to slant.

Color enhancement is often applied to ABI bands to provide better resolution of low-contrast areas. Enhancement schemes have been standardized as far back as the 1970s, when they were often referred to as "enhancement curves". Due to great flexibility in assigning color palettes on end-user computers, enhancement curves are no longer standardized.

Bands 1 through 6 are only available during the daytime. Fires however may appear at night on bands 4 through 6. Bands 7 through 16 are available continuously.

The pre-2017 GOES 5-channel imagers used the following bands:

nm	Purpose
0.65	Visible (VIS)
3.9	Shortwave infrared
6.7	Water vapor (WV)
10.7	Standard infrared (long wave)
12	Split window IR component

ABI equivlaent channels are:
VIS   Band 2 (Red)
IR   Band 13 (Clean Longwave IR)
WV   Band 9 (Mid WV)

real-time Internet sources

star.nesdis.noaa.gov/goes
weather.cod.edu/analysis
weather.ndc.nasa.gov/GOES
weather.ral.ucar.edu/satellite
www.ssec.wisc.edu/data

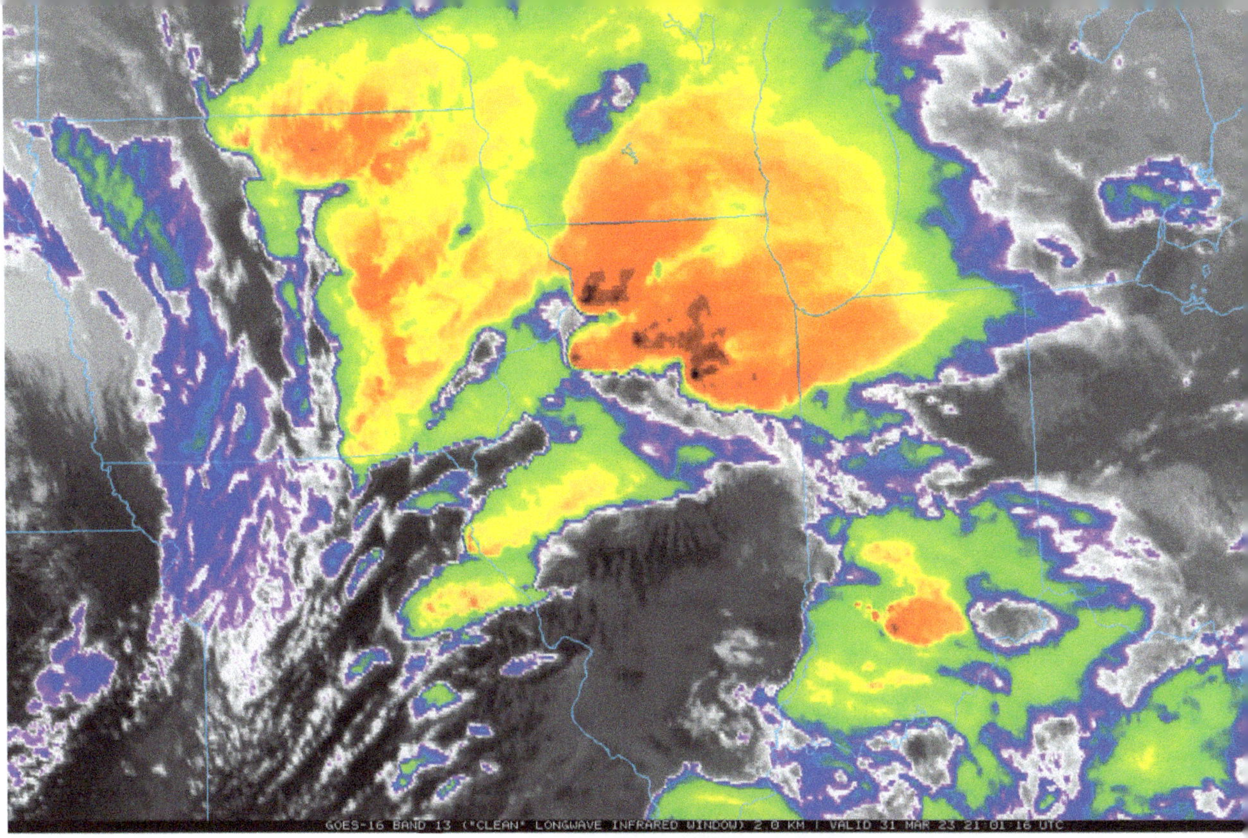

Above: ABI band 13, "Clean LW IR", during the 31 March 2023 tornado outbreak in Iowa, Illinois, and Missouri. Color enhancement schemes like this one are very common and are simply a replacement of the specific shades of white and gray with more colorful bands to provide enhancement. The very small updraft towers appear as black dots. The original ABI products are grayscale images.

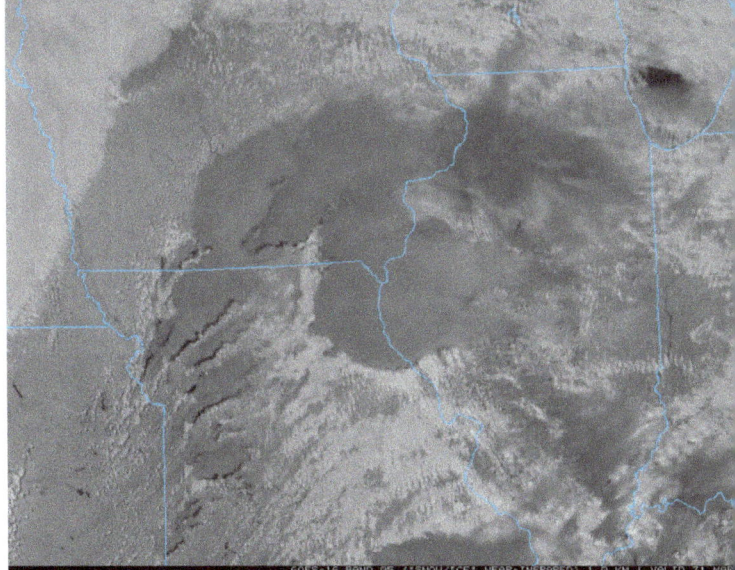

Right, above: ABI band 5, the "Snow/Ice" band, for 31 March 2023 as storms develop in southern Iowa and Missouri. The storm anvils show up as dark, smoke-like plumes due to strong absorption by ice at this wavelength.

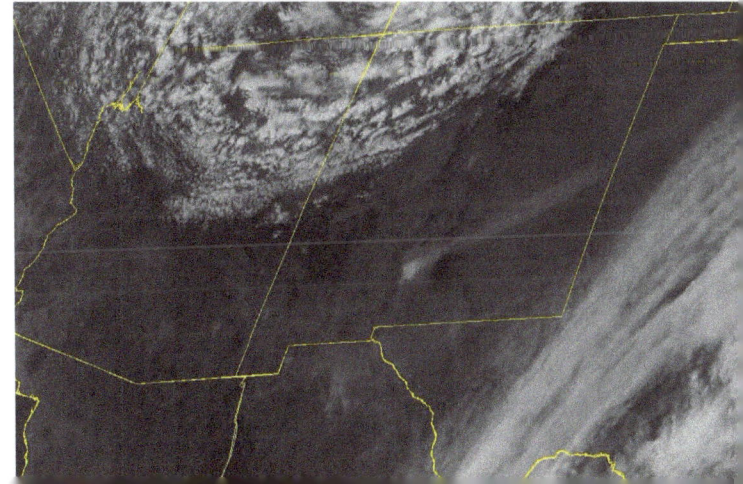

Right, below: ABI band 1, "Blue", provided the best contrast for this dust plume extending from White Sands, New Mexico at center on 4 April 2023. The multispectral dust product was somewhat contminated during this event. *(NOAA/NESDIS)*

Multispectral imagery

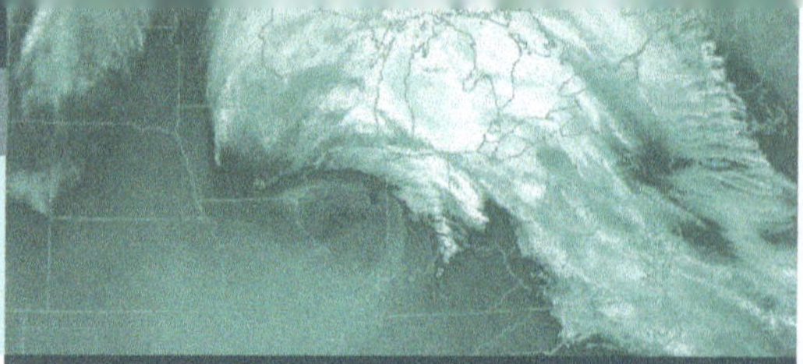

The ABI bands provide scalar fields that are generally displayed as simple grayscale images. However by combining information from multiple bands, it's possible to create enhanced graphics that detect specific types of weather phenomena and cloud characteristics. This is known as multispectral processing. These bands can also be split into red, blue, and green components (RGB) to provide color imagery, which provides richer detail on the resulting image.

Multispectral imaging is not specific to GOES-R technique and was developed during the earliest days of forecasting. By the 1990s it was in limited use at forecast offices. With the expanding capabilities of GOES-R, the use of these images have been growing rapidly.

There are not "official" sets of multispectral products. Most of these are generated locally by data centers, websites, and university data providers. However several well known multispectral algorithms have entered widespread use. Shown here are some of the ones that are most commonly encountered (bands used are in brackets):

- **True color.** Provides a pseudo-color image as it would be seen in the visible spectrum. Since there is no sensor for green, data from the Veggie channel is substituted. [ABI Bands 1, 2, 3]

- **Natural color.** Constructs a pseudo-color image using part of the NIR spectrum. Snow and ice take on bluish tints, while vegetation takes on an intense green color. A variation of these scheme uses   [ABI Bands 1, 3, 5]

- **Nighttime (NT) Microphysics.** Provides excellent indications of visible clouds and fog at nighttime. A suitable a de-facto replacement for visible imagery. Offered at the College of DuPage weather site. A "fog" band product also exists using only Band 7 and 13. [ABI Bands 7, 13, 15]

- **Air Mass.** Channel differences are mapped into red, green, and blue colors, providing some differentiation of air masses. Tropical air appears greenish, while continental air appears reddish. [ABI bands 8, 10, 12, 13]

- **Day Cloud Phase.** This combination tends to differentiate liquid water into bluish or positive tints and ice clouds into reddish or negative tints. Due to the use of visible bands it can only be used during the day. [ABI bands 2, 5, 13]

- **Sandwich.** A hybrid visible image with an infrared overlay of the coldest clouds, generally those at the cirriform levels. [ABI bands 2, 13]

key details

Multispectral imagery combines data from multiple bands (wavelengths) using addition, subtraction, and other types of processing. There are usually specific algorithms for the red, green, and blue output channels, which provides an RGB product.

Multispectral imagery differs from color enhancement, which is simple color replacement on a single ABI channel. In multispectral imagery, color shades originate from the algorithm rather than directly from the raw image.

During the nighttime hours the NT Microphysics or "Fog" products offer some of the most useful detail. These products are not suitable for dense overcast or precipitation, these product will provide little useful detail, and other infrared products are more appropriate such as ABI Channel 13.

Multispectral fog detection is the same as low stratus detection. To the satellite, there is no perceptible difference. To discriminate between the two, the forecaster must evaluate surface observations and check realtime cameras.

The Air Mass product is biased toward red by dry upper-level conditions, toward green by a high tropopause and ozone, and toward blue by moist upper levels. Purple colors, which are a mix of red and blue, are associated with cold temperatures and low slant angle near the edge of the Earth's disc.

Detailed information on many more multispectral techniques may be found at: rammb.cira.colostate.edu/training/visit/quick_guides

real-time Internet sources

star.nesdis.noaa.gov/goes
weather.cod.edu/analysis
weather.ndc.nasa.gov/GOES
weather.ral.ucar.edu/satellite
www.ssec.wisc.edu/data

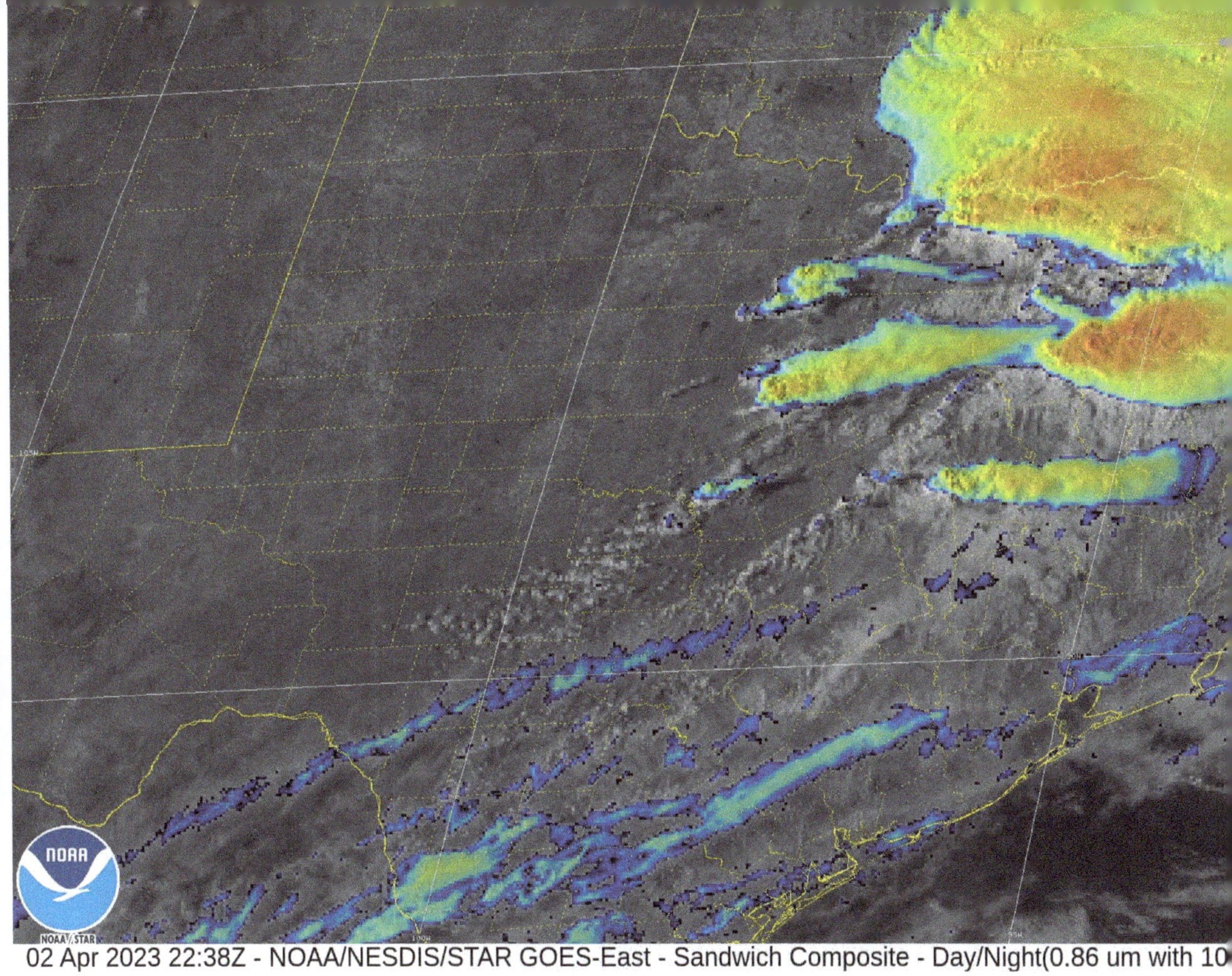
NOAA
NOAA/STAR
02 Apr 2023 22:38Z - NOAA/NESDIS/STAR GOES-East - Sandwich Composite - Day/Night(0.86 um with 10

Above: Sandwich multispectral prduct during a severe outbreak on 2 April 2023. This shows excellent structure in the thunderstorm anvil tops. *(NOAA/NESDIS)*

Right: NT Microphysics imagery of Northwest Mexico and south Texas during the pre-dawn hours of 2 April 2023. This shows extensive stratus and fog as bluish-white patches. Altocumulus layers appear as reddish-yellow, while cirrus appears dark red and black. *(College of DuPage)*

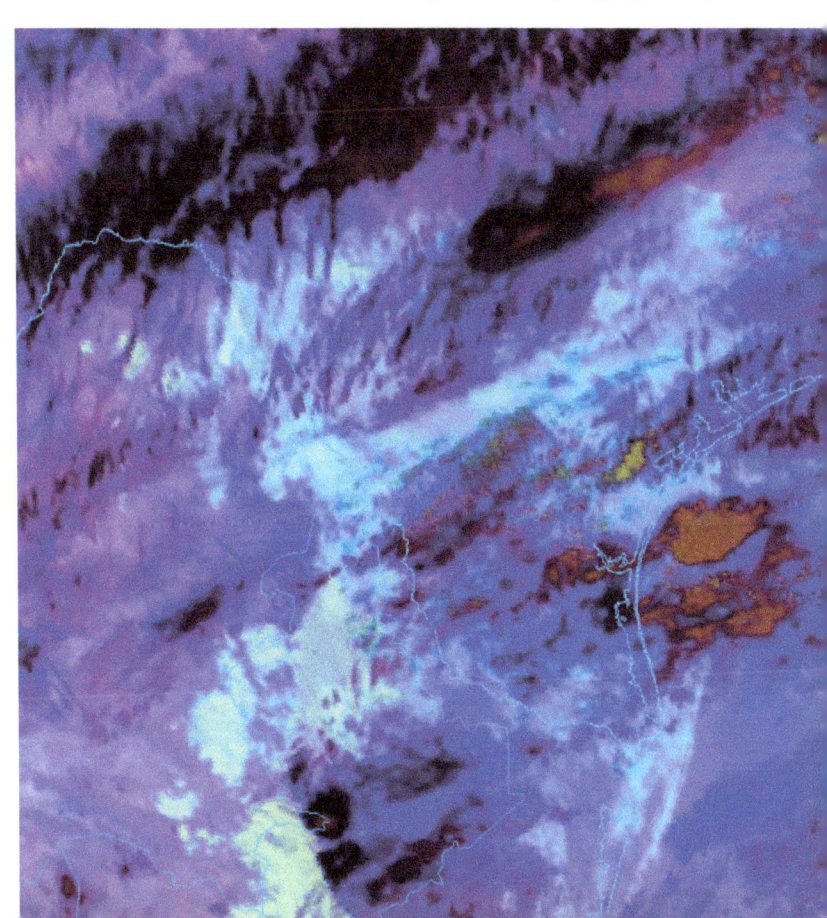

Full-disc image from the Feng Yun FY-2A geostationary satellite manufactured, launched, and operated by China. This image shows Asia on 23 December 1997 in a combination of visible and infrared. Until 1997 nearly all geostationary imagery of Asia was obtained from the Japanese GMS satellite. *(CMA)*

CHAPTER 3
RADAR IMAGERY

Radar

The entire radar section of this book is dedicated to the United States WSR-88D radar network. Though this admittedly does not help most international readers, the underlying principles are similar and can be used to understand the slowly emerging products from radar programs offering public data, such as those in Australia, Canada, Germany, Mexico and South Africa.

History

The roots of the Next Generation Radar (NEXRAD) Program go back to 1977. At this time, a network of WSR-74C and WSR-57 radars had been in operation for about 20 years. Research was beginning to prove the value of velocity data in storm detection. Studies began to see if the radars could be upgraded. The Joint Doppler Operations Project was established in 1979 at the National Severe Storms Laboratory by the National Weather Service (NWS), the Federal Aviation Administration, and the U.S. Air Force. The task would be to oversee the development of a next-generation radar.

During much of the 1980s, engineering was performed by Paramax, a division of Unisys, with algorithm development largely the responsibility of the National Severe Storms Laboratory. Radar units were delivered between 1990 to 1998, gradually converting the entire national radar network. In 1996 the last of the old-guard WSR-57 radars was retired in Charleston, South Carolina, marking the end of an era.

Engineering design

The WSR-88D consists of two major subsystems: the RDA (Radar Data Acquisition) and the RPG (Radar Products Generator). Older equipment included the PUP (Principal User Processor), which was the weather office graphical workstation, and the UCP (Unit Control Position), an interface to the RDA kept at the weather office. These latter two systems were integrated into AWIPS as part of the transition to open architecture in the 2000s.

The RDA (Radar Data Acquisition) unit consists of the antenna, transmitter, receiver, and signal processor. This produces a raw Level II data stream consisting of reflectivity, velocity, spectrum width, and dual-polarization data. The RPG (Radar Products Generator) converts the raw data stream into Level III data, which provides numerous enhanced data products.

The power output of the WSR-88D radar is 750,000 watts. This is comparable to the transmitter of a large television station. Its klystron tube delivers a frequency of 2700 to 3000 MHz (10 to 11.1 cm; in the S-band) using a 28-foot dish that produces a beam 0.95° in width. A pulse length of 1.57 microseconds (1545 ft) is possible with the radar system.

The WSR-88D radome is 39 feet in diameter and made of rigid fiberglass. It is meant to protect the antenna from wind, lightning, hail, and problems like ice and corrosion. Since the dome only causes a 0.6 dB signal loss it is almost transparent to the radar unit. The radome is placed on a tower anywhere from 20 to 98 ft in height, depending on the terrain and obstructions.

Meteorological design

The WSR-88D is designed to operate in two radically different modes: *clear air* and *precipitation*. It can only operate in one of these modes at any given time. The main difference between the two is that clear air mode offers the advantage of greater sensitivity due to a slower antenna rotation rate, which allows more energy to be returned back to the radar. This comes at the cost of poor temporal resolution, with products generated half as often and sampling too slow to keep track of precipitation and especially storms.

Above: WSR-88D reference map for the conterminous United States, with radar station identifiers. *(NOAA)*

The radar operating modes are further subdivided into a number of *volume coverage pattern* schemes, or VCPs. For a number of years there were only four VCPs: 11, 21, 31, and 32. VCP 11 was for convective precipitation, 21 for stratiform precipitation, 31 for long-pulse clear air in weak winds, and 32 for short-pulse clear air in high winds. Starting in 2003 additional VCPs were introduced, all of which offer more choices for elevation, pulse length, and so forth but are based on the four basic VCPs. For detailed information, it's strongly recommended to look through the Warning Decision Training Branch modules at <*training.weather.gov/wdtd/courses/rac/outline.php*>

The WSR-88D system is built on a set of *base products*: base reflectivity, base velocity, and spectrum width. The new polarimetric upgrade fielded in 2012 adds differential reflectivity, differential phase, specific differential phase, and correlation coefficient. All of these are direct measurements produced by the RDA.

The RPG works with all these base products to create additional datasets, such as vertically integrated liquid (VIL), enhanced echo tops (EET), and tornado vortex signature (TVS). Each of these draws from the base data to produce the desired output. Composite reflectivity, for example, is built from Level II base reflectivity at different tilts. Tornado vortex signature data is based on an the tornado detection algorithm (TDA) which examines Level II reflectivity and velocity measurements.

RADAR VIEWER WEBSITES, APPS, AND SOFTWARE OF GENERAL INTEREST

WEBSITE-BASED DISPLAYS
College of DuPage - weather.cod.edu/satrad/
MRMS - mrms.nssl.noaa.gov/qvs/product_viewer
CIMMS viewers - cimss.ssec.wisc.edu/severe_conv
NCEI - www.ncei.noaa.gov/maps/radar
WeatherTap RadarLab Local ($) - www.weathertap.com

DESKTOP SOFTWARE
GRLevelX
RadarOmega
Digital Atmosphere (Level III only)

MOBILE APPS
RadarOmega   RadarNow!
RadarScope   MyRadar

Base reflectivity

Standard base reflectivity (Z, or Z_{HH}) is energy reflected from water droplets and ice particles back to the radar receiver. The stronger the reflection, also known as *power*, the more "intense" the echo. The WSR-88D radar specializes in detecting only precipitation, not clouds and fog, which are primarily composed of microscopic droplets that are too small for the radar to see, except when droplets grow or ice crystals form. Base reflectivity can be contaminated by ground clutter and anomalous propagation. Insects and birds can produce large patterns of faint reflectivity around the radar site.

The finest-grained WSR-88D reflectivity image has a resolution of 0.5 deg x 0.25 km, but some products are provided at a coarser resolution. Images before the 2008 Super Resolution upgrade were provided at 1.0 deg x 1 km, thus, the pre-2008 years of NEXRAD tend to show coarse data. Intensity is described in decibels of power reflected, or dBZ. The scale is logarithmic, so an increase of 3 dBZ is a doubling of power returned.

The technique of analyzing strong thunderstorms is a science in itself. Strong winds can shape the precipitation core into unique patterns, such as the hook echo, which is indicative of a tornado. A strong thunderstorm core which shifts more toward the edge of the cell, rather than remaining centered, is indicative of a severe thunderstorm. The biggest threat of severe weather is at the strong reflectivity gradient.

A feature called a "bright band" is often observed in stratiform winter precipitation situations. This is a ring that appears on the base reflectivity product, centered on the radar site. It occurs when the radar beam intersects a layer of snow melting into rain as it falls, which creates enhanced radar reflectivity. As the range and height to this feature is similar across the region, it appears at a constant range from the radar, and thus appears as a ring or a partial ring. The height of the feature can be easily estimated.

It must be remembered that the sweep of the radar beam is conical in shape; it is near the ground at the radar site and increases to higher heights at increasing distance from the radar site. Therefore echoes far from the radar are being sampled at a much higher elevation — as high as 15,000 ft at 100 miles from the radar. This makes it impossible to directly sample low-altitude features such as hook echoes at such ranges. Also a much larger volume is being sampled, so key storm features may be smeared out at these distant ranges.

key details

The base reflectivity product has a strong correlation to rainfall intensity, and especially to the presence of hailstones.

Base reflectivity polarization is both transmitted and received in the horizontal plane. The WSR-88D actually transmits in both planes at once (at a 45-degree angle) and measures the vertical and horizontal components. By using characteristics of both channels, we get polarimetric information (covered in other sections).

Most "local radar" views seen on television are base reflectivity. However anything covering an entire state or region will usually be a composite reflectivity image.

In clear air mode, reflectivity shows echoes spanning -28 to 28 dBZ. In precipitation mode, echoes range from 5 to 70 dBZ. Web sites typically assign different color sets to each mode so that it is immediately obvious which color set and radar mode is in effect.

Always be alert for suspicious echoes, such as chaff, birds, and solar spikes. They are surprisingly common.

Radar intensity relationships:
10 dBZ - Very light rain / light snow
20 dBZ - Light rain / heavy snow
30 dBZ - Moderate rain
40 dBZ - Heavy rain / some thunder
50 dBZ - Heavy rain / thunderstorm likely
60 dBZ - Thunder; extreme rain and/or hail
70 dBZ - Rare; usually indicates large hail.

Product code: 94/DR (RPG), N0Q (WMO).

real-time Internet sources

weather.cod.edu/satrad
www.weathertap.com ($)
weather.gc.ca

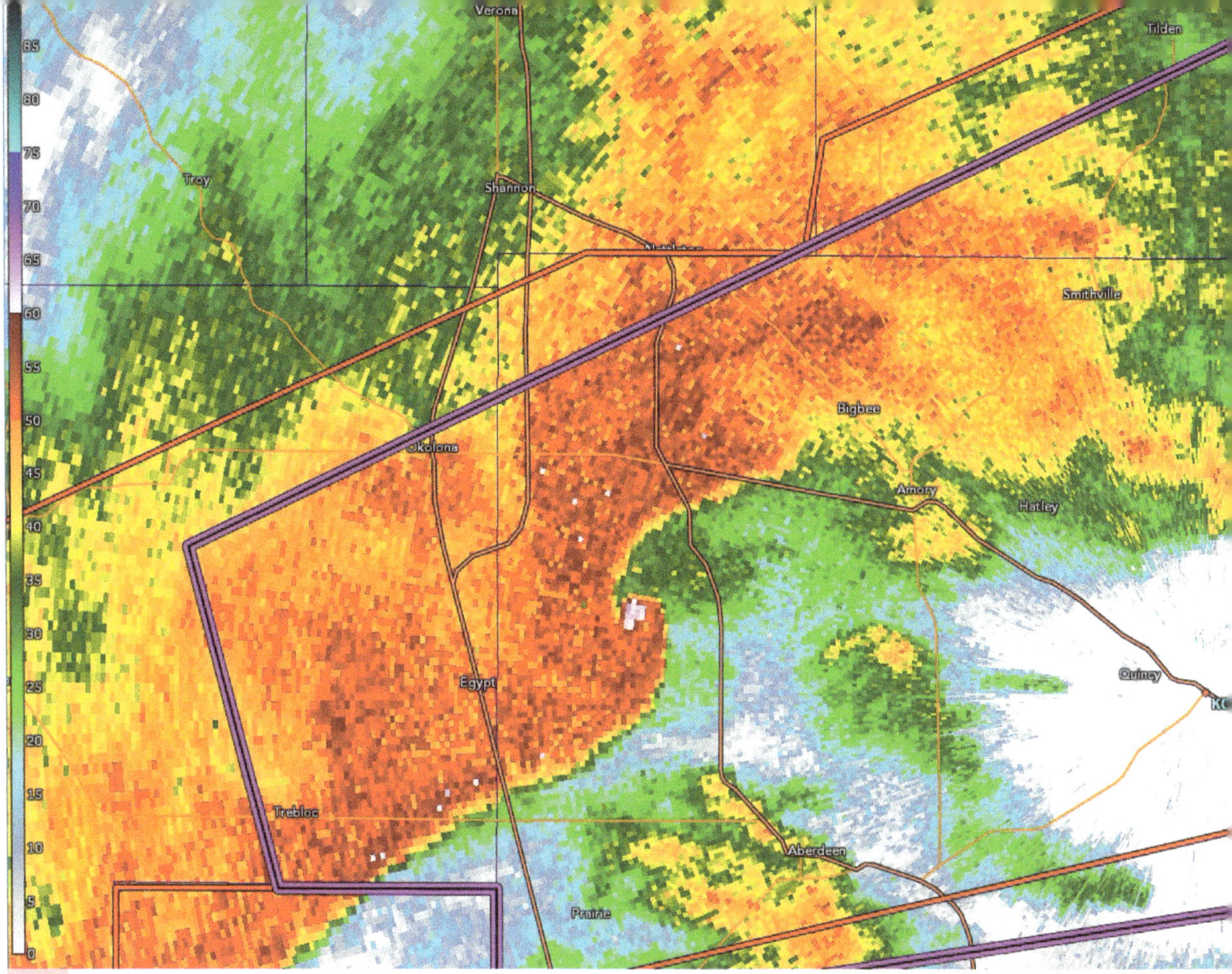

Above: Base reflectivity at 10:45 pm CDT on March 24, 2023 showing a destructive nighttime tornado approaching Amory, Mississippi. At least two fatalities occurred in the town. This was sampled at a range of 10 miles from the radar (lower right), allowing excellent detail to be captured. The tornado shows a "debris ball" of high intensities, caused by metal debris entering the circulation and increasing backscattering of radar power. *(GRLevel2)*

Right: Naperville, Illinois tornado of 20 June 2021 at 11:06 pm CDT (circled). This was an unusual presentation for such a storm, in which development of a significant tornado (EF3) occurred along an MCS. It's a reminder that forecasters must always be on guard. The large inflow notch ahead of it formed the classic "line echo wave pattern" (LEWP) producing more of a broken than solid cell structure which decreased cell competition. *(GRLevel2)*

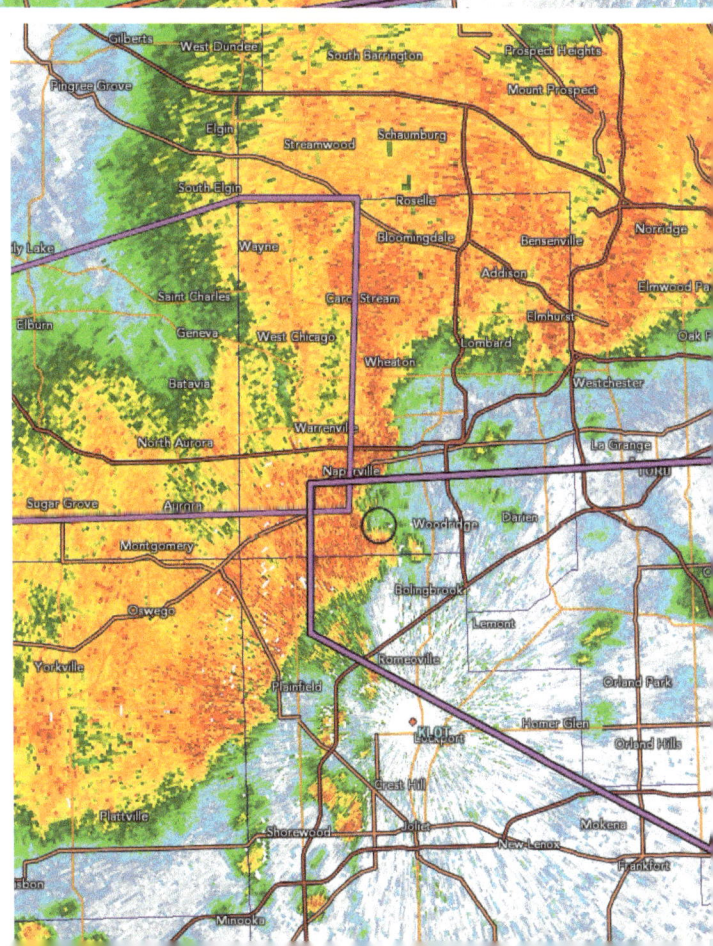

Composite reflectivity

Composite reflectivity (CR) works exactly the same as base reflectivity (see previous pages) However composite reflectivity examines not just one scan elevation but all of them. From this it displays the maximum reflectivity found in any of the "bins" vertically above a given location. In short, it displays the highest detected reflectivity value at any given location. This makes composite reflectivity suitable for displaying of elevated precipitation, droplets suspended high in an updraft (such as during storm initiation), and even cloud decks.

One of the best uses of composite reflectivity is in detecting the first signs of thunderstorm development. The first detectable echoes within a towering cumulus transitioning to a cumulonimbus cloud will typically occur at a height of 10,000 to 20,000 ft. Within roughly 75 nm of the radar, this is usually *above* the lowest base reflectivity scan. Therefore the first echoes will typically show up on one of the higher base reflectivity scans and also appear on the composite reflectivity product.

Since composite reflectivity merges echo information at numerous levels, important structures seen on one base reflectivity frame can be completely lost. Hook echoes, bow echoes, weak echo regions, rear-inflow notches, bright bands, and other interesting features will be smeared out or lost. Therefore base reflectivity should always be monitored during periods of severe weather.

Also it is not possible to draw conclusions about three-dimensional structure using composite reflectivity. Three-dimensional assumptions can be easily estimated using base reflectivity since the height is obvious to us as a function of scan height and range. For example, an intense signature close to the radar site on the 0.5° scan implies it is close to the ground. However this assumption cannot be made using composite reflectivity. Investigation of other products is necessary.

The composite reflectivity product is limited by the scan height. The radar antenna only rises to only 19.5° above the horizon, so the upper portion of storms close to the radar site are not sampled. This produces a void on various products known as the *cone of silence*.

There is also a cost in terms of time. Base reflectivity products are available immediately after one full sweep of the radar beam. However, since composite reflectivity is a volumetric product, it will not be available until the entire volume scan is complete. While you are looking at a five-minute old composite reflectivity product, an updated reflectivity scan is often already available.

key details

Composite reflectivity is a depiction showing the highest reflectivity detected at a given location, regardless of the elevation slice being used. It uses the entire volumetric scan.

Precipitation areas look bigger than they really are at the surface. This is especially true of large thunderstorms and in dry air masses. Since many radar mosaics (maps of multiple radars) use composite reflectivity, it is important to be aware of whether you are looking at base or composite reflectivity at all times.

Always use composite reflectivity when expecting precipitation, since monitoring just one base reflectivity elevation may cause developing echoes at another elevation level to be missed. Storm echoes develop initially at the 15 to 20 thousand ft AGL height, which is above most of the 0.5 deg elevation.

Do not use composite reflectivity to examine severe storm features and boundaries, since fine detail at one particular elevation will get blurred out by other elevations. Hook echoes are severely degraded in composite reflectivity since the concave area of the hook is usually topped at higher levels by the storm's overhang.

The limitations of composite reflectivity include masking of features by echoes at other levels, inability to determine structure, scan height limitations, and time considerations since this is one of the last products generated in a volume scan. Its best use is to provide a quick overview of conditions throughout the volume scan.

Product code: 38/CR (RPG), NCZ (WMO).

real-time Internet sources

weather.cod.edu/satrad
www.weathertap.com ($)

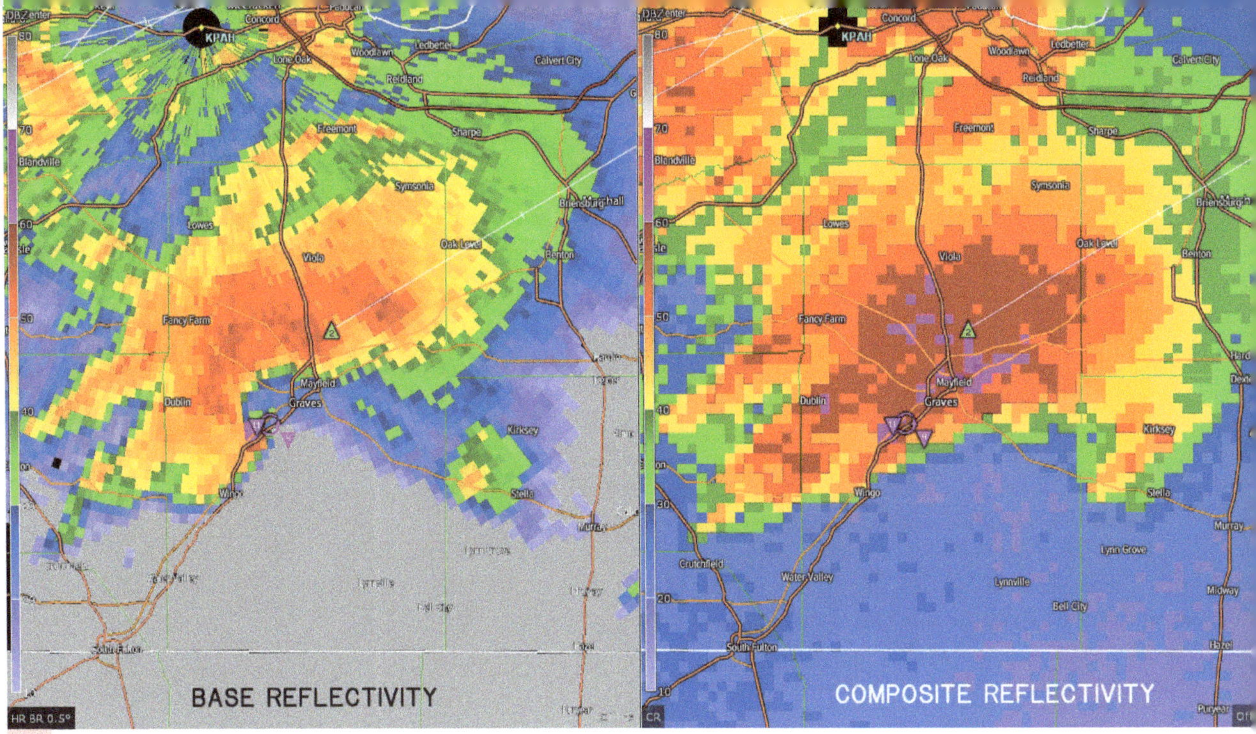

BASE REFLECTIVITY **COMPOSITE REFLECTIVITY**

Above: Comparison of base reflectivity (left) and composite reflectivity (right) for the December 10, 2021 Mayfield, Kentucky supercell. It is immediately obvious that the hook echo does not appear on composite reflectivity. This depiction however does show the enhanced reflectivity due to the presence of very heavy rain and large hail.

Below: Composite reflectivity is useful for monitoring convective initiation. Below we see a comparison of 0.5° base reflectivity (left) and composite reflectivity (right) for 2128 UTC on 31 May 2013. Higher reflectivities only appear on the composite reflectivity chart at this stage since initial precipitation development typically occurs in the middle troposphere. This shows initial development of the storm clusters west of El Reno, one of which led to a well-publicized storm chaser disaster about two hours later. It took about 10 minutes for similar intensities to appear on base reflectivity. Proper use of tools enhances the forecaster's situational awareness by providing additional leadtime. *(Both images GRLevel3)*

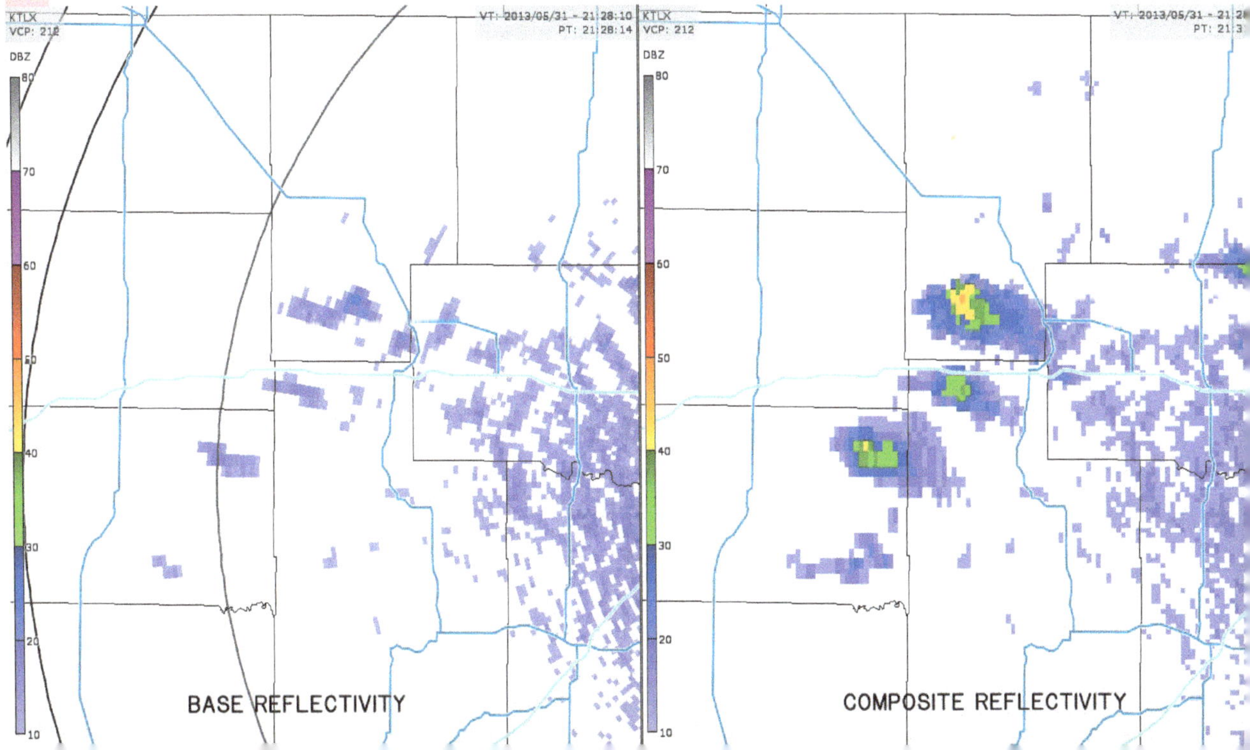

BASE REFLECTIVITY **COMPOSITE REFLECTIVITY**

Velocity

Radial velocity (V_R) describes the radial velocity (along the beam) of scatterers within a given bin. The radar determines this by measuring the Doppler shift of the reflected energy. Since velocity is only available radially, this makes the product challenging to interpret. Tangential and pure two-dimensional motion cannot be measured; only assumed.

The overall shape of the broad-scale negative and positive shading tends to outline the tropospheric wind flow. The radar swath defines a cone within the troposphere, so all heights are included in the scan. For example a spiral appearance to the broad-scale coloring suggests change in wind direction with height. A lack of a spiral appearance suggests unidirectional winds with height.

Small-scale circulations are analyzed by finding a couplet; in other words, a pair containing positive and negative velocity. The exact orientation of this couplet relative to the radar determines whether convergence, divergence, cyclonic rotation, or anticyclonic rotation is present (see illustration). When peak velocities of a couplet touch each other, velocity is expressed in terms of "gate-to-gate shear"; this usually only occurs with the tight rotation or convergent rotation signature of a tornado.

When possible, the Storm Relative Motion (SRM) product should be used to analyze a storm. This attempts to balance velocity data by compensating for the drift of features with the prevailing flow, and can help features stand out much better. The motion is automatically derived from an average of all storm velocities as determined by the Storm Tracking Algorithm.

Range folding is an artifact that occurs when the distance to a storm exceeds the maximum unambiguous range. Beyond this range the echo arrives after the next pulse has been transmitted and a false echo occurs. Echoes that are range-folded are usually shaded gray or purple to show that they have no usable data.

Aliasing is another problem. The velocity of a scatterer may exceed the maximum unambiguous velocity (the Nyquist co-interval): the highest velocity observable by the radar given its current pulse settings. This can mask tornadic couplets by giving strange speed readings in the couplet. Dealiasing schemes do exist. Velocity products on the web are generally not dealiased. Radar viewing programs like GRLevel2 have a dealiasing toggle which should always be used in severe weather situations.

key details

Positive velocity is movement away from the radar. Negative velocity is movement toward the radar. These are key concepts!

There is a de-facto shading scheme in use for velocity. Positive (away) velocity usually gets a warm shade like red, while negative (toward) velocity usually gets a cool shade like green or blue. This rule of thumb can quickly help get one's bearings when the product legend is not available.

The Storm Relative Motion product should not be used by itself to make assumptions about surface winds. The SRM product uses a storm-relative frame of motion, while the Radial Velocity product uses a ground-relative frame of motion. In a downburst situation, the SRM product can help locate mechanisms for high winds, while the radial velocity product suggests the actual winds experienced at the surface (relative to the radar, of course).

If you trace the border formed between the broad-scale negative and positive areas on the velocity scan, it can be used to judge the wind direction throughout the entire troposphere. At any given point along this line, the wind is perpendicular to the radar beam, blowing toward the positive velocity area. By tracing this line from the radar site to the outer range area, you can follow the wind direction from the ground to the middle or upper troposphere. The wind speed is traced in a like manner by going outward from the radar site to the maximum range and finding the maximum wind velocity at each particular range.

Product code: 99/DV (RPG), N0U (WMO).

real-time Internet sources

weather.cod.edu/satrad
www.weathertap.com ($)

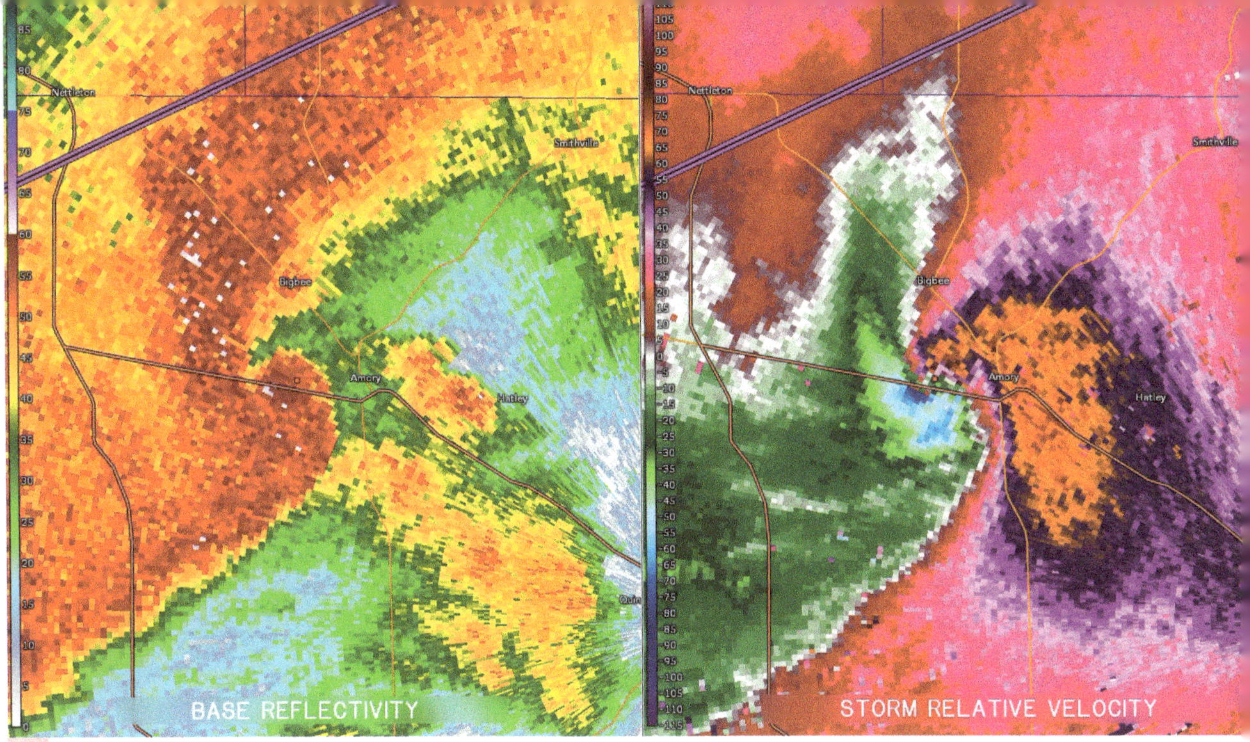

Above: Storm-relative velocity image for the Amory, Mississippi tornado at 10:52 pm CDT on 24 March 2023. Radar is at lower right; reds are outbound and greens/blues are inbound. *(GRLevel2)*

Right: Fundamental couplet models for different types of small-scale circulations. *(Tim Vasquez)*

Below: Hurricane Laura on 26 August 2020 just as the radar (at center of ring on right) was being destroyed. This shows the intense wind field within the hurricane. *(GRLevel2)*

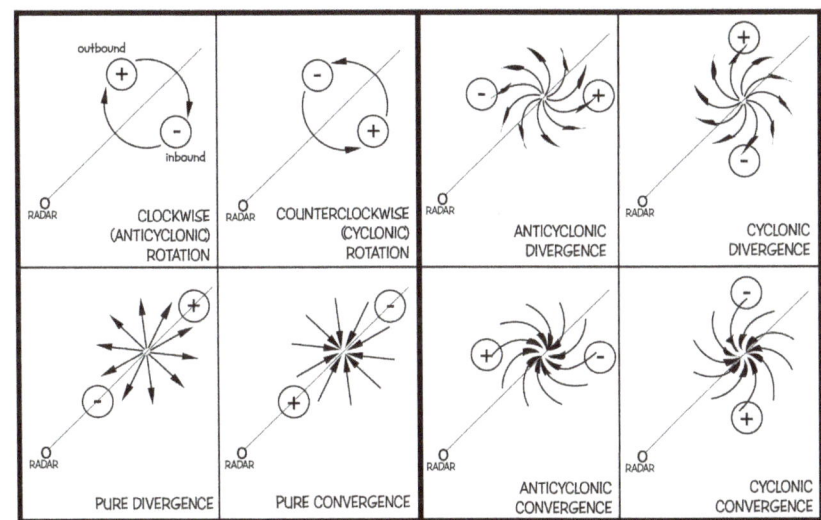

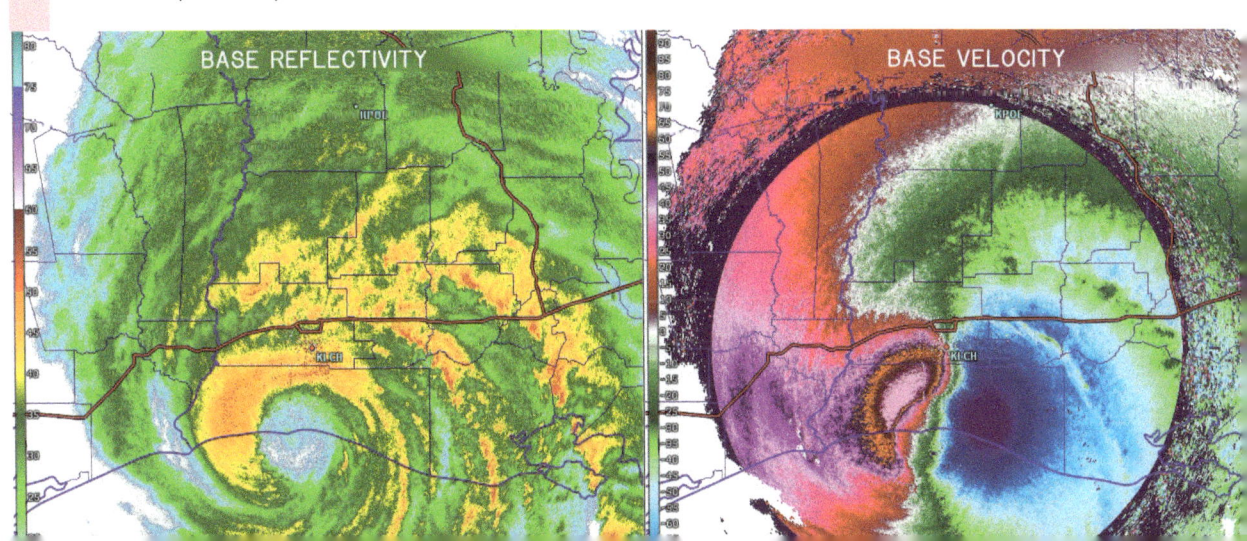

Spectrum width

Spectrum width (σ, or sigma) measures the variance in velocity within a given radar volume. Particles in any given volume are almost never moving cohesively at a single velocity. Turbulence, varying particle sizes, and different forces at work within the volume can all combine to cause varying trajectories and velocities. The velocity estimated by the radar in one specific bin is actually an average of all of these motions.

The variance in the motion corresponds to the spectrum width. This can be measured by examining the "width" of the Doppler shift that is returned to the radar. Low width will cause the reflected energy to peak at one very specific frequency, while high width will cause a broad dispersion of the echo at different frequencies. A spectrum width is expressed as low (narrow) or high (broad).

Unfortunately spectrum width has been largely neglected in the study of operational meteorology. One of the more interesting investigations into spectrum width was a paper by Keith Browning, drawn upon by Leslie Lemon. It suggested that an updraft core normally contains air that ascends uniformly and smoothly, with turbulent flow largely dampened out. This implies that the updraft may be found by locating a core of narrow spectrum width within the cloud.

There is some evidence that spectrum width products can help locate small tornadoes, gustnadoes, and waterspouts. Within a radar bin, a small tornado may not contain enough fast-moving scatterers to trigger a shear signature. However these scatterers would produce a wide range of velocities within that particular bin. The spectrum width product would return a high value. Spectrum width can help locate and measure areas of turbulence.

Another use of spectrum width is to identify suspected three-body scatter spike (TBSS) signatures, which are artifacts that seem to project behind a hail core along the beam radial. The presence of a high spectrum width can help suggest that the artifact is indeed a TBSS artifact.

In more quiescent weather patterns, spectrum width data can be used to help locate fronts and outflow boundaries. Often these features are easily identified, particularly in clear air mode, but at times they may be masked. Spectrum width can help confirm the validity of radial velocity data. A large spectrum width may indicate that the averaged radial velocity for that bin is not reliable. Finally, spectrum width may also help locate initial convective development.

key details

Spectrum width gives information on the variance of velocity within a particular radar bin. It's an indirect measure of turbulence and gustiness as expressed by the motion of rain, ice, and other particles.

Spectrum width tends to increase with range from the radar. This is a natural consequence of beam broadening: as the beam widens, there is bound to be an increasingly large range of particle motions.

Spectrum width values are not affected by VCP changes, though clear-air modes provide better sampling and more accuracy.

Spectrum width is one of the three base products of the WSR-88D, in addition to reflectivity and velocity.

The idea of using spectrum width to identify small vortices was pioneered in 1995 by Joseph Golden and Carin Goodall-Gosnell, and in another 1995 paper by Waylon Collins. Both papers studied waterspout events in the southeastern United States and related them to WSR-88D spectrum width data.

WSR-88D spectrum width data has been the focus of several studies by UCAR and other institutions designed to help reduce the risk of clear air turbulence to aircraft.

Weak reflectivity bordering on the noise threshold for the radar will cause erratic spectrum estimates and noisy spectrum width returns.

Product code: 30/SW (RPG), NSW (WMO).

real-time Internet sources

None are known to exist. Software and apps should be used instead for this product.

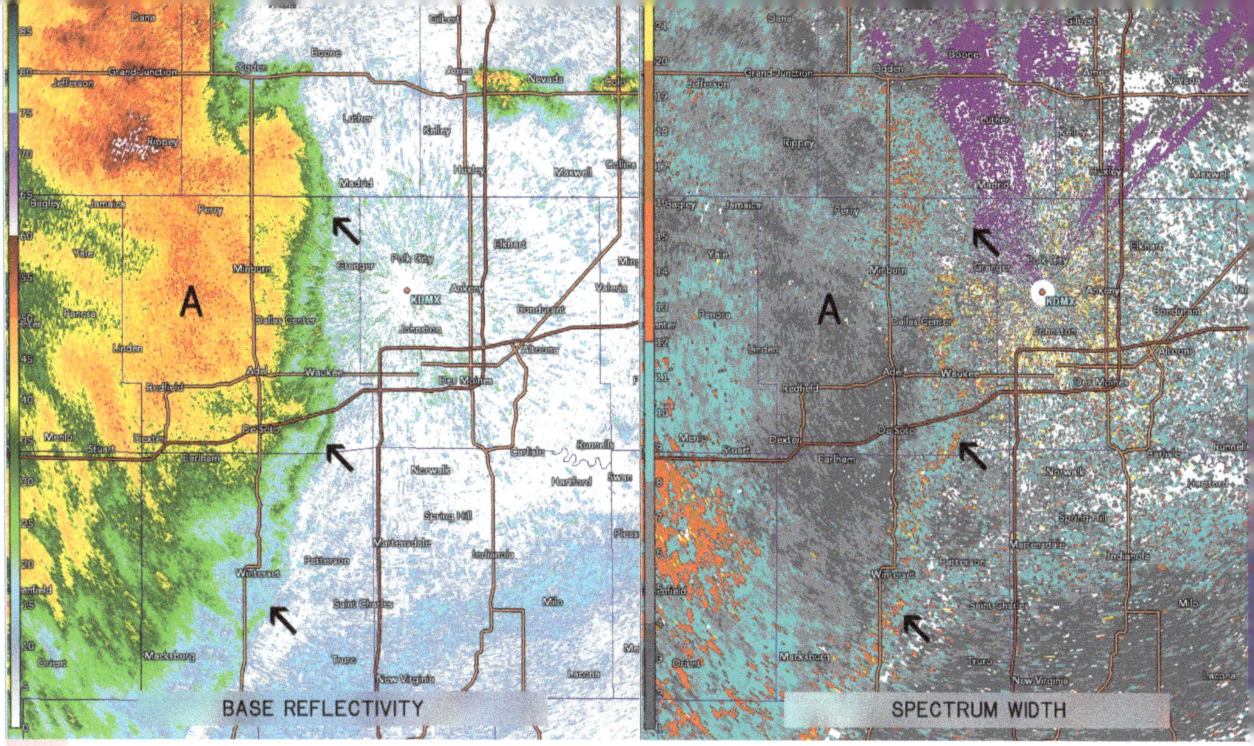

Above: Base reflectivity and corresponding spectrum width image for 10:34 am CDT for the 10 August 2020 Iowa derecho as it approached Des Moines. The outflow boundary is distinctly visible (arrows) and is sometimes easier to identify on spectrum width, especially when rain is falling onto it. The downdraft region (A) often has low values of spectrum width while the inflow region often has a noisy texture.

Right: Spectrum width for the 14 April 2012 Wichita tornado (0319 UTC) before it went on to produce a 13-mile track with EF3 damage.

Lower right: The corresponding 0.5° reflectivity image with important spectrum width features overlaid as black lines. The products complement each other and help reinforce insight into what is going on. Quite often when hailstones or rain drops are falling into the inflow region, reflectivity pattern definitions become smeared out. Spectrum width can often uncover the actual locations of outflow boundaries and air mass contrasts. *(All images from GR2Analyst)*

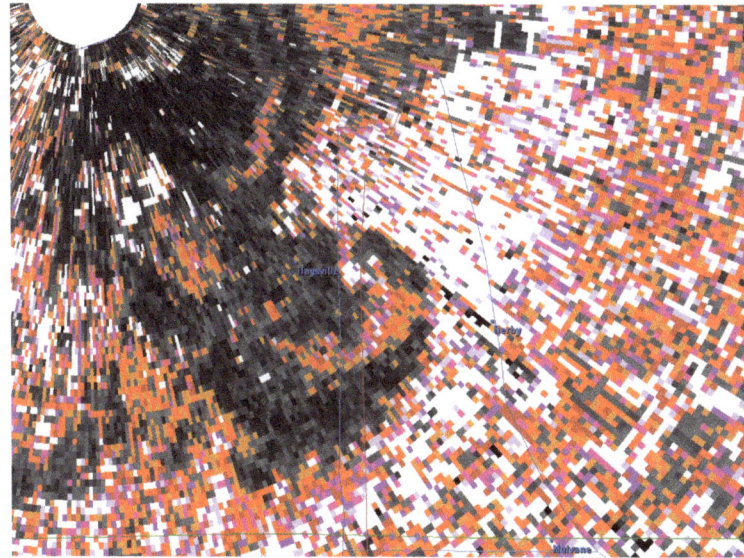

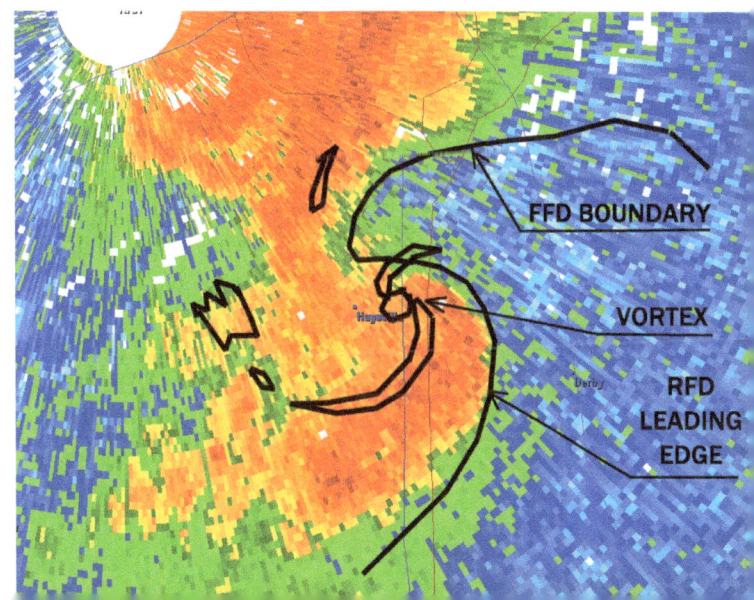

Differential reflectivity

Differential reflectivity (Z_{DR}) is one of the base products of polarimetric radar. Consider that most "normal" radar data is sampled in the horizontal plane, with electromagnetic energy oscillating from side to side. Next-generation radar technology, which is now available on the United States NEXRAD network, adds sampling of energy in the vertical plane: up and down. By examining the ratio between the horizontal and vertical reflectivity, we get a sense of the shape of scatterers in a given volume. If we take a pancake-shaped droplet, for example, and orient it broadside to the ground, it will return considerably more power in the horizontal axis than the vertical.

When we talk of ratios, 1 means that both terms are equal. Differential reflectivity however is expressed in a manner where zero indicates equivalence. At this value, the scatterers can be presumed to be spherical. Negative values indicate more power is being returned in the vertical plane, and suggests scatterers are vertically elongated (prolate). Positive values are more common, suggesting more power is being returned in the horizontal plane and that scatterers are horizontally elongated (oblate).

Let's take the simplest example: a raindrop. The smallest raindrops, drizzle, are extremely small and fall very slowly, so they remain fairly spherical and have a differential reflectivity of about zero. But large falling raindrops are shaped by drag as they fall. This causes them to flatten out like a pancake broadside to the ground. On radar, this yields a strongly positive differential reflectivity signature. On the other hand, a hailstone tends to takes on prolate characteristics as it grows and begins falling, with its shape elongated toward the ground and sky. This yields a negative differential reflectivity.

It is important to remember that although thunderstorm cores often show high positive values, differential reflectivity is not an indicator of intensity or rainfall rate. It is more sensitive to shape and size of the scatterers. For shape discrimination with sensitivity to intensity, KDP (specific differential phase) is a better tool.

What is differential reflectivity useful for? Hail cores in thunderstorms are one noteworthy example, because the differential reflectivity signatures are almost diametrically opposite. What may be seen is a core of negative Z_{DR} (hail) within a broad area of positive Z_{DR} (rain). As a result, the differential reflectivity product can give the forecaster a better idea of whether higher base reflectivities are the result of very heavy rain or hail.

key details

Differential reflectivity (Z_{DR}) gives information on the shape of precipitation particles within a particular radar bin.

Negative values indicate prolate shapes, i.e. elongated up and down. This is associated with hailstones.

Neutral values indicate spherical shapes. This is associated with drizzle, graupel, and rain/hail mixes.

Positive values indicate oblate shapes, i.e. elongated side to side. This is associated with rain, snow, and wet graupel.

Differential reflectivity is an expression of Z_H/Z_V, but its exact definition is $10\log_{10}(Z_H/Z_V)$. Equal values of Z_H and Z_V will yield a ratio of 1 (1 divided by 1), but the differential reflectivity equation shows that this will actually produce a value of zero. This simplifies prolate vs. oblate to a negative vs. positive relationship.

Very large raindrops tend to have high values of positive Z_{DR}. The larger they are, the higher their terminal velocity and the more they are spread out by drag as they fall.

Polarimetric processing was retrofitted to all United States WSR-88D radars in 2012-2013. The better radar data websites offer polarimetric images, and software programs like GRLevelX handle them too.

Insects have high values of Z_{DR} because they prefer to fly with their bodies oriented horizontally. They often make up the bulk of clear air scatterers late in the day.

Product code: 159/DZD (RPG), N0X (WMO).

real-time Internet sources

weather.cod.edu/satrad
www.weathertap.com ($)

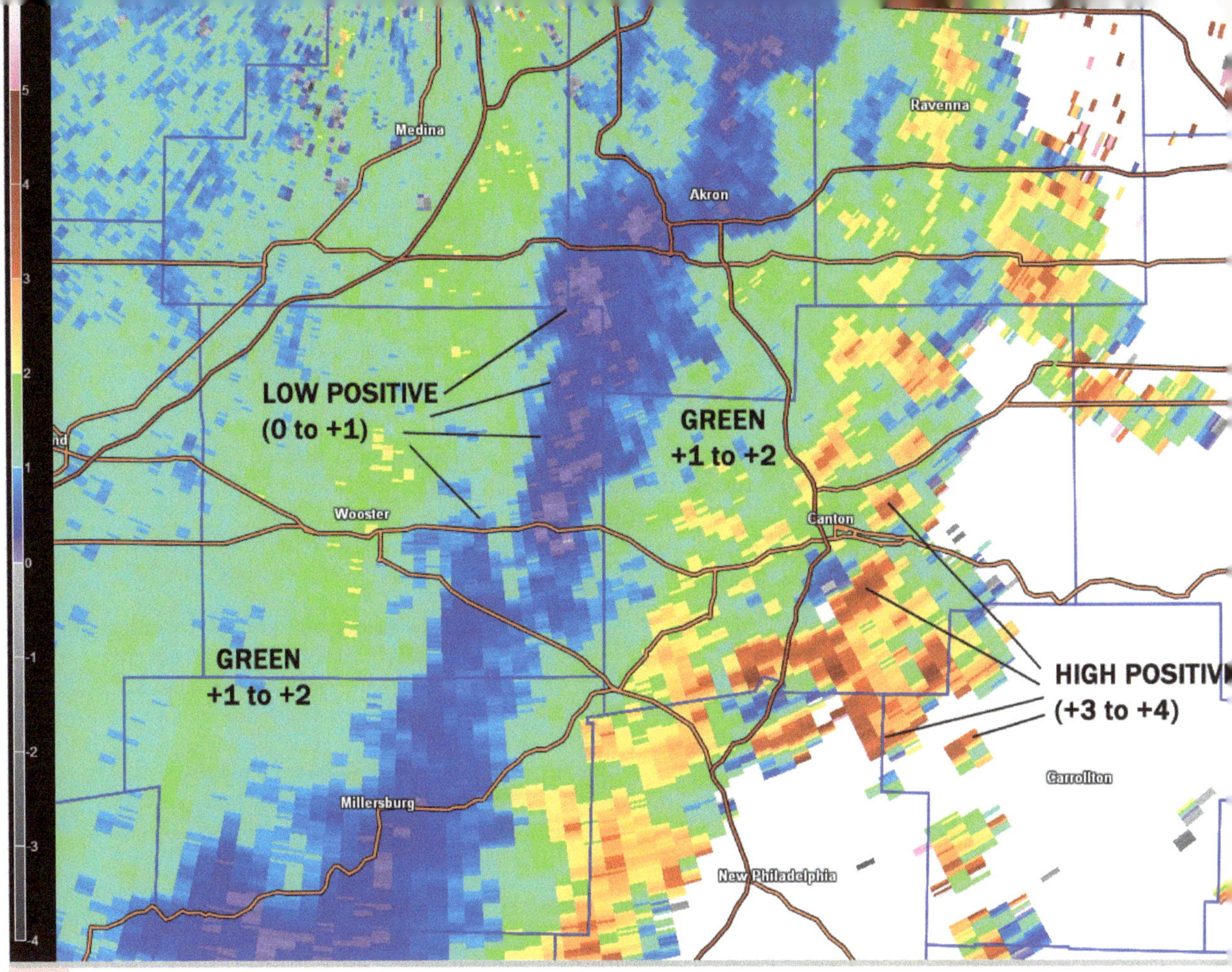

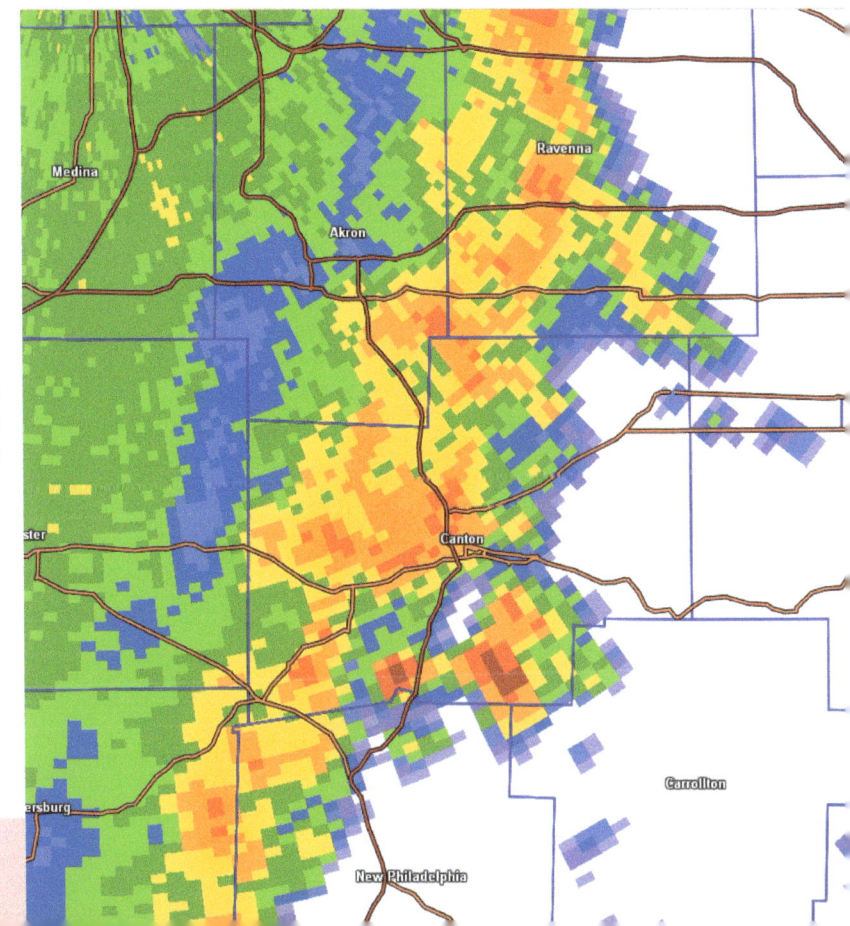

Above: Differential reflectivity image for a derecho storm in Ohio for June 13, 2013 at 0544 UTC. Compare this with the reflectivity image below. This shows higher Z_{DR} (more oblate shapes) in heavier precipitation, with lower Z_{DR} (more spherical shapes) in the weather precipitation. The lack of low Z_{DR} in the heavy cores indicates that the threat of large hail is low. The primary hazard in this storm is wind. *(GRLevel3)*

Right: The base reflectivity 0.5° reflectivity image corresponding to the image above. *(GRLevel3)*

Correlation coefficient

Correlation coefficient (also known as CC, rho, or ρ_{HV}) compares the consistency in the backscattered power between the vertical and the horizontal planes. Whereas spectrum width gives a measure of diversity of velocity within the sample volume, correlation coefficient yields the diversity of shapes in the volume. However, unlike spectrum width, higher values of CC indicate lower diversity. Correlation coefficient is affected strongly by oscillation, wobbling, and canting of the scatterers. The range of values is 0 to 1.05, with 1 indicating very high consistency. Values below 0.8 to 0.9 indicate a low consistency, implying a large and irregular assortment of shapes.

Most meteorological targets have low shape diversity (high values of CC, but there are some noteworthy exceptions. For example, birds and insects have a very high diversity of shapes and movements and reflect a variety of signals back to the radar in the vertical and horizontal planes. This will result in low CC values. Wet snowflakes and large, wet hail have similar characteristics.

A higher diversity of energy (i.e. low CC) is also produced in mixed phase precipitation regimes. For example, regions with only rain and only snow will each have CC values of above 0.95. But in a volume that has mixed wintry precipitation, the CC values will drop below 0.95. Since hail shafts usually contain a mixture of rain and hail, they will also show relatively low CC values.

Likewise, sharp gradients of CC values in a winter weather situation, rather than just the absolute value in each zone, can be used to define locations where a change in precipitation type is occurring. At the transition line from sleet and freezing rain to snow, a CC gradient will exist.

One particular noteworthy example of low CC values is in the "bright band" surrounding stations during the cold season: the ring surrounding a station where the radar beam is penetrating the melting level. Here, there are mixed phases which produce a high diversity of shapes and sizes and produce a highly visible ring of low CC values on horizontal displays and an elevated horizontal axis aloft.

In the 2010s there has been renewed interest in the detection of tornadic debris. The diversity of shapes and oscillations in a debris cloud will yield sharply lower CC values than the precipitation areas. This technique is more useful for tornadoes embedded in rain areas. With some classic supercells and LP storms, spectrum width may be a better discriminator.

key details

Correlation coefficient gives information on the diversity of oscillation, wobbling, and canting of scatterers within a particular radar bin.

The simplified formula for correlation coefficient is
$$CC = S_{VV}S_{HH} / (|S_{HH}^2||S_{V}^2|)^{1/2}$$
where S_{V} is the vertical phase power return and S_{HH} is the horizontal phase power return.

The Greek letter rho is a standard variable for representing statistical correlation coefficient. This is the linking of two variables, one on the X axis and the other on the Y axis.

High values (above 0.96) are associated with low diversities and correspond to typical precipitation areas.

Marginal values (around 0.95) are associated with mixed phase precipitation regimes.

Low values (below 0.95) tend to be associated with non-meteorological scatterers, along with wet snow and large, wet hail.

A decrease in correlation coefficient with time indicates a shift away from homogenous shapes. This may be due to development of mixed-phase precipitation, the appearance of hail. This makes it an important indicator in winter weather forecasting situations.

Be aware of the conical nature of the beam, especially in winter weather situations. Low CCs close to the radar and higher CCs at longer distances may indicate melting in the low levels and solid ice crystals aloft.

Product code: 161/DCC (RPG), N0C (WMO).

real-time Internet sources

weather.cod.edu/satrad
www.weathertap.com ($)

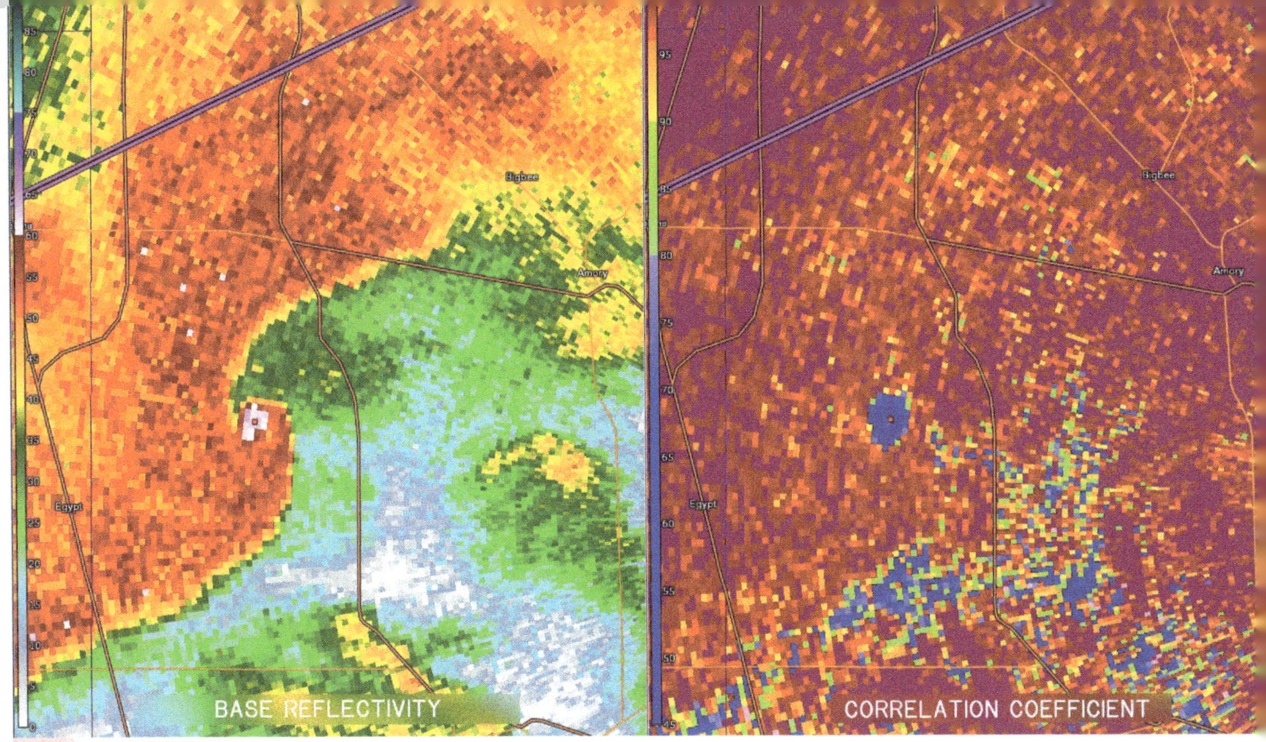

BASE REFLECTIVITY

CORRELATION COEFFICIENT

Above: Comparison of base reflectivity (left) and correlation coefficient (right) for the 24 March 2023 Amory, Mississippi tornado. The tornado vortex contains a large area of correlation coefficient values of 40% or less, showing as blue. This reflects the wide distribution of backscatter polarity probably due to the vortex filling with tornadic debris.

Right: Reflectivity (upper panel) and correlation coefficient (lower panel) for Buffalo at 1220 UTC 23 December 2022 during the onset of a historical lake effect snowstorm. At the time of the radar image, the surface precipitation is rain, while aloft it is snow. The bright band shows as a ring on reflectivity, while on correlation coefficient it shows as a ring of decreased values, indicating mixed phase precipitation. It is important to remember the radar beam is conical, so the further the distance from the center, the higher in the atmosphere we are sampling. *(GRLevel2, all images)*

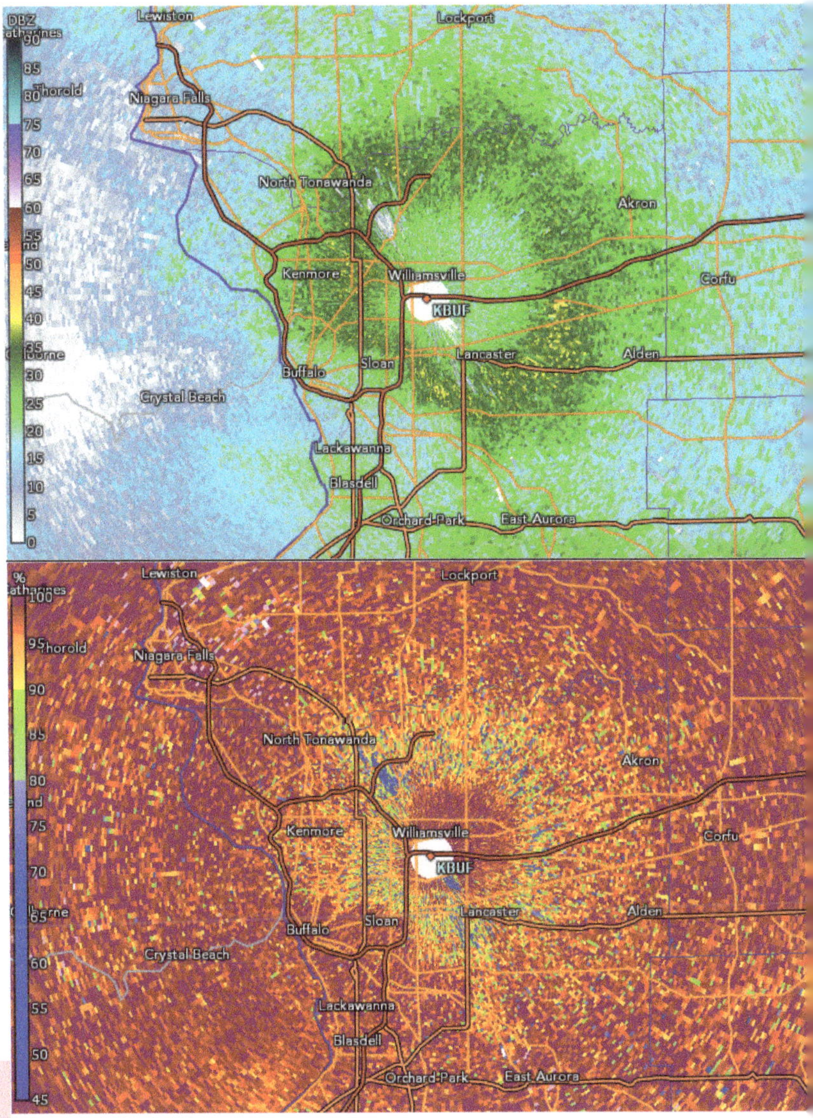

Specific differential phase

Specific differential phase (KDP, or K_{dp}) identifies locations where the phase difference between the vertical and horizontal wavetrains are changing. Radar data arrives from distant targets in the form of a continuous electromagnetic wave, one for the horizontally-polarized axis and one for the vertical. Each of these individual waves making up the wavetrain do not arrive with perfect regularity but become shifted slightly with respect to time due to the presence of water, ice, and other scatterers. Some of the greatest changes in phase are caused by liquid droplets, which cause a relatively large change in phase between the vertical and horizontal signals. A single thunderstorm, for example, can desync the wavetrains by about 90° (one quarter of a complete sine wave).

The basic differential phase product is simply called differential propagation phase, or ϕ_{dp}. This indicates what the difference in phase is between the horizontal and vertical wavetrains at a given instant or at a given range bin. This information is useless because differential phase is the result of phase changes that have already occurred in bins between the target and the radar, not what is in the range bin being examined. Plotting differential propagation phase would simply show charts with radial streaking. What is more important is to highlight the *phase change* along a radial from one bin to the next, instead of the *phase*. This gives us *specific differential phase* (KDP), identifying exactly where phase changes are occurring.

Specific differential phase is most sensitive to anisotropic particles, such as oblate raindrops, whereas isotropic particles such as small round drizzle drops, spherical hailstones, and snow produces no significant phase shift. As a result, KDP is sensitive mostly to rain. This has important forecasting implications. For example, KDP is more strongly bound to rainfall rates than reflectivity. In a wintry precipitation regime, changes in KDP can help differentiate rain and mixed phase areas from snow.

One unique advantage of specific differential phase is that it is not affected by attenuation. This can be a problem when the radar beam is looking lengthwise down a squall line, with backscattered energy from distant cells passing through cells that are much closer. Phase change is a property that cannot be "blocked" by closer cells, so specific differential phase will yield a more accurate representation of intensity, compared to the base reflectivity product.

key details

Specific differential phase (KDP) is expressed in degrees per kilometer (° km^{-1}).

Strong positive values of KDP are seen in anisotropic particles, most notably rain drops, where values may be as high as 10° km^{-1}.

Low values of KDP are seen in hail (-1 to 1° km^{-1}) and in snow (0 to 2° km^{-1}). Spherical or tumbling hail has a KDP near 0. Drizzle and cloud droplets will also produce very low KDP values.

Negative KDP values are often associated with vertically-oriented ice crystals where strong electrical fields produce a strong canting response. This phenomena may be seen in the tops of thunderstorm clouds and aloft in trailing stratiform regions, though the ice crystal response is brief and the WSR-88D is not adapted to this type of detection.

KDP columns may be seen in storm updraft regions where hailstones shed liquid drops.

KDP has similarities to ZDR in that oblate scatterers such as raindrops are particularly effective at increasing values. However KDP is sensitive to shape and the number of drops, while ZDR is mostly a function of drop size and shape.

For hail detection, comparison of reflectivity to KDP can uncover locations where hail is occurring. A difference in the shape of the core or a difference in the core location between the two products is the most obvious clue that the KDP and reflectivity product do not match each other.

Product code: 163/DKD (RPG), N0K (WMO).

real-time Internet sources

weather.cod.edu/satrad
www.weathertap.com ($)

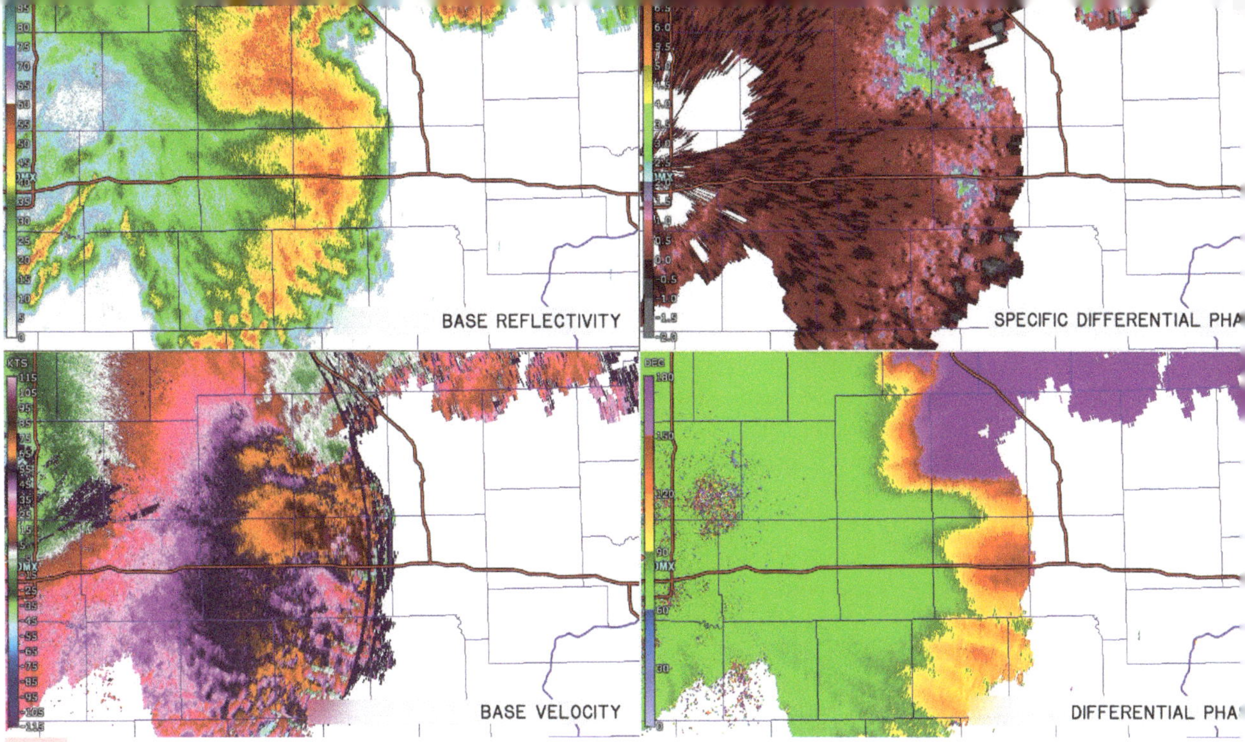

Above: Base reflectivity, specific differential phase (KDP), base velocity, and differential phase (phi), all for 0.5° at the same date and time, for the Iowa derecho of 10 August 2020 at 12;24 pm. The patterns shown on base reflectivity and KDP show strong similarities. The KDP product is built from the differential phase product (below it) which by itself is probably not useful. The velocity product at lower left shows high outbound velocities (in orange, the radar is to the west) from the intense downbursts. *(GR2Analyst)*

Below: Comparison of base reflectivity and specific differential phase, similar to above, for Hurricane Harvey as it made landfall near Port Aransas, Texas on 25 August 2017 at 7:03 pm CDT. The product is fairly similar to base reflectivity but may reveal additional detail in the spiral bands since it uses a different method of acquiring information from backscattered energy. *(GR2Analyst)*

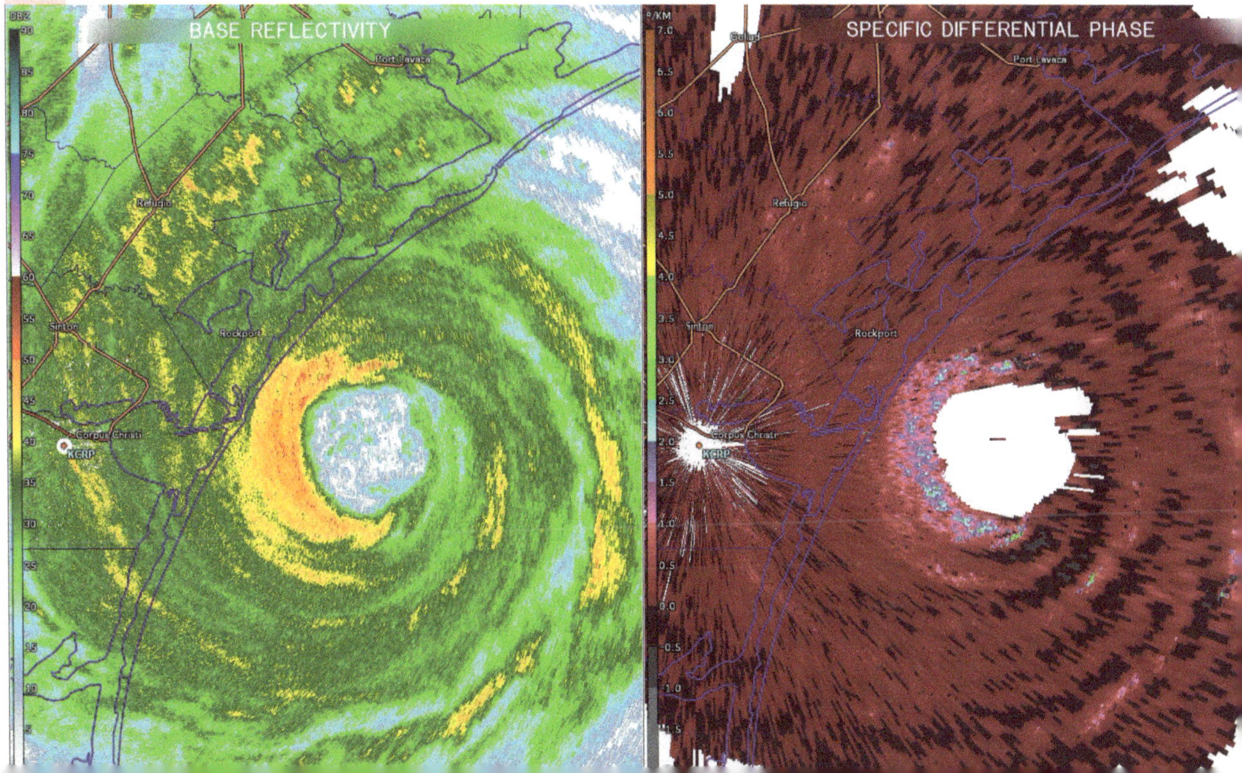

Hydrometeor classification

The Hydrometeor Classification Algorithm (HCA) use dual-polarization signatures to determine the type of precipitation in each sample bin (the smallest "pixel" resolvable by the radar). The HCA output is divided into two main products: Hydrometeor Classification (HC) and Hybrid Hydrometeor Classification (HHC).

HC products can be obtained for any tilt showing possible hydrometeors at each of these levels. HHC (hybrid HCA) outputs a single product valid for the surface. HHC is used as input into the radar's precipitation measurement algorithms, but is also perhaps the most useful for identifying surface precipitation. HHC however lacks the "large hail" and "giant hail" categories.

The HCA output is not designed to be used alone, but provides a first glance at the precipitation areas for the forecaster and a starting point for using other polarimetric products. It can also help less experienced radar users get a rough estimate of where hazards might be happening.

Most of the categories like rain, hail, and snow are fairly self explanatory. Wet snow (WS) is the "clumpy" form of snow with high water content. Dry snow (DS) is a fine powdery snow that often occurs in drier regimes such as cold advection zones, and due to its low density it rapidly produces accumulation and tall drifts. The category "graupel" (GR) is the name given to snow pellets, which form when a snowflake becomes coated with supercooled water droplets. The result is a pellet that looks like soft hail with a fine, fluffy coating.

The category "big drops" (BD) is category many longtime professionals and amateur forecasters may find to be quite unusual, especially since it does not fit reportable METAR and SYNOP precipitation types. This is the name given to very large drops that fall in relatively small numbers, and are most common along the leading edge of thunderstorms.

Ice crystals (IC) might be mistakenly thought of as snow, but a snowflake is actually an aggregation of ice crystals. Ice crystals are very fine, almost invisible particles, and form in a wide variety of shapes such as columns, plates, and hexagons depending on the temperature and moisture. Ice crystals are usually never seen at the ground unless the temperatures are below about 0°F to -10°F.

The HCA product is built on a mix of empirical and physical assumptions, and there will always be a weather situation that does not fit the algorithm's expectations. Also there is no indication of confidence. Forecasters should always crosscheck this product with other sources.

key details

Hydrometeor Classification Algorithm (HCA) provides the radar's best estimate of precipitation type based on a number of techniques and algorithms.

HCA signatures may not be representative at the surface since the radar beam is always some distance above the ground.

HCA is based on a chain of assumptions and can fail in unusual weather situations.

The algorithms have difficulties in specific situations due to crossover of favored Z, ZDR, and CC values. For example, weak light rain and dry snow are poorly differentiated.

Freezing rain is a function of ground-level temperature, which is not considered by the radar and is for the forecaster to decide. It will be observed by the HCA as rain (RA, HR) or mixed-phase precipitation (WS, etc). There is also no specific category for ice pellets, which may show as UK, GR, RA, or other categories.

Product code:
Dig HCA: 165/DHC (RPG), NOH (WMO)
Hybrid HC: 177/HHC (RPG), HHC (WMO)

HCA categories
BD: Big drops: low numbers of large drops.
BI: Biological. Insects, birds, bats, etc.
GC: Ground clutter. Buildings, cars, hills, etc.
DS: Dry snow: fluffy and light.
GH (HC only): Giant hail
GR: Graupel: snow pellets
HA: Hail or rain/hail mix
HR: Heavy rain. Large numbers of drops.
IC: Ice crystals. Plates, columns, etc.
LH (HC only): Large hail
ND: No data / below threshold
RA: Rain. Liquid drops.
RF: Range folded (corrupted data).
UK: Unknown scatterer.
WS: Wet snow: sticky, high water content.

real-time Internet sources

weather.cod.edu/satrad
www.weathertap.com ($)

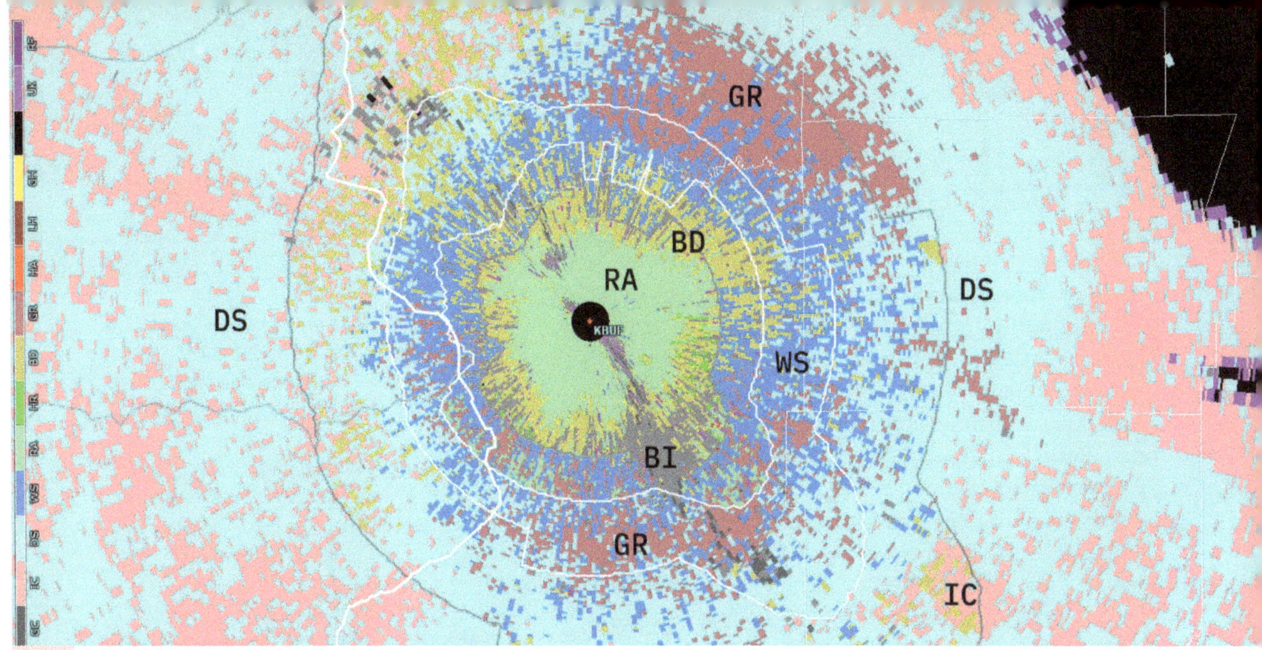

Above: Hydrometeorological classification algorithm (HCA) on 23 December 2023 at 1200 UTC at the time a massive winter storm with heavy lake effect snow was beginning at Buffalo. This image uses the 1.5° tilt to better sample vertically close to the radar site. It shows a large area of snow at altitudes above 2000 ft AGL, but in the 1500 to 2000 ft AGL layer (closer to the radar) we see a ring of wet snow. Closer to the radar and and lower it transitions to big drops (from melting snow) and finally to rain at the surface. Due to very strong cold air advection, the outer HCA patterns closed in on the inner ones shortly after this image as rain transitioned to snow. *(GRLevel3)*

Right: Structure of one of the core techniques in the WSR-88D Hydrometeor Classification Algorithm. The X (horizontal) axis shows base reflectivity and the Y (vertical) axis shows differential reflectivity. Data from a specific range gate is mapped into this table and the results provide an initial guess at precipitation type.

Below: HCA indications for the Andover, Kansas tornadic storm of 29 April 2022 at 8:19 pm CDT. This is one of several tools that can be used to differentiate rain from hail areas. The tornado vortex is mostly misclassified, as would be expected, but its location shows prominently with a previous TVS mark close by. *(GR2Analyst)*

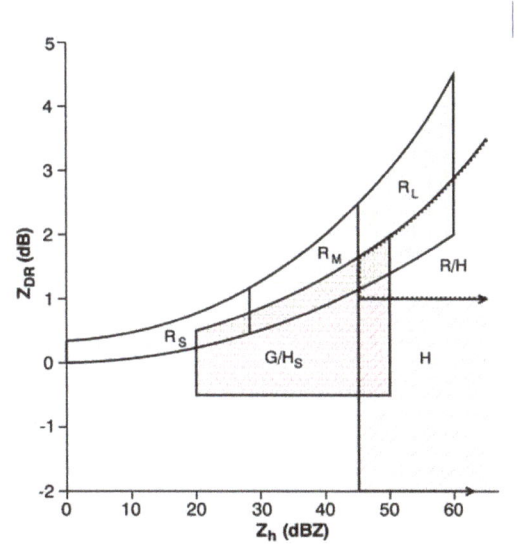

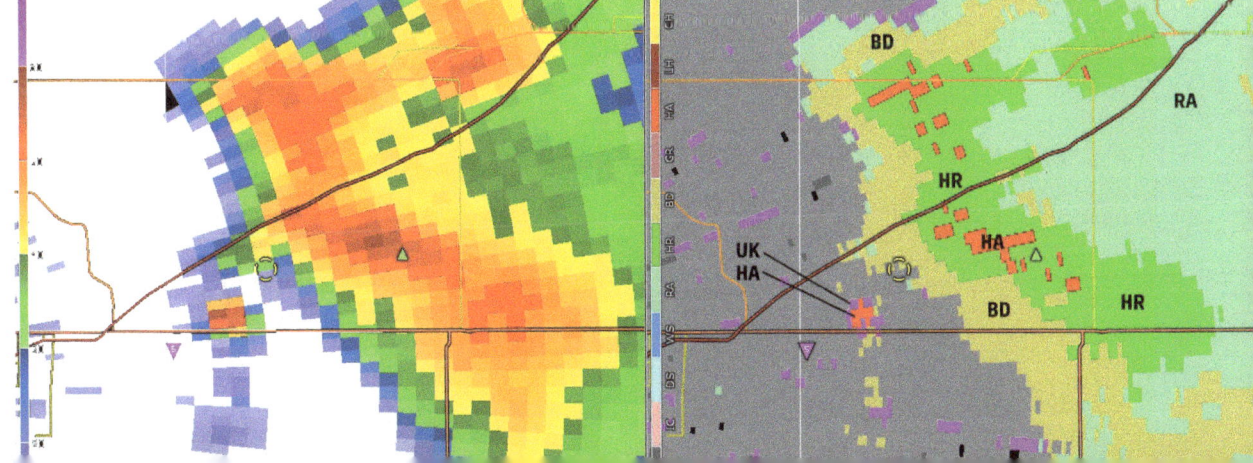

Precipitation total

The Precipitation Total product actually refers to three specific products generated by the NEXRAD site. These are 1-hour precipitation, 3-hour precipitation, and storm total precipitation. These provide a graphic estimate of precipitation totals within 124 nm of the radar site.

The Precipitation Total product is excellent for monitoring areas that have had excessive precipitation. These areas may be subject to flooding. It can also reveal where grounds are saturated, which can further compound flooding and can feed moisture back into a convective situation through evaporation. The product can determine where heavy snow has fallen. Hydrologists will use Precipitation Totals to find where basins are approaching saturation and where streams might reach flood stage.

If no precipitation, as determined by the precipitation totals algorithm, has fallen within 124 nm of the site after more than an hour, the storm total precipitation (STP) product is automatically reset and all plots show a zero precipitation total. However, during long rain events the storm total period may exceed 24 hours.

As with many other products, precipitation total estimates are limited by the tilt of the radar antenna, which reaches no higher than 19.5°. Not all of a precipitation shaft which is very close to the antenna will be sampled. The problem usually occurs within 20 nm of the site for convective precipitation and within 10 nm of the site for stratiform precipitation.

The data is easily contaminated by ground clutter and anomalous propagation. Chaff (from military aircraft) will produce false totals and produce plots that look unusually like precipitation. When bright bands occur during cold weather events, these will distort the totals.

Fast-moving systems will distort the precipitation total pattern, as the storm will move across multiple bins between scans. This will produce a herringbone pattern in the precipitation total "trails" left by storms.

Hail will cause overestimation of rainfall amounts, as hail particles reflect much more power back to the radar than water droplets do. The three-body scattering spike signatures can also "lay down" precipitation values behind the hail core, where no precipitation actually exists.

It should be noted that there is no quality control of the precipitation product at any stage. Users have to evaluate the results against their knowledge of radar principles, climatology, and empirical knowledge of the precipitation total accuracy.

key details

The depiction of precipitation totals was changed from Cartesian 1.1 nm blocks to standard polar coordinate format effective with NEXRAD Build 9 in late 1996.

Precipitation total estimates may be degraded by the presence of ice, snow, and especially hail. The algorithm works best with warm, slow-moving precipitation.

It is useful to conduct a study of your local area and determine whether the radar tends to underestimate or overestimate precipitation. Be sure to use official day-to-day readings at an airport weather station for representative results, rather than using extreme rainfall amounts that could be difficult for the radar to estimate.

Storm total precipitation values are reset when no precipitation totals are detected by the algorithm for at least one full hour.

Storm total precipitation is not confined to a specific time limit.

Plots near military operating areas are frequently degraded by chaff released by military aircraft during mock dogfights. Notorious chaff regions are the area north of Las Vegas, the area west and southwest of Phoenix, and the region southwest of Salt Lake City.

Precipitation total estimates are corrupted by precipitation too close to the radar antenna, by ground clutter or chaff, by fast moving storms, and by hail cores.

Product codes:
  169/OHA (RPG), OHA (WMO) (1-hr)
  138/DSP (RPG), DSP (WMO) (storm total).

real-time Internet sources

weather.cod.edu/satrad
www.weathertap.com ($)
water.weather.gov/precip (daily radar estimates)

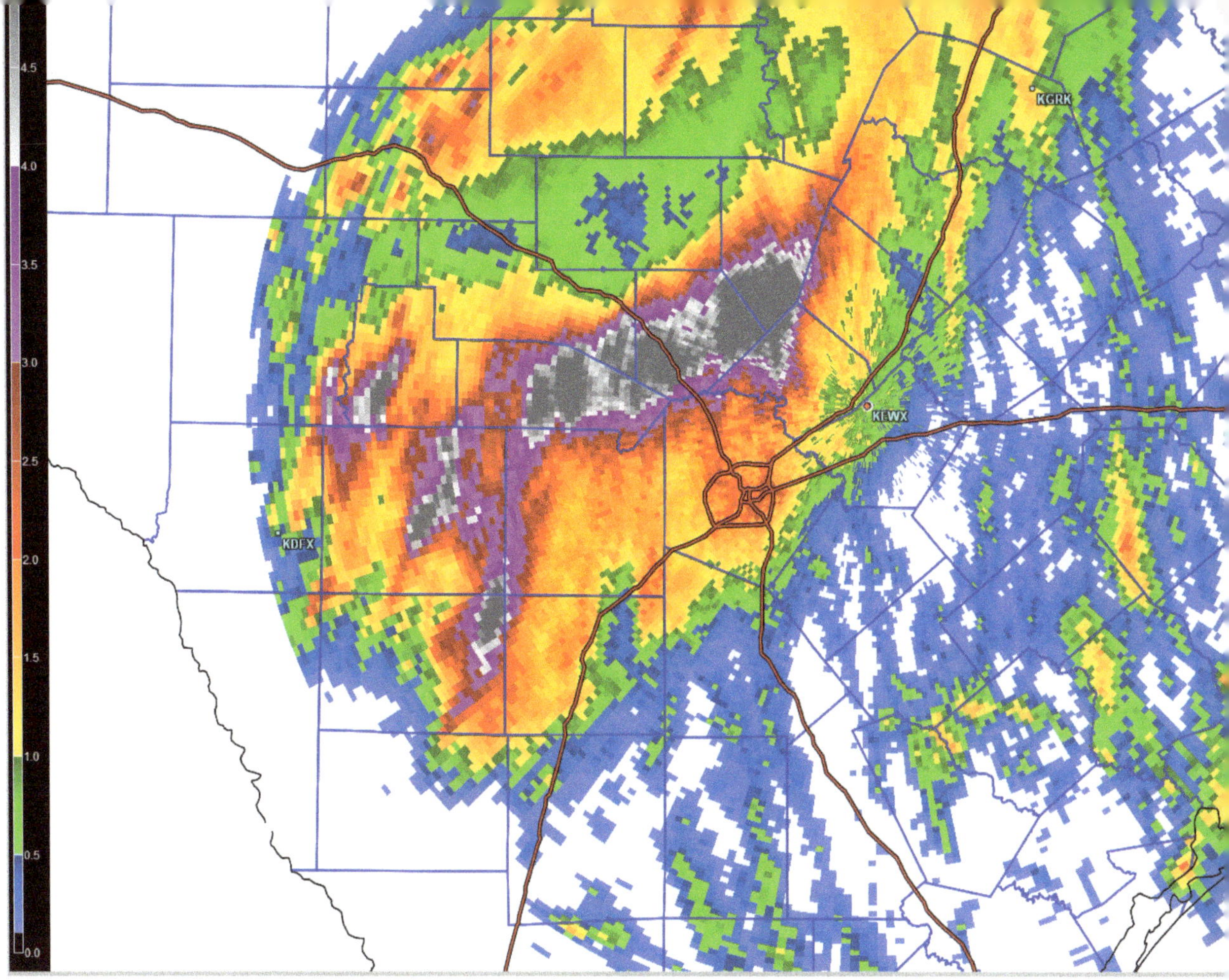

Above: Storm total precipitation product for the damaging 2015 Memorial Day weekend flooding in the Texas Hill Country. *(GRLevel3)*

Right: A National Weather Service hydrologist adjusts a 1-hour precipitation total estimate from the Fort Worth WSR-88D before feeding it into a local model. This quality control procedure makes for better runoff and soil saturation estimates, improving the quality of flood forecasts. *(Tim Vasquez)*

Vertically integrated liquid

Vertically Integrated Liquid, or VIL, indicates the average mass of liquid water per cubic meter within a column above the earth's surface. It examines all of the bins above a given location and uses empirical relationships to estimate the total liquid water content, using a technique formulated by Greene and Clark in 1972.

VIL is calculated in units of kg m^{-2}. In the WSR-88D, inputs into VIL beyond 56 dBZ are disregarded in order to filter out hail. VIL values have traditionally been capped at 80 kg m^{-2} but this upper limit can be circumvented by various display programs.

The quantity *VIL density* (VILD) divides VIL by the echo top to produce a value in kg m^{-3}, or multiplied by 1000 to give g m^{-3}. The largest values will be produced by low-topped storms with high VIL values. As a general guideline, VILD values of 3.5 comprise a threshold for severe hail, and as VILD passes 4.0, 1-inch hail is likely.

There is also digital VIL (DVIL) in which contaminated radials are eliminated. The algorithm includes reflectivity below 18 dBZ, making it more useful in winter weather situations. The 56 dBZ filter is not imposed, so DVIL output is much more sensitive to hailstorms.

The most important strength of VIL is that, being an additive total of all echoes in the vertical, it is comparable to composite reflectivity and allows immediate assessment of which storms are most important. A cyclic increase and decrease in VIL values indicates a multicell thunderstorm structure, though this could also indicate a storm that is close to the radar and is undergoing the "stadium effect" due to clipping of its tops above the maximum tilt. On the other hand, persistent high VIL values are associated with supercells.

VIL has formed one of the strongest links to observed hail size, and in the early 1990s it was recommended as a first-line technique for identifying severe hail. However the 56 dBZ hail filter described above defeats this type of analysis. Still, though, VIL remains closely correlated with hail since numerous vertical bins containing hail will produce reflectivity at the upper bounds, enhancing the VIL value. DVIL also eliminates this hail filter.

VIL is affected by any process that distorts reflectivity values. A storm that is tilted or moving rapidly (changing position between individual scans within a VCP cycle) will produce artificially low VIL values. Both of these spread the storm footprint across multiple vertical bins, thinning out the VIL response. Chaff, bright bands, and three-body scatter spikes also corrupt VIL values.

key details

VIL is an indirect estimate of the instantaneous liquid content above a given spot. It is analogous to a summation of reflectivity in the vertical. VIL was often used in the 1990s to evaluate hail risk, however the hail algorithms such as POSH and MEHS are considered to be superior for this purpose.

VIL is closely correlated to updraft strength and can be used to quickly identify cells that require monitoring for severe weather or flooding.

There is no magic number for VIL values. They vary according to the season and type of weather system. Trends over time and local differences between neighboring cells are more meaningful.

VIL values are the least reliable in VCP 21 due to the wide elevation gaps present above 4.3°.

VIL is adversely affected by storm tilt, fast storm motion, elevation gaps, proximity to the radar, and operation in VCP 21. All of these will produce erroneously low VIL indications.

GRLevelX products use HRDVIL (DVIL) for "Vertically Integrated Liquid", and if not available will fall back on traditional VIL.

VIL will not be correctly represented within 20-30 nm of the radar, as the radar antenna is limited to a tilt of 19.5°, clipping storm tops.

Product code:
Standard VIL: 57/VIL (RPG), NVL (WMO)
High-res VIL: 134/DVL (RPG), DVL (WMO)
Note: High-res VIL or HRVIL is also known as "Digital VIL"

real-time Internet sources

weather.cod.edu/satrad
www.weathertap.com ($)

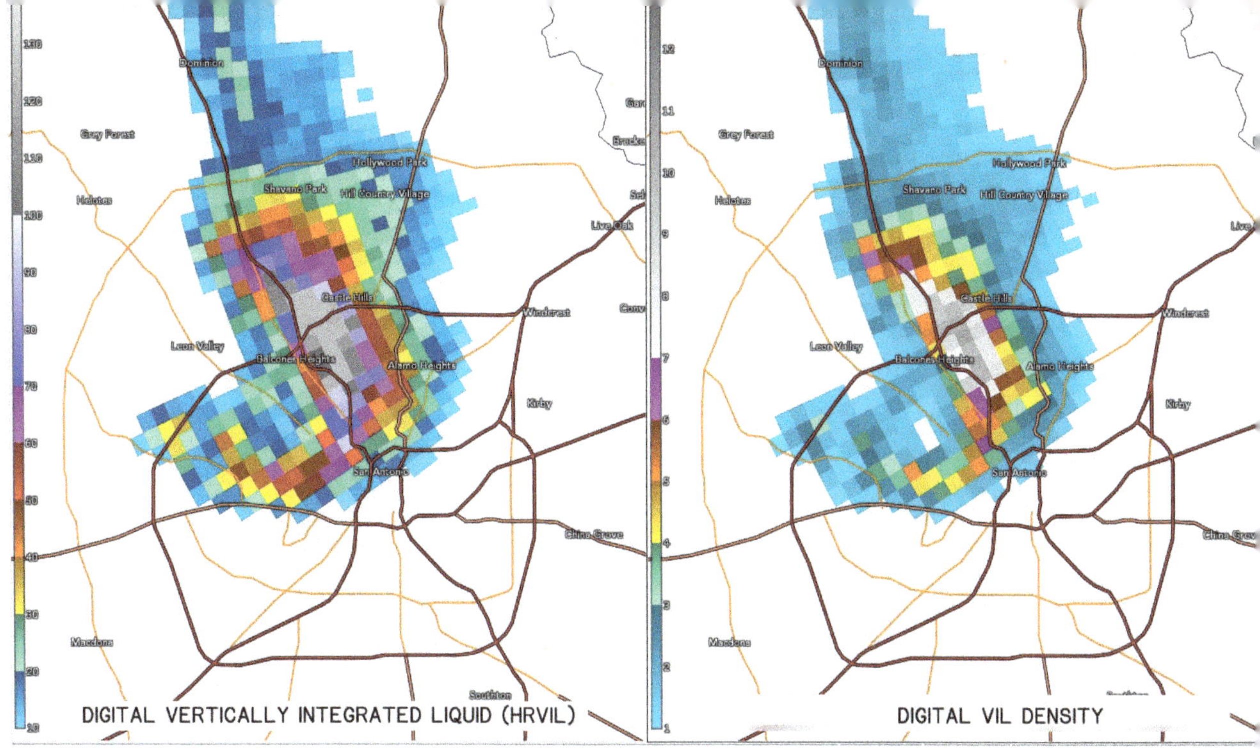

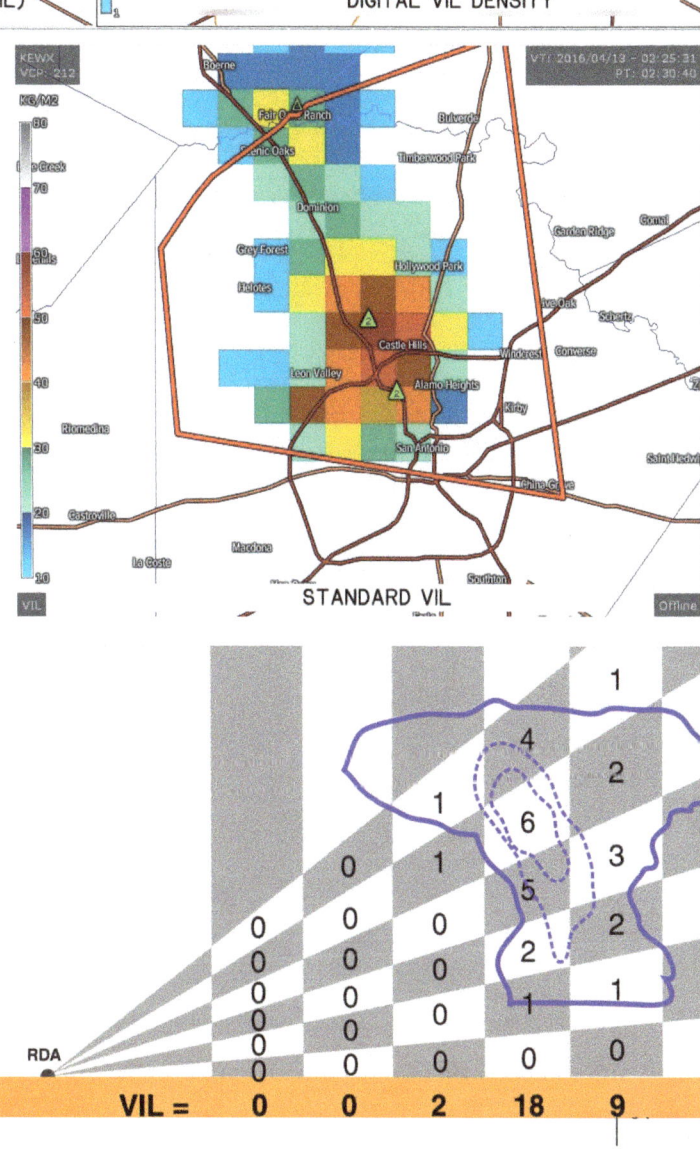

Top: HRVIL (left) and HRVIL-based VIL density (right) for the severe San Antonio hailstorm of 12 April 2016, which caused $1.5 billion in damage. Peak values reached the algorithm cap of 127 kg m^{-2}, while peak HRVIL density reached 10 kg m^{-3} at the time of this frame. *(GR2Analyst)*

Right: Classic VIL depiction (NVL) of the same data above. Due to filtering of reflectivity exceeding 56 dBZ, indicated values do not exceed 58 kg m^{-2}. Though this example is contaminated by hail, the classic VIL product is more suited for rainfall estimates.

Below: A cross-section of a storm shows the simplified concept of VIL. Arbitrary reflectivity values are shown within the storm cloud (outlined). The values are summed to produce a total. Notice how the lowest reflectivity scans miss any significant precipitation, but VIL detects the higher core. Composite reflectivity is similar in principle and would show the highest value in each vertical stack.

Echo tops

The Echo Tops product is a processed type of image. All bins in a vertical column are analyzed to find the highest grid box that exceeds 18.5 dBZ. The center beam height for the highest scan meeting the reflectivity threshold is the echo top.

Technically the term "Echo Tops" refers to the Enhanced Echo Tops (EET) or High Resolution Enhanced Echo Tops product, which is a revision of the legacy Echo Tops (ET) product. The EET update appeared in RPG Build 4 in late 2003. It has effectively replaced the legacy echo tops product, which used a Cartesian (rectangular grid) that plotted data at 2.2 x 2.2 nm resolution.

The biggest problem with this product is that the echo tops are not always sampled, or may be sampled too coarsely. Another type of problem is related to the antenna's maximum tilt, which is 19.5°. As a result, when storms exceed 30,000 ft within 15 nm of the radar or 50,000 ft within 25 nm of the radar, the radar is unable to scan clear air above the storm and the echo top is underestimated.

Elevation gaps are a significant problem, in which are slices "skipped" by the scan strategy. This is a significant problem in VCP 21 which has only four scans between 4.3° and 19.5°. For example, imagine a storm with a top of 55,000 ft located 55 nm from the radar. The 6° elevation intersects this range at 60,000 ft and the 4.3° elevation intersects at 37,000 ft. Unfortunately it is only the 4.3° elevation that detects the storm, so the elevation is counted as 37,000 ft. As an echo gets closer, its elevation seems to drop as it descends the beam until it is detected on the next highest elevation slice and jumps to another height. This behavior causes the "stairstep" pattern often seen on Echo Tops products. It is most prominent when stratiform tops are present.

Beam width is also a detrimental factor. At 120 nm the width of the radar beam spreads to about 13,000 ft. This means that a storm registering a top of 48,000 ft top may actually be topping out at anywhere between 38,500 and 51,500 ft.

The threshold value of 18.5 dBZ means that an echo top with a reflectivity of less than 18.5 dBZ will be cut off. An echo top, which includes an overshoot or anvil top, may be actually higher than what is depicted.

Finally, the echo top product does not compensate for three-body scatter spikes and sidelobes, which are artifacts that appear due to scattering in the extreme reflectivity near hail cores. This may lead to the product *overestimating* echo tops.

key details

The Echo Tops product indicates the highest altitude at which significant radar reflectivity exists.

Thunderstorm tops are likely to be higher than what the echo top product indicates, since the product cuts off detection below the 18.5 dBZ threshold value. Radar climatology from the 1950s to the early 1990s used human interpretations of the range-height scope, where the top of the echo constituted the maximum top.

The echo top product may occasionally overestimate storm tops in a severe weather situation when hailstorms cause three-body scattering spikes (TBSS). These are spurious false echoes that extend behind and above the storm. Increasing echo top heights may be caused by both strengthening of a storm and the production of large hail.

Echo top values cannot be compared between different days to measure severity. The variation of tropopause height significantly influences thunderstorm height.

An echo top that is increasing in height may indicate a strengthening storm, with the opposite true for a weakening storm. An echo top decreasing very rapidly may indicate a collapsing updraft, with the possibility of downburst damage or a tornado at the surface. TBSS's may corrupt these values.

Echo tops products are often not very accurate as the radar can only select from one of a limited number of radar tilts to determine the top, yielding an accuracy of roughly 5000 ft. Beam broadening is a limiting factor. The "stadium effect" curtails use at close range.

Product code: 135/EET (RPG), EET (WMO).

real-time Internet sources

weather.cod.edu/satrad
www.weathertap.com ($)

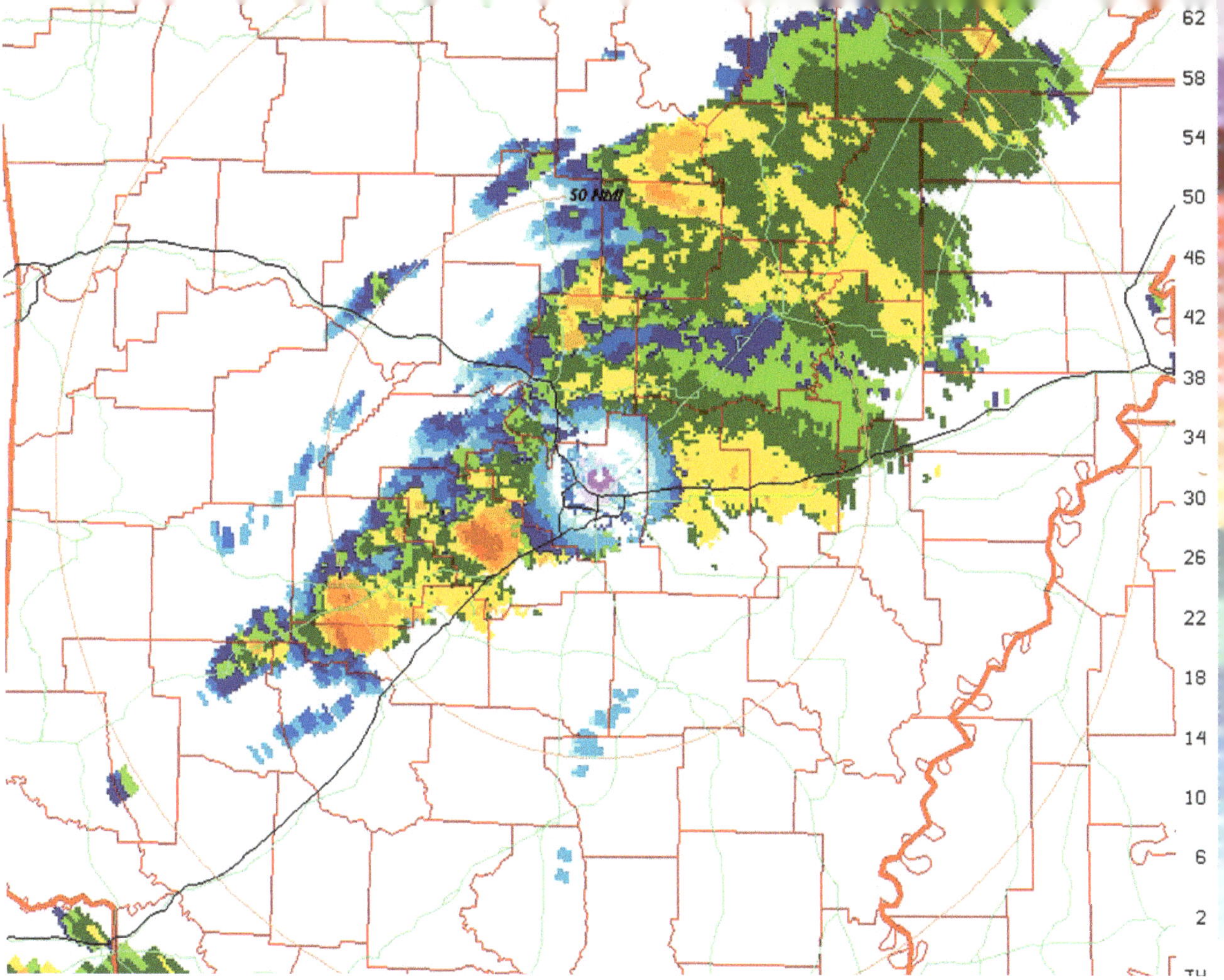

Above: Echo tops product at 1850 UTC about 30 minutes before the 31 March 2023 Little Rock / NLR tornado. The storm that is responsible is just to the southwest of the radar site and is entering the cone of silence, where vertical measurements will become degraded. *(College of DuPage)*

Right: This diagram demonstrates the problems caused by elevation gaps as a storm draws closer, using three thin beams for emphasis to judge each cloud top. As a result, the storm seems to get shorter, then jump to 40,000 ft as it approaches. This produces a stadium effect on echo top products. The sampling problem is worst in VCP 21.

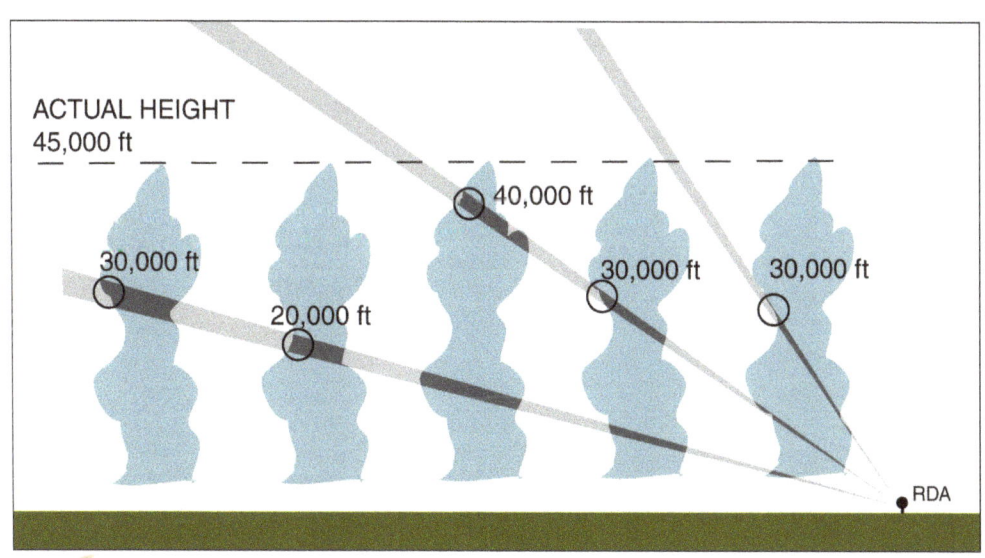

Storm tracking information

The purpose of the Storm Tracking Information product is to identify storm centroids and display their past, current, and projected locations. It uses the Storm Cell Identification and Tracking (SCIT) Algorithm, which replaced the old Storm Series Algorithm that was phased out with NEXRAD Build 9 in late 1996.

First the Storm Segments algorithm is run. One-dimensional storm segments are located by searching along a radial (i.e. at different azimuths at a constant distance) for one-dimensional runs of contiguous high reflectivities. If a series of high reflectivities has sufficent radial length to meet or exceed a parameter called the "overlap threshold", it is classified as a storm segment.

The Storm Centroids algorithm then runs. It gathers all storm segments and attempts to build two-dimensional representations of each storm. This produces a storm component. Each storm component is checked to see if it meets the minimum accepted size. Generally a size of 2.2 nm is considered sufficient. If it does not meets the size criteria, it is deleted. Each storm component is measured and assigned a "disk" representing its location and radius.

When this is complete, storm components are resolved into three dimensions by looking at all elevations, measuring overlap of components, and determining an overall centroid for the storm.

At this point, the Storm Tracking Information algorithm executes. The algorithm's job is to relate each storm found in a current volume scan to a storm found in a previous volume scan. The algorithm starts with the largest storm centroid and works down to the smallest ones, linking current storm centroids to those from previous volume scans. This builds a track history. There are also sanity checks that are performed. A storm motion change over time must not exceed 90 degrees from the last scan. A storm centroid may also not have changed mass by more than one order of magnitude.

Finally the Storm Position Forecast algorithm is run. Each storm's anticipated motion during the upcoming hour is made by a simple extrapolation of its average direction and speed of movement. The algorithm also checks its work; if it has performed poorly, no forecasted positions are displayed. However if it has done a good job, forecast centroids for the next four volume scans are displayed. No projection is made for new storms.

When storms organize into clusters or linear structures, storm tracking will become degraded, but the algorithm will still do a good job tracking dominant cells.

key details

The Storm Cell Identification and Tracking (SCIT) algorithm identifies thunderstorm cells and plots a history of paths and a projected track for each detected cell.

A key concept in the storm tracking algorithm is its dependence on precipitation signatures, and thus downdrafts. Updrafts cannot be "tracked", though the mesocyclone detection algorithm provides some insight.

The SCIT algorithm was formerly known as the Storm Series Algorithm. This older algorithm focused exclusively on cells that were 30 dBZ or greater and lacked a sophisticated identification scheme.

Unusual structures, especially linear configurations seen in squall lines, will cause problems with the SCIT algorithm and may make the output cluttered or unusable.

Storm position forecast (SPF) tracks are based only on previous movement. The algorithm does not forecast changes in intensity, path, or speed, and will not predict deviant movement. By convention a 1-hour SPF is shown with 15-minute ticks.

A history track that shows a curve deviating from the mean tropospheric flow or from other cells may signal that the storm has transitioned to a severe phase.

Graphics display workstations such as WDSS often have a product available called "Cell Trends". This shows, for a given cell, its cell top, cell base, height of the storm centroid, its maximum reflectivity, probability of hail, probability of severe hail, cell-based VIL, and maximum reflectivity.

Product code: 58/STI (RPG), NST (WMO).

real-time Internet sources

cimss.ssec.wisc.edu/severe_conv/psv3.html
www.weathertap.com ($)

Desktop software and mobile apps are extensively used for viewing this type of data.

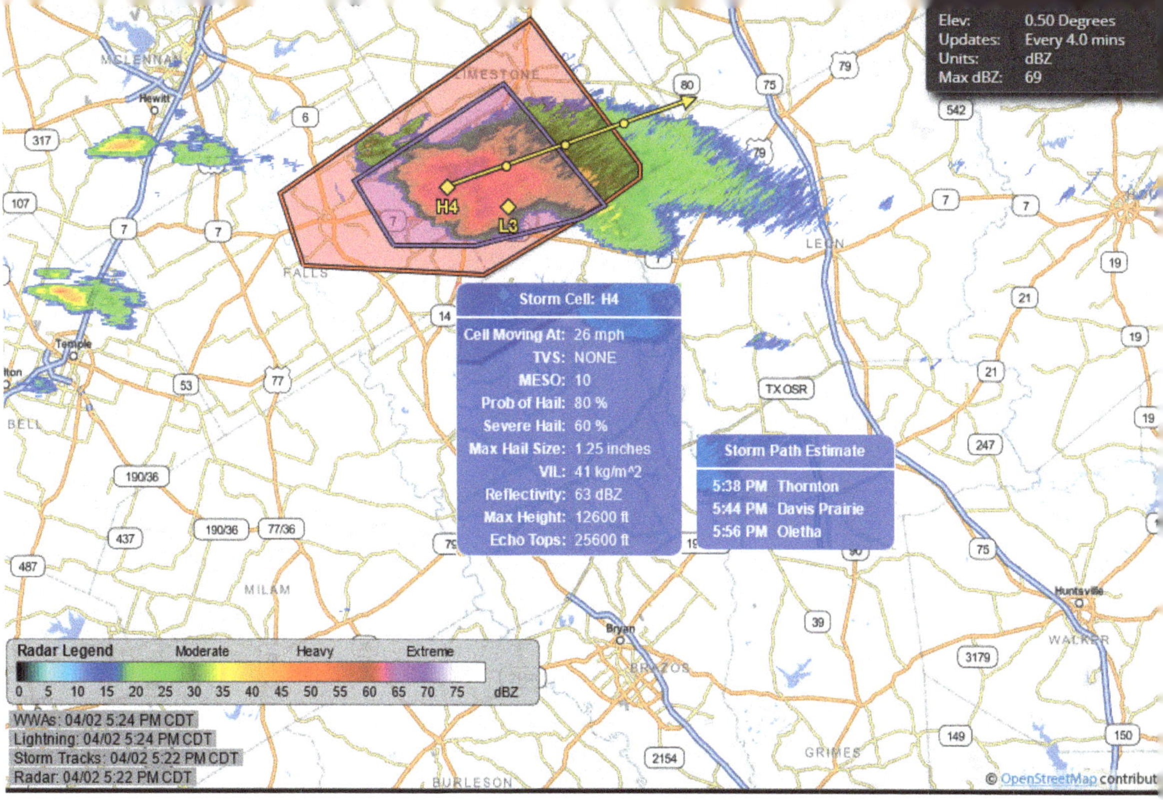

Above: Weathertap.com offers RadarLab Local, probably the most sophisticated web-based interface for radar data (i.e. requiring neither software nor a mobile app). It includes extensive storm tracking information. *(Weathertap)*

Right: Storm attributes as displayed in the standalone radar application GRLevel3. Cell X6 at center has been clicked on, and an attribute popup window appears. This radar scan also appears in the image below. *(GRLevel3)*

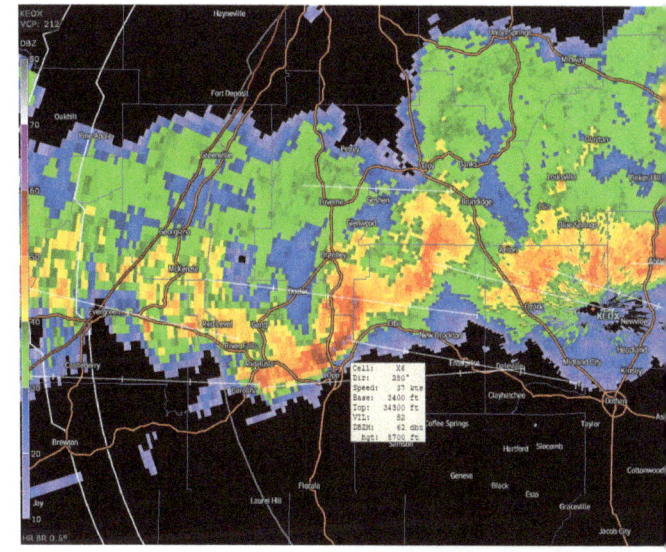

Below: Same radar scan as previous but viewed on the AWIPS system used by the NWS. An attribute table appears at the top with the same basic information. *(AWIPS)*

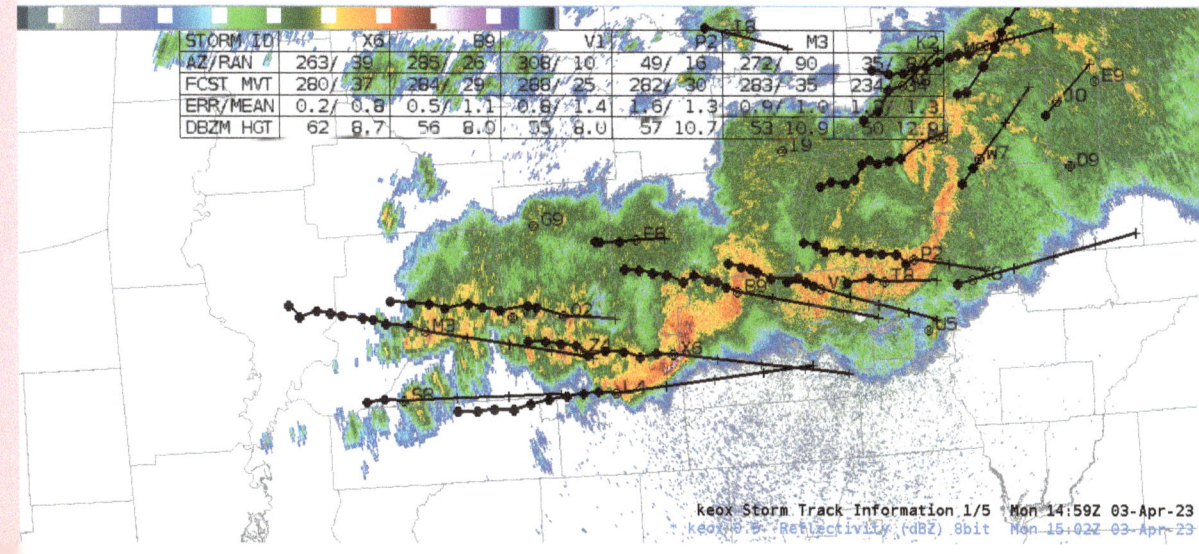

Hail algorithm

The Hail Index is designed to display whether a storm's structure is conducive to hail formation. It incorporates the new Hail Detection Algorithm (HDA), which replaces the old Hail Index Algorithm that was phased out with NEXRAD Build 9 in late 1996. The new algorithm provides more robust detection capability, along with a new set of statistics for the end user such as hail size. An upgrade in 1998 allowed the generation of hail probability. Its input is storm centroids and components from the Storm Cell Identification and Tracking (SCIT) algorithm, along with user-defined melting level information.

The HDA produces three quantities for each location where hail is flagged: Probability Of Hail (POH), Probability Of Severe Hail (POSH), and Maximum Expected Hail Size (MEHS). In all instances, the difference in elevation between the melting level and the radar are factored in, so results are valid for the radar's elevation. Storms in higher terrain may require special evaluation.

The Probability Of Hail (POH) parameter determines the chance of any hail of any size reaching the Earth's surface. It looks for the height of the 45 dBZ echo above the melting level. According to this algorithm, if the 45 dBZ echo is 2 km above the melting level the POH is about 20%, and if it exceeds 6 km the POH rises to 100%.

The Probability Of Severe Hail (POSH) runs a process called the Severe Hail Index (SHI). This process uses storm components from the SCIT algorithm as input "objects" and relates reflectivities to melting level data. It outputs a SHI value in joules per meter per second. From this, a POSH probability value is determined.

The Maximum Expected Hail Size (MEHS) uses only the SHI data, and applies a simple function that increases the expected hail size as the severe hail index grows. It is considered to be the most difficult part of the forecasts output by the HDA, since it cannot forecast unusual hail shapes and minor elevation differences.

How does the algorithm perform? So far it has received excellent reviews. A study done at NWS Wichita suggested that the HDA tends to overestimate hail size, but works very well in strong mid-level flow scenarios.

It must also be pointed out that the hail algorithm may fail in unusual structures, which include highly sheared storms and deviant movers. It may also fail in squall lines, since the Storm Tracking algorithms have trouble with linear structures. Furthermore the algorithm needs to be able to measure the full depth of a storm for best results; not good for storms right over a radar site.

key details

The Hail Detection Algorithm provides a robust technique for measuring the hail potential of individual storm cells, calculating its probability of producing large hail and hail in general, along with maximum hail size. Hail size is an indirect indicator of updraft strength.

By convention, the symbol for hail is a triangle that points upward. Probable hail is identified by a hollow triangle. Positive hail is identified by a filled triangle.

The original NEXRAD hail detection algorithm (HDA) was based on Leslie Lemon's 1978 storm structure concept and defined by Pio Petrocchi, John Smart, and Ron Alberty in 1982-85.

The old HDA was based on seven weighted questions, in order of increasing importance: does the highest storm component reach 8 km or more; does the storm's maximum reflectivity exceed 55 dBZ; is the low-level storm component north of one at a higher level; does the storm exhibit tilt; is the mid-level reflectivity (5 to 12 km) greater than 50 dBZ; does overhang of more than 4 km exist; and does the highest storm component exist above an overhang. The resulting score was flagged as either probable or positive.

The MEHS is calculated by the HDA as:
  MEHS = 2.54 x SHI$^{0.5}$
where MEHS is in millimeters and SHI is the severe hail index in joules per meter per second.

Product code is 59/HI (RPG), NHI (WMO).

real-time Internet sources

www.weathertap.com ($)

Desktop software and mobile apps are extensively used for viewing this type of data.

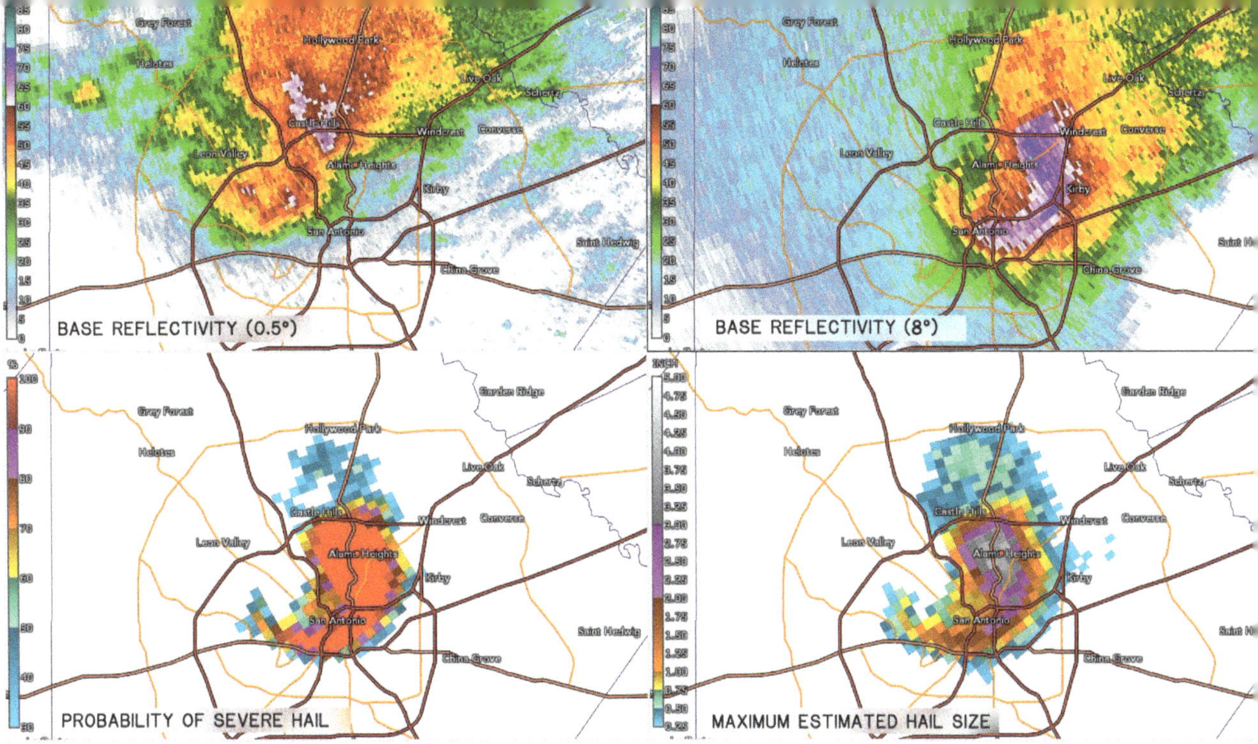

Above: Key NEXRAD Level II hail indicators for the 12 April 2016 hailstorm in San Antonio, Texas, which produced $1.5 billion in damage. Image is for 9:35 pm. The top right image shows reflectivity at 8.0°, corresponding to 21,000 ft AGL, indicating 75 dBZ reflectivities being suspended aloft, caused by wet rimed large hailstones at that level. Very large hail was indicated by the MEHS algorithm. Storm movement was from WNW to ESE and exhibited forward tilt, so hail indicators do not match with the low-level scan, though much of this hail did make it to the surface. Maximum tops were 52,000 ft.

Right, above: MEHS product on 3 June 2018 at 7:08 pm MDT, in which an American Airlines A319 bound from San Antonio to Phoenix turned into a storm over Carrizozo, New Mexico at 34,000 ft and suffered a broken cockpit window and a damaged nose. The flight landed safely in El Paso with no injuries. Indicated MEHS in this storm, which is calculated for the surface (not at flight level), was 2.2 inches. However since hail has always undergone some degree of melt when it reaches the surface, hailstone sizes are always largest within the cloud. *(GR2Analyst, both images)*

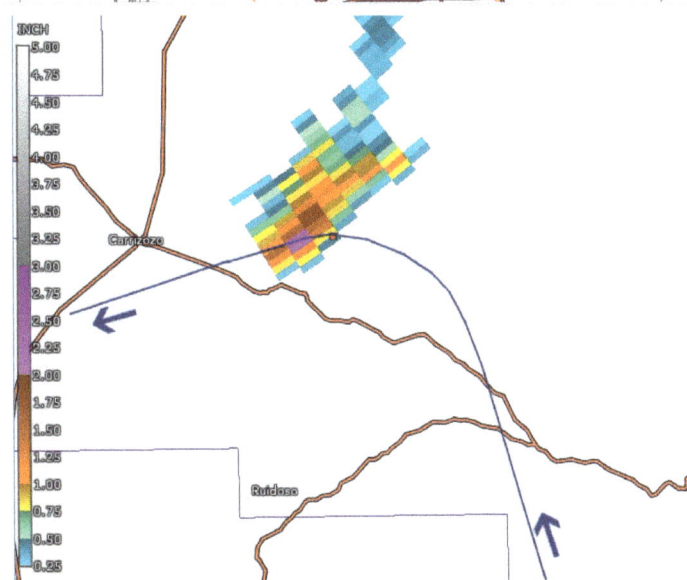

TABLE OF HISTORICAL HAILSTORM VALUES

Event	Location	POSH	MEHS	HRVIL	HRVILD
2010-10-05	Phoenix AZ	100%	4.5"	127-M	13.0-S
2012-04-28	St Louis MO #1	100%	4.3"	127-S	13.0-S
2012-04-28	St Louis MO #2	100%	3.4"	127-S	13.0-M
2012-06-13	Dallas TX	100%	2.9"	104	10.2
2016-04-11	Denton/Wylie TX	100%	3.2"	127-S	13.0-S
2016-04-12	San Antonio TX	100%	4.1"	127-L	13.0-M
2017-05-08	Denver CO	100%	3.1"	121	13.0-S
2017-06-11	Blaine MN	100%	2.8"	127-S	11.8

Maximum values for 20-minute window ending at peak hail event. Values from GR2Analyst, which does not compute values exceeding HRVIL of 127 and HRVILD of 13.0, so they are described with a suffix. S is small areal coverage at this intensity (about 2 nm² or less), M is medium coverage, and L is large coverage (about 8 nm² or more).

Mesocyclone detection

The NEXRAD Mesocyclone Detection Algorithm (MDA) pairs base velocity information with processing power to determine the location of thunderstorm mesocyclones. The idea is not to take the responsibility away from the forecaster, but to provide the means to identify a potentially dangerous storm that might otherwise go unnoticed. The algorithm also identifies areas of uncorrelated shear which alerts the forecaster to rotation in a suspect area that is below the physical thresholds for a mesocyclone.

A mesocyclone is a region of strong, consistent rotation within a thunderstorm. It typically measures several miles in diameter and may have tangential winds of over 40 mph. A mesocyclone signifies a very strong, organized updraft, and is often a precursor to tornado development. The vast majority of tornadoes are spawned within a mesocyclone, and perhaps all supercells contain a mesocyclone. Mesocyclones are typically cyclonic but a few may be anticyclonic. Most develop in the mid-troposphere, and may descend to the surface and itdensify with time.

The processing algorithm starts from scratch with the base velocity product. At each elevation, and range, it examines all of the bins throughout the entire azimuth sweep to find groups of adjacent bins that show an increase or decrease in velocity from one end to the other. The change must span a large enough number of bins to be counted. This value is called the "pattern vector threshold" and can be modified by the radar operator. If the change exceeds the pattern vector threshold, typically ten bins, and meets a threshold value of shear and momentum, the change is classified as a pattern vector.

The MDA tries to link all pattern vectors to form two-dimensional features. When this is completed, it attempts to resolve these features into three-dimensional circulations by linking the features to others at higher elevations.

Based on the success of this three-dimensional correlation, the feature is then identified as one of three things: three-dimensional uncorrelated shear, uncorrelated shear, or a mesocyclone. Three-dimensional uncorrelated shear has vertical but not horizontal consistency. Uncorrelated shear is the opposite; it has horizontal consistency but none in the vertical. A mesocyclone, however, contains both. Finally the MDA creates an attribute table showing information about it.

The algorithm has undergone a series of refinements over the years, with the most significant ones added to the WSR-88D RPG in the late 1990s and 2000s.

key details

The mesocyclone detection algorithm (MDA) provides an automated method for identifying areas of storm-scale rotation and judging whether the rotation has vertical consistency. A mesocyclone is one of the key features of a strong, persistent thunderstorm updraft.

All Mesocyclone Detection Algorithm features are dependent on the accuracy of the base velocity data. Range folding, dealiasing problems, and other anomalies can produce false or spurious indications.

Typical attributes:
- **CIR, CircID**: Unique mesocyclone ID
- **STMID, Cell**: Unique storm ID
- **SR**: 3D strength rank (SR) (1 to 25). SR corresponds to specific levels of DV or shear, thus circulation diameter is considered. 3D SR is the highest 2D value met through the entirety of a 3 km deep column up to 5 km.
- **AZ (AWIPS)**: Azimuth RDA to meso (deg)
- **RAN (AWIPS)**: Distance RDA to meso (nm)
- **LLRV**: Mesocyclone base RV.
- **LLDV**: Mesocyclone base DV.
- **BASE**: Height of mesocyclone base.
- **DPTH**: Depth of mesocyclone.
- **REL_DEPTH, STMREL**: Ratio of mesocyclone depth to storm top (%).
- **HGT (AWIPS)**: Height of MXRV (below)
- **MXRV, MaxRV**: Max RV at all mesocyclone tilts, with this height ("HGT" on AWIPS).
- **MSI**: Mesocyclone Strength Index. An integrated measure of 2D strength ranks throughout the mesocyclone depth, multiplied by 1000.

Notes: DV is delta velocity, the absolute difference between maximum and minimum velocity. RV is rotational velocity, i.e. DV divided by two, thus it measures tangential velocity. If the vortex reaches either 0.5° or 19.5°, a < or > symbol is added.

Product code: 141/MD (RPG), NMD (WMO).

real-time Internet sources

www.weathertap.com ($)

Desktop software and mobile apps are extensively used for viewing this type of data.

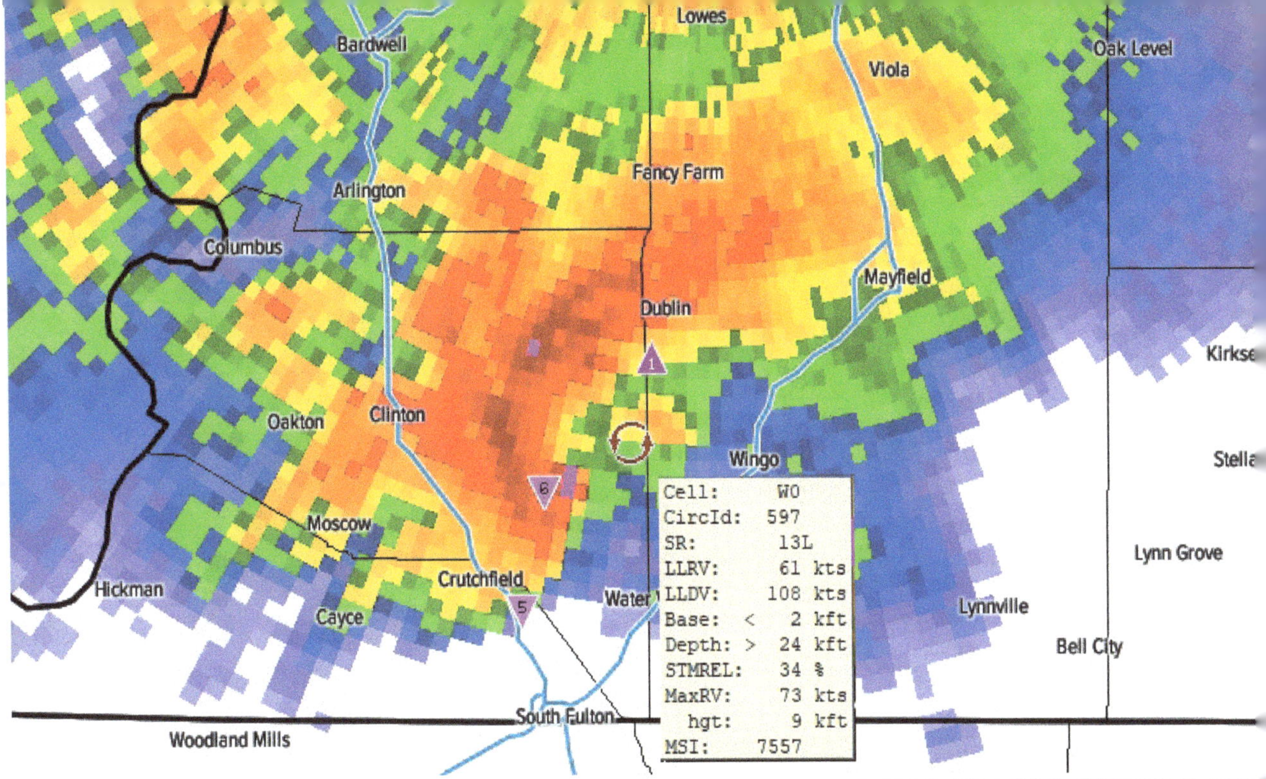

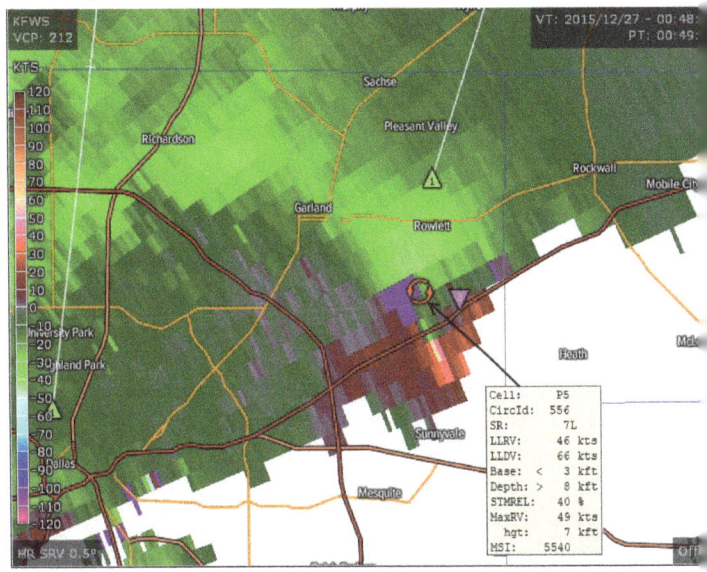

Above: Mesocyclone symbol (circle) with attribute table for the 10 December 2021 Mayfield, Kentucky supercell. It shows an exceptionally strong 3D strength rank of 13 and an MSI of nearly 7600. Strong mesocyclones such as these are often accompanied by TVS icons, and it may be necessary to turn off the TVS overlay temporarily to query for mesocyclone data.

Right: Garland tornado of 26 December 2015. The mesocyclone algorithm plots as a red icon in GRLevel3, which indicates a strength rank of 7 to 8. *(Both images GRLevel3)*

GRLEVEL3 SYMBOLOGY
Solid ring - Low level meso (base at lowest tilt)
Broken ring - Elevated meso
Light green - Very weak meso (SR 1-2)
Green - Weak meso (SR 3-4)
Yellow - Moderate meso (SR 5-6)
Red - Strong meso (SR 7-8)
Purple - Extreme meso (SR 9+)

AWIPS SYMBOLOGY
Circle diameter scaled with meso diameter
Thin circles - SR 1-4
Thick circles - SR 5+
Spikes - Base below 1 km

TABLE OF HISTORICAL MESOCYCLONE VALUES
Maximum values for 60-minute window ending at peak tornado event.

Event	Location	SR	MSI	LLRV	MXRV
2011-04-27	Phila. MS EF5	9	7354	74 kt	88 kt
2011-05-22	Joplin MO EF5	16	10439	86 kt	86 kt
2013-05-20	Moore OK EF5	13	9464	74 kt	91 kt
2013-05-31	El Reno OK EF3*	17	14040	94 kt	104 kt
2015-04-09	Fairdale IL EF4	8	5679	50 kt	52 kt
2015-12-26	Garland TX EF4	7	5540	46 kt	49 kt
2019-10-20	Dallas TX EF2**	8	6655	42 kt	54 kt
2021-12-10	Mayfield KY EF4	13	8171	65 kt	91 kt

* Path over open country precludes higher EF estimates.
** Peaked at late stage of tornado at SR 11, LLRV 52 kt, and MXRV 66 kt.

Tornado detection

Tornado vortex signature (TVS) identification algorithms were developed as early as 1970 at the National Severe Storms Laboratory (NSSL). These algorithms continued to be refined by NSSL over the following decades, culminating in their implemention in the new WSR-88D NEXRAD system where it was known as the TVS Detection Algorithm (TDA). This emphasized that Doppler radars do not detect tornadoes but can identify shear patterns, and can attempt to resolve them into features. In spite of this, a somewhat controversial name shift to "Tornado Detection Algorithm" took place around 2000.

The NEXRAD TDA system was extensively revised in RPG Build 10, released in November 1998. The TDA came with its own logic to scrutinize the entire radial velocity product, freeing it from dependence on the Mesocyclone Algorithm.

Much like the Mesocyclone Detection Algorithm, the TDA builds one-dimensional pattern vectors, then two-dimensional features, then three-dimensional circulations. It then uses altitude, depth, and shear criteria to identify possible tornadoes. TDA actually examines the pattern vectors to calculate shear, rather than finding couplets.

A TVS requires a 3D feature depth of at least 1.5 km, a 3D feature base at 0.5 deg or below 600 meters AGL, and a low-level velocity differential (LLDV) of 25 m s^{-1} (49 kt) or a maximum velocity differential (MAXDV) of 36 m s^{-1} (70 kt).

An Elevated TVS (ETVS) is the same in principle, but exists above both 0.5 deg and 600 meters and requires only a low-level velocity differential of 25 m s^{-1} (49 kt). It generally indicates sharp rotation aloft and is useful for identifying possible developing tornadoes.

When scrutinizing a TVS, consider whether the atmosphere is capable of providing the instability and shear neccessary for a tornado. False alarms do occur. Also note that at ranges beyond 35 nm the TDA is often triggered by strong mesocyclones.

In closing, it must be emphasized that the TDA product, as well as other algorithms, *were never intended to serve as a primary tool for the forecaster* but to supplement what is usually a complex interpretation process. The forecaster must interrogate all available data, including radar images and spotter reports, and produce a coherent picture of what is happening. This must be integrated with one's own experience and understanding of meteorological and radar detection principles.

key details

The Tornado Detection Algorithm (TDA) identifies velocity patterns that correlate with tornado development. The output of this algorithm is a TVS (tornadic vortex signature). Thus, TDA is more correctly described as a TVS detection algorithm.

The standard symbol for a non-elevated tornadic vortex signature (TVS) is a filled triangle that points downward. An elevated signature is hollow.

The TDA algorithm uses gate-to-gate velocities (i.e., connecting gates between adjacent radials) which exceed a predetermined threshold for cyclonic rotation. These are known as pattern vectors. A minimum of three adjacent pattern vectors exceeding a required threshold velocity produces a 2D feature. This pattern matching is done with a decreasing series of threshold velocities. Then adjacent 2D features are combined vertically into 3D features.

Typical attributes:
- **Number on triangle** (GRLevelX): Tens digit of LLDV in knots, a measure of strength.
- **AVGDV**: Average DV through TVS depth.
- **LLDV**: DV at base of TVS.
- **MDV, MAXDV**: Maximum 2D DV, any level.
- **DPTH**: Depth of TVS.
- **BASE**: Height of TVS base.
- **TOP**: Height of TVS top.
- **MXSHR** (GRLevelX): Level of maximum shear divided by TVS diameter.

Notes: DV is delta velocity, the absolute difference between maximum and minimum velocity. If the vortex reaches either 0.5° or 19.5°, a < or > symbol will be indicated with dimensional measurements.

Product code: 61/TVS (RPG), NTV (WMO).

real-time Internet sources

www.weathertap.com ($)

Desktop software and mobile apps are extensively used for viewing this type of data.

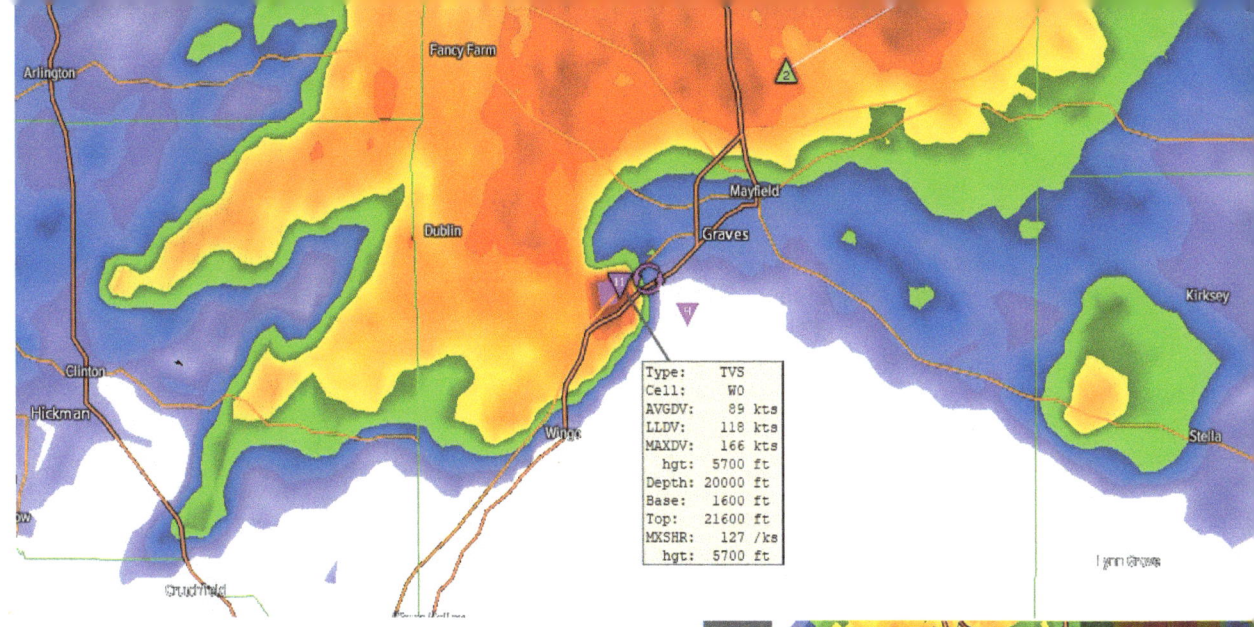

Above: TDA plot and TVS signature (purple triangle) for the Mayfield, Kentucky tornado on 10 December 2021 at 9:23 pm CST. It struck at night from a fast-moving supercell, destroying the town. Technical TVS parameters are displayed in a popup panel, providing differential shear values and dimensions of the circulation. A weaker secondary satellite tornado is indicated 2 nm to the southeast.

Right: Mayflower, Arkansas EF4 tornado, 27 April 2014.

Below: TDA plot for the 31 May 2013 El Reno tornado at 6:19 pm CDT, when it was near its peak. The TDA attributes show differential velocities ranging from 131 to 190 kt. The broad-scale couplet is the mesocyclone and is centered on the round mesocyclone ID mark. *(All images by GRLevel3)*

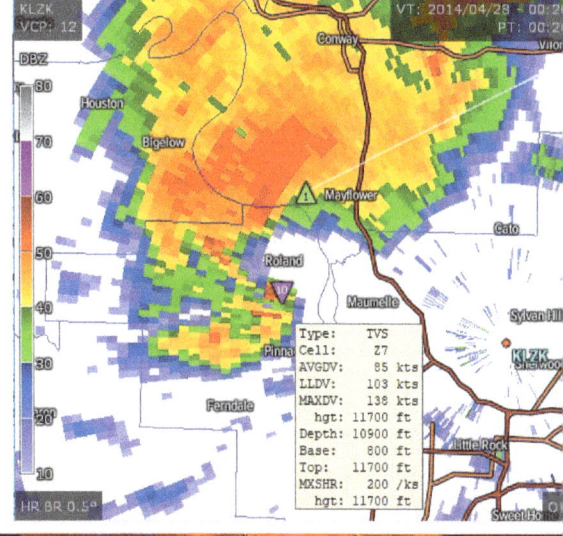

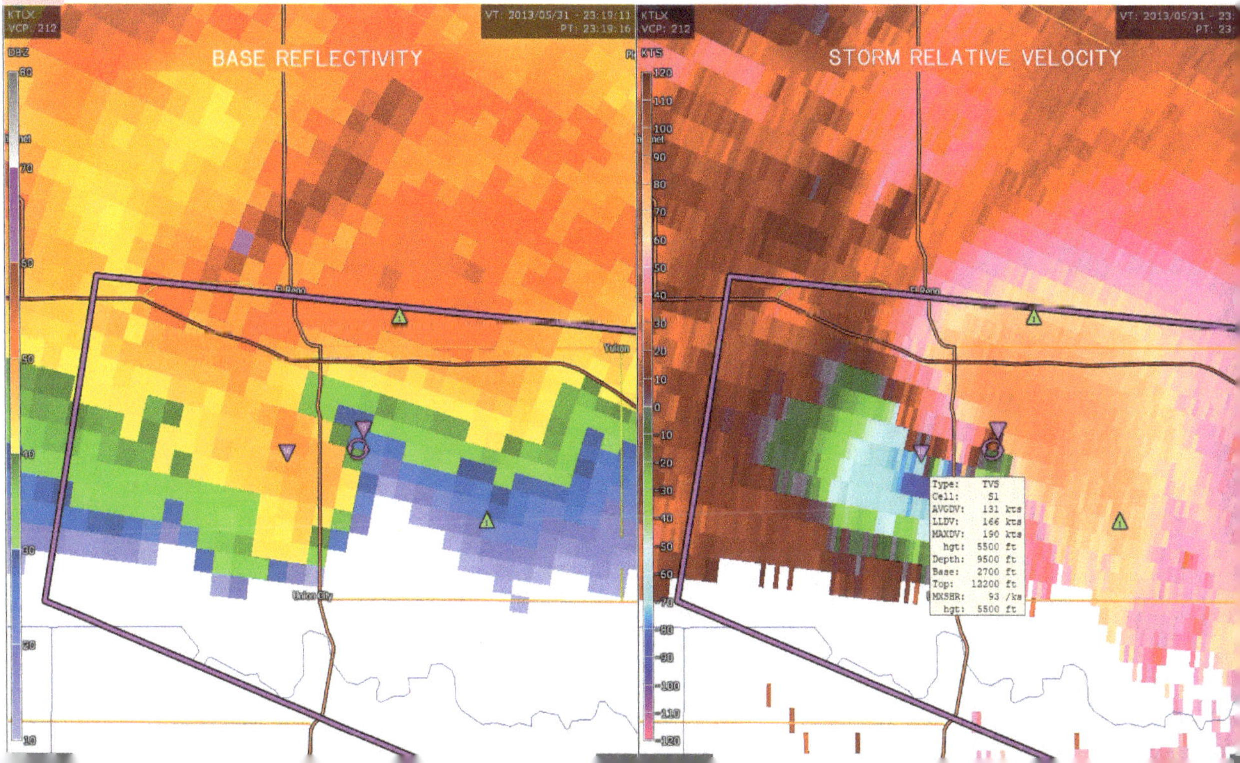

VAD wind profile

The Velocity Azimuth Display (VAD) Wind Profile (VWP) displays a time-height diagram of wind direction and speed. The data is derived from base velocity data. VWP data is provided every 1,000 ft from 1,000 to 70,000 ft MSL. The data is color coded to show the root mean square error in knots, which is inversely proportional to the reliability of the data at that height.

The VAD algorithm first waits for all reflectivity and velocity data to be processed, dealiased, and range unfolded. It then sequentially figures a wind speed at each height level. It does this by finding the scan elevation and range closest to a distance called the VAD Analysis Range or VAD Optimum Slant Range (usually 18.6 nm), which is a direct distance to the scatterer rather than a horizontal range. The algorithm analyzes all of the scatterers at that range and elevation on a sweep of all azimuths.

If 25 scatterers are detected, the algorithm attempts to fit a sine wave to the azimuth and velocity of all the data points. The sine wave configuration is used because an ideal plot of winds that are constant throughout the scan volume will resemble a sine wave if the velocity is plotted with respect to azimuth. Thus the data will fit this curve.

The processor checks the fit of the data to the sine wave by computing a root mean square (RMS) error. This determines the reliability of the entire sample. A symmetry analysis is then done by comparing the departure of the fitted curve's baseline (zero-velocity line) from that of a standard sine wave. If the data fails the analysis by not meeting either check it is discarded and the data is considered void at that level.

During convective weather situations, changes in the low levels of the atmosphere may show important trends in storm-relative helicity. Also strengthening of upper-level flow may hint at upper-level dynamics and increasing bulk shear moving into the threat area.

The VWP may indicate coupling or decoupling of the boundary layer, especially at night or during the morning. This is most obvious when lower-tropospheric winds remain constant but winds in the lowest 1 to 2 thousand feet drop to calm after dark or increase to match the lower-tropospheric winds during the morning.

Also frontal inversions can be assessed using the VWP product, signified by two layers with markedly different wind regimes. The VWP can reveal the depth and character of the cold air mass. If the front itself is too close to the radar site, though, the radar may have trouble finding a representative wind figure in the lowest level.

key details

VWP shows a profile plot of winds aloft very much like wind profiler systems. However the sample is volumetric, covering a region nearly 200 miles wide, and so important features may be smoothed over. Another significant difference is the VWP requires many more scatterers than wind profilers.

VWP requires a significant number of scatterers to work. The best scatterers are produced by dust, insects, and cloud droplets. If scatterers are not present, a data void will occur. This is quite common, especially during good weather.

Bad data will be produced by birds, air mass boundaries, and thunderstorms. This will reflect velocity signatures that differ from the mean wind at that level. It may increase the RMS error at that level or cause it to be rejected altogether.

The VWP product is not used as often as it should be for identifying trends in low-level shear that could increase the potential for severe thunderstorms.

A RMS error of over 9.7 kt or a symmetry error of over 13.6 kt will invalidate the data for that level.

For a given level, the highest amplitude of the VWP sine wave is considered to be the wind speed, while the phase of the highest inbound amplitude is considered to be the wind direction.

A void will occur when any data sample fails the VWP algorithm.

Product code: 48/VWP (RPG), NVW (WMO).

real-time Internet sources

weather.cod.edu/satrad
www.weathertap.com ($)

VAD Wind Profile (VWP). This graph shows wind speed and direction as a function of altitude (Y) and time (X). In this example, the column on the far right represents the most current observation. Winds are southwesterly at about 15 kt near the surface and 40 to 50 kt in the upper troposphere. Note the color coding used for the RMS error. It is important to examine the scale, shown here at the bottom, to understand what colors are being used for good data.

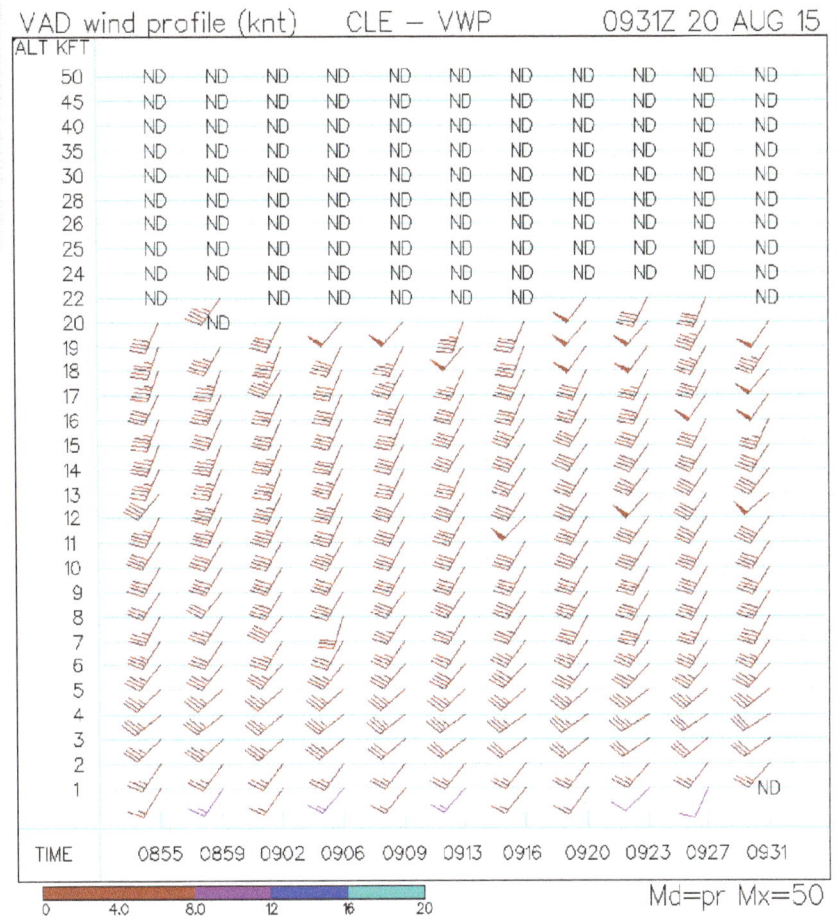

Velocity Azimuth Display (VAD). Not to be confused with a Velocity Wind Profile (VWP), this is what a velocity azimuth display really looks like, as obtained from a hardcopy at a WSR-88D site. It is a graph of azimuth (X) versus velocity (Y) at a given height, in this case 13,000 ft. This is a primitive radar product and is unlikely to be found on the Internet since a single volume scan could generate dozens of these graphs. However it serves a useful example by showing the sine wave and data points from which a VWP product is constructed.

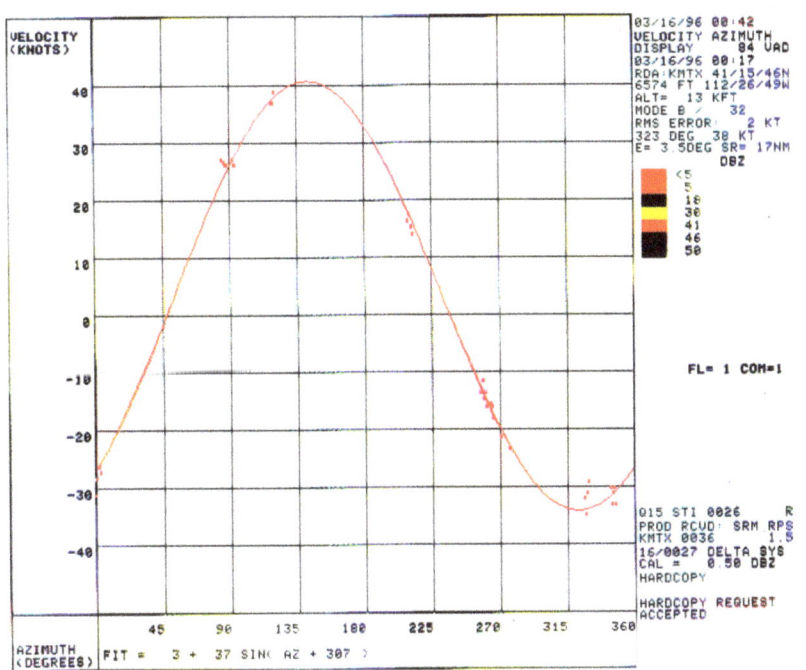

Free text message

FTMFWS

MESSAGE DATE: AUG 14 2014 23:51:00

THE KFWS 88D RADAR HAS A WIDEBAND COMMUNICATIONS PROBLEM. THE TELECOMMUNICATIONS COMPANY IS AWARE OF THE PROBLEM...HOWEVER THE ISSUE WILL LIKELY NOT BE RESOLVED UNTIL FRIDAY MORNING. THE RADAR IS EXPECTED TO BE DOWN THROUGH 8/15/2014 AT 18Z. UPDATES WILL BE PROVIDED AS TECHNICIANS TROUBLESHOOT THE COMMUNICATIONS FAILURE.

When the radar goes offline, it's helpful to be able to quickly locate information on what the problem is and when the radar will be back up. The WSR-88D designers had a plan in mind: the Free Text Message (FTM). This is transmitted over normal data distribution circuits.

The product is available for all WSR-88D radar sites, and it gives the local NWS Forecast Office the ability to post public information pertinent to the radar. This reaches the hands of not only hobbyists and media forecasters, but also air traffic control centers, other weather offices, and the NOAA Radar Operations Center.

```
==================
NOUS63 KIND 231240
FTMIND

WSR-88D OUTAGE NOTIFICATION
NATIONAL WEATHER SERVICE INDIANAPOLIS IN
840 AM EDT THU MAR 23 2023

THE NATIONAL WEATHER SERVICE INDIANAPOLIS (KIND)
RADAR WILL BE OUT OF SERVICE FOR SLEP GENERATOR
REPLACEMENT UNTIL MARCH 29TH.   WE APOLOGIZE FOR
ANY INCONVENIENCE.

ALTERNATE NEARBY RADARS INCLUDE...KILX...KIWX...
KLVX...KVWX... KILN AND TIDS.

$$


LASHLEY

==================
TTAA00 KBTV 120259
WSR-88D OUTAGE NOTIFICATION
NATIONAL WEATHER SERVICE BURLINGTON, VT
952 PM EST SAT MAY 11 1996

TO: ALL KRMX USERS

FM: BTV

KRMX REMAINS OFF LINE AS OF 03Z. FORT DRUM
ADVISES THAT THEY HAVE NOT BEEN ABLE TO REACH
ANY MAINTENANCE PERSONNEL AND SO RADAR LIKELY
DOWN OVERNIGHT AT LEAST...THEY WILL HOWEVER
CONTINUE TO TRACK SOMEONE DOWN.

Y
```

key details

Free Text Messages (also sometimes called Status Messages) reveal important information about outages, defective equipment, and other conditions that may affect the radar unit. It is sent over the standard NOAAPORT text feed.

When a radar is not generating any products or the data looks erroneous, look at the Free Text Message to find what is causing the problem and to determine when the site will be back online.

NWS Instruction 10-2201 designates the use of free text messages. It states: "Should the radar become nonoperational, the back up office will send a free text message notifying users of the outage and expected time of return to service. In this case, the backup office will use neighboring radars, including supplemental WSR-88Ds (Air Force), which have overlapping coverage to cover the shortfall."

Free Text Messages are aggregated at NCEP, which operates a website showing the nationwide radar status: radar2pub.ncep.noaa.gov

During much of the 1990s FTM messages could only be entered on a UCP (Unit Control Position) terminal that was co-located with a forecast office. Radar tasks such as these have now been integrated with the AWIPS workstation system.

Product code: 75/FTM (RPG), FTM (WMO).

real-time Internet sources

www.weather.gov/nl2/NEXRADView
radar2pub.ncep.noaa.gov (replaces ROC status page)
forecast.weather.gov/product_types.php
tgftp.nws.noaa.gov/data/raw/no/ (search FTM)
textwx.com/list/product/FTM

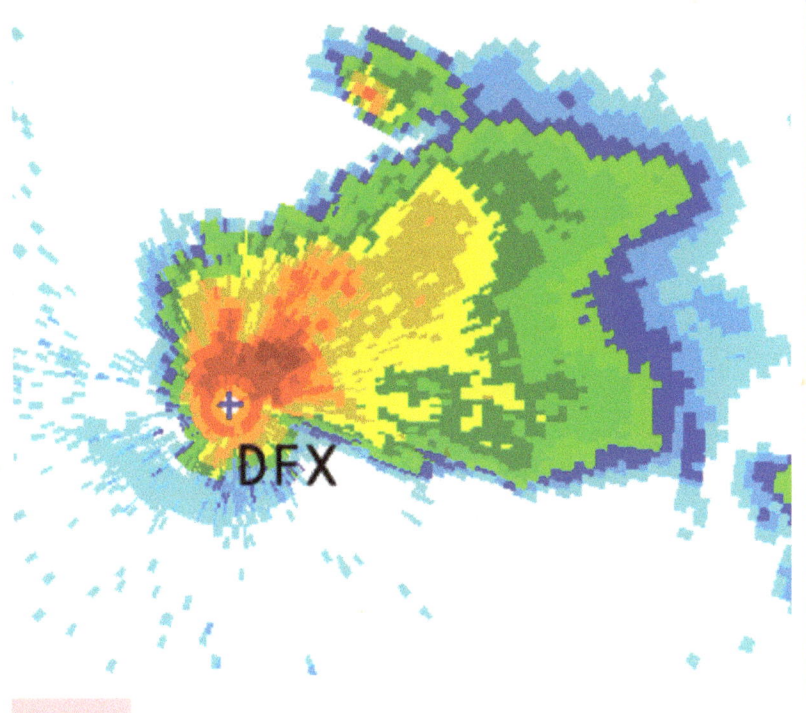

```
..........................................
000
NOUS64 KEWX 260612
FTMDFX

THE RADAR WILL BE DOWN UNTIL FURTHER NOTICE DUE TO AN UNEXPECTED
OUTAGE. SORRY FOR THE INCONVENIENCE. 26/7Z


..........................................
000
NOUS64 KEWX 260843
FTMDFX

THE KDFX RADAR IS OUT OF SERVICE FROM A DIRECT HIT BY A SEVERE
THUNDERSTORM.   RADOME DAMAGE HAS OCCURRED.   IT IS UNKNOWN HOW LONG
THE RADAR WILL BE OUT OF SERVICE BUT IT WILL TAKE SEVERAL DAYS TO
ASSESS THE OVERALL DAMAGE AND MAKE REPAIRS.   345 AM CDT MAY 26 2001
JDW


..........................................
000
NOUS64 KEWX 130723
FTMDFX

THE KDFX RADAR IS BACK IN SERVICE.         13 JULY 2001   PASHOS, FIC


..........................................
```

On May 26, 2001 at 1:10 am CDT the Del Rio, Texas WSR-88D radar was hit with 80 mph wind and hail, destroying the 39-foot diameter fiberglass radome. It took 49 days to bring the site back online. The Free Text Message product shows a sequence of messages during this event. *(NWS photo)*

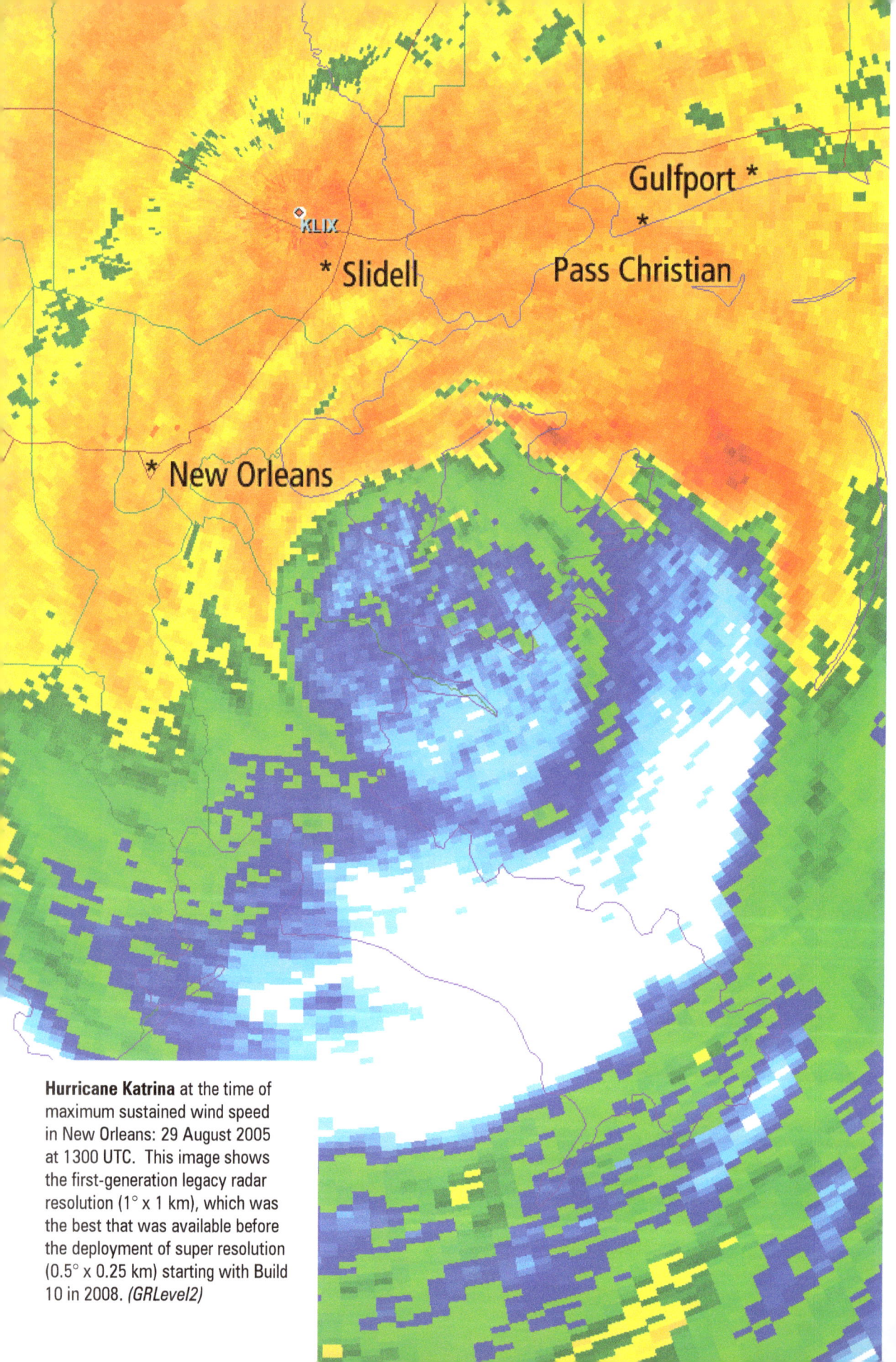

Hurricane Katrina at the time of maximum sustained wind speed in New Orleans: 29 August 2005 at 1300 UTC. This image shows the first-generation legacy radar resolution (1° x 1 km), which was the best that was available before the deployment of super resolution (0.5° x 0.25 km) starting with Build 10 in 2008. *(GRLevel2)*

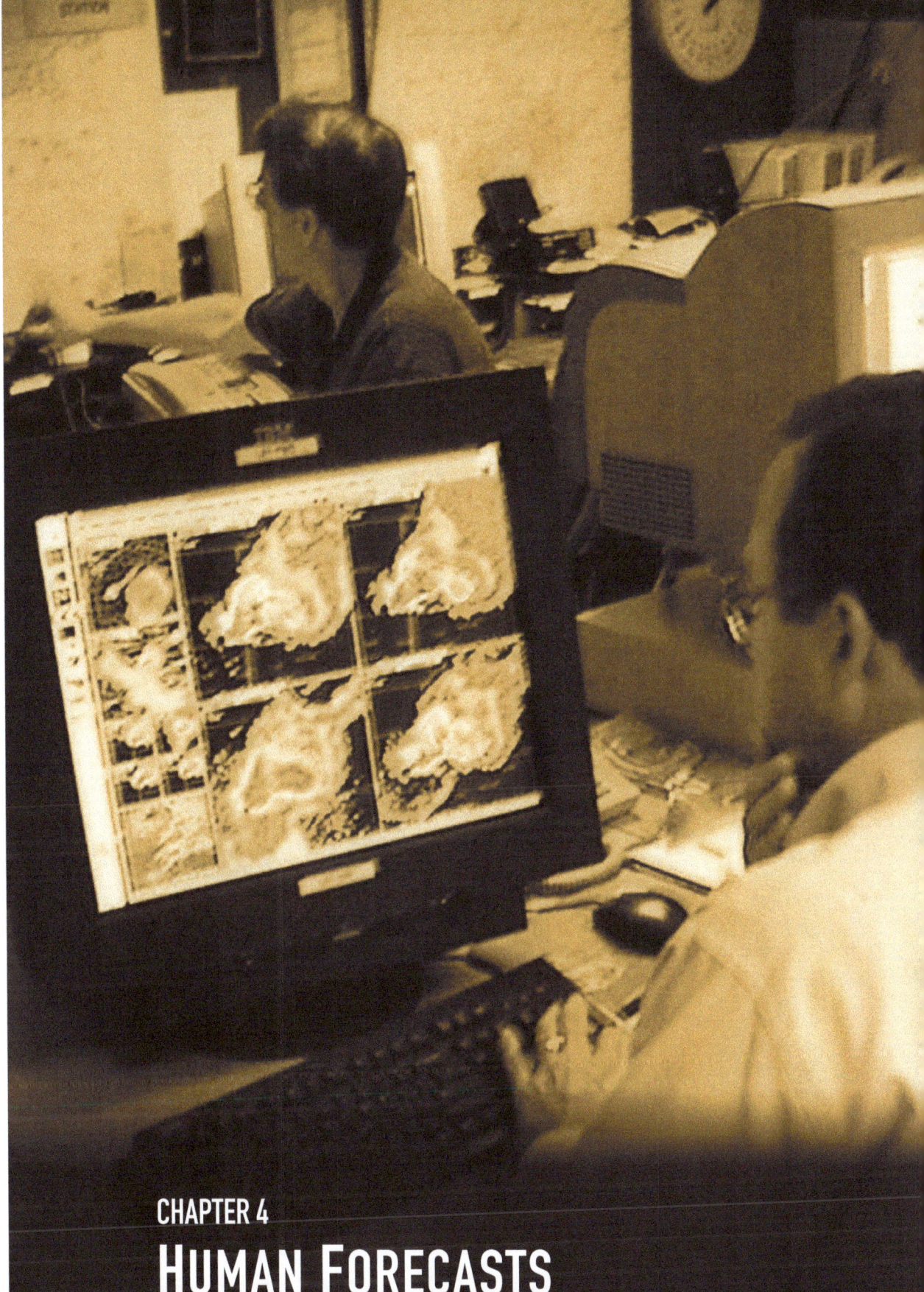

CHAPTER 4
HUMAN FORECASTS

Storm Prediction Center

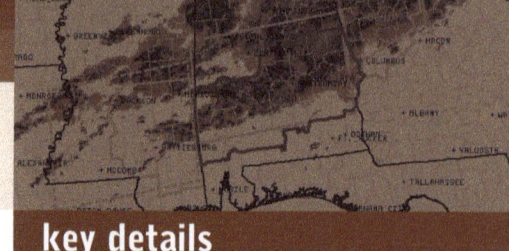

The Storm Prediction Center (SPC) is a NOAA agency that was originally created at Kansas City in 1952 as the Severe Local Storms Unit (SELS), renamed to the National Severe Storms Forecast Center (NSSFC) in 1966. It then took on its current name in October 1995 when it moved to Norman, Oklahoma. The agency is staffed 24 hours a day by a team of highly qualified severe weather forecasters.

SPC's most recognizable product is its tornado and severe thunderstorm watches, which have been issued to the public since 1952. These identify enhanced potential for these hazards and are intended for the general public.

To seasoned forecasters as well as storm chasers, a highly recognized product is the convective outlook. This is an official forecast transmitted five times a day which establishes the threat of severe thunderstorms across the United States and divides them into several risk categories. Originally there was only a Day One (0-24 hour) and Day Two (24-48 hour) outlook, but the Day Three product was added in 2001 along with the Day 4-8 products following in 2007.

A vast array of observational data and models, along with rich forecast experience, are used to develop the convective outlook, and they include a detailed technical discussion. This makes them highly valuable for assessing severe weather risk, and they are referred to by NWS forecast offices nationwide.

SPC also provides numerous other useful products. The Mesoscale Discussion (MCD) identifies areas of concern that may warrant a severe thunderstorm or tornado watch within the next few hours. It is usually transmitted when there are significant indicators of impending thunderstorm activity or an increased threat. Storm reports provide an ongoing tabulation of confirmed severe weather. Past reports may be accessed on the website going back to June 1999; earlier reports are available at the National Climatic Data Center (NCDC) in the Storm Reports publication.

All forecasters should explore the "Forecast Tools" tab. This offers access to an upper air data viewer, mesoscale data fields from the Rapid Refresh model, special fields from the High Resolution Rapid Refresh model, short-range ensemble model displays, a detailed sounding climatology, short range ensemble plumes, and composite weather maps. An archive page at <www.spc.noaa.gov/archive> provides access to historical products.

key details

The Storm Prediction Center (SPC) was known as the National Severe Storms Forecast Center (NSSFC) until 1995. This closely coincided with its move from Kansas City, Missouri to Norman, Oklahoma in 1997.

SPC issues only watches and outlooks. It does not issue warnings. These are the responsibility of local forecast offices.

The SPC website should be explored as it contains extensive storm climatology and a number of experimental forecasting tools.

RISK DEFINITIONS

General. There is a 10-percent or better chance of thunderstorms.

Marginal (MRGL). Organization, longevity, coverage and/or intensity of any storms wil be limited, Replaces the SEE TEXT category.

Slight risk (SLGT). Well-organized severe thunderstorms are expected but in small numbers and/or coverage.

Enhanced (ENH). There is a greater risk of severe weather than would be indicated by SLGT, but not warranting MDT.

Moderate risk (MDT). Great concentration of severe storms. Verification is made up of either 30 or more reports of 1-inch or larger hail, and/or 6 to 19 tornadoes, and/or 30 or more wind events.

High risk (HIGH). A major severe weather outbreak. At least 20 tornadoes are expected with at least two producing F3 damage or an extreme derecho event with widespread wind damage.

real-time Internet sources

www.spc.noaa.gov/products/outlook
kamala.cod.edu/spc

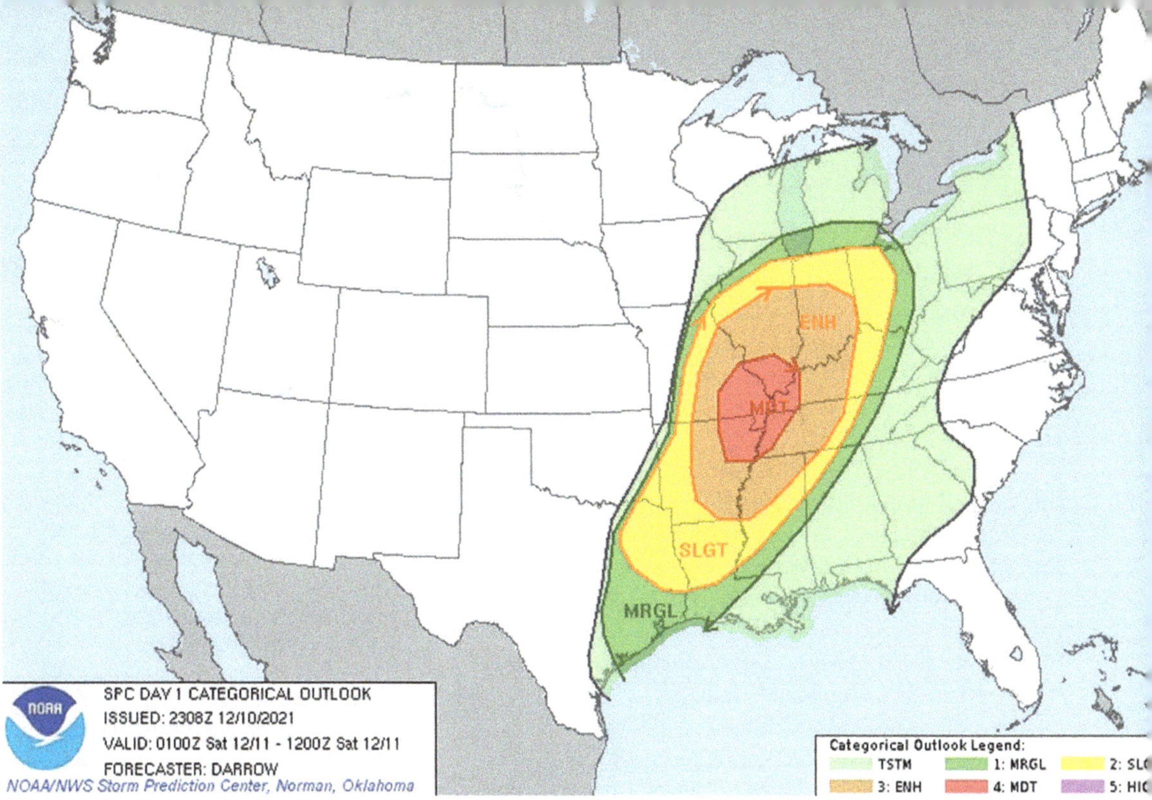

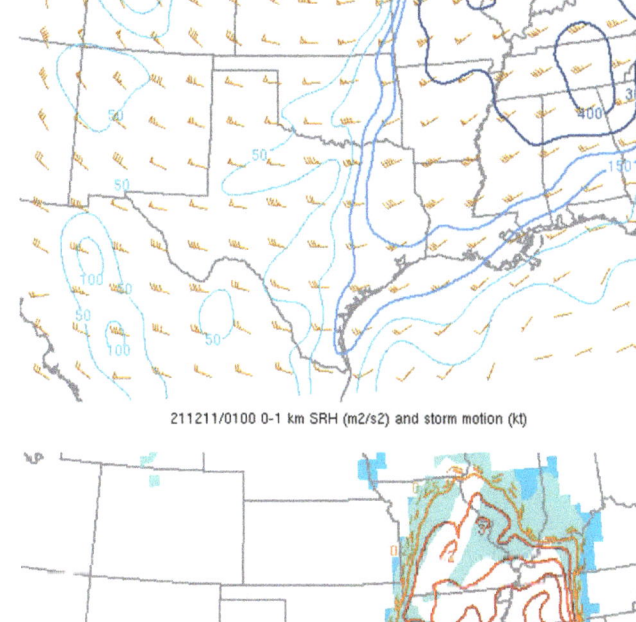

Top: Day 1 Convective Outlook preceding the central Mississippi River and Midwest tornado outbreak of December 10, 2021. This brought several long-tracked tornadoes, including one that extended from near Jonesboro, Arkansas to north of Bowling Green, Kentucky. Another long-track tornado moved through much of southern Kentucky and tracked through Bowling Green. Mayfield, Kentucky suffered catastrophic damage. Other tornadoes ripped through parts of northern Tennessee, Missouri, and southern Illinois. *(SPC)*

211211/0100 0-1 km SRH (m2/s2) and storm motion (kt)

Middle: 0-1 km storm relative helicity product from the SPC experimental products page for the above date, generated from RAP model data. *(SPC)*

211211/0100 Significant Tornado Parameter (eff layer) and MLCIN (J/kg, shaded at 25 and 1

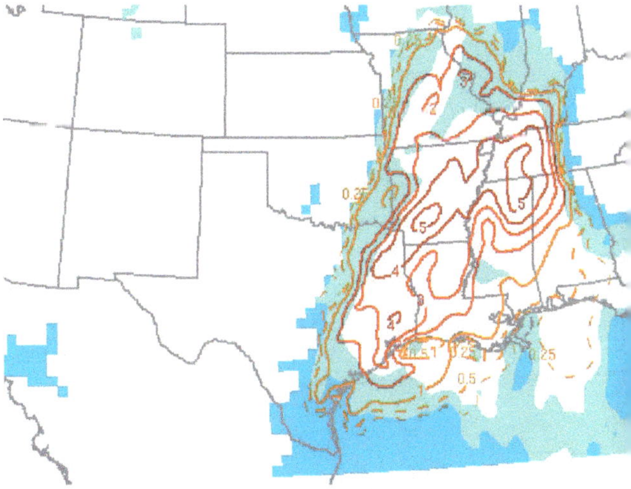

Bottom: Significant tornado parameter product from the SPC experimental products page for the above date. This identifies the enhanced environment across much of Arkansas and neighboring states to the east. *(NOAA/SPC)*

Weather prediction center

The Weather Prediction Center (WPC) in Camp Springs, Maryland assumes much of the centralized forecast guidance for United States weather offices. It was known as the Hydrometeorological Prediction Center (HPC) in Camp Springs until a March 5, 2013 name change.

The department came into being in late 1995 with the reorganization of the National Meteorological Center (NMC) into the National Centers for Environmental Prediction (NCEP). The main objectives of WPC are to provide quantitative precipitation forecasts, medium-range public forecasts, numerical model diagnostics and discussions, surface analysis responsibilities, and an international desk for visiting meteorologists.

One of the most visible functions of WPC is the production of rainfall and flooding forecasts. All NWS offices refer to the products from WPC when assessing these risks. There is also a heavy emphasis on winter storm guidance; numerous products are available here for estimating the potential for snowfall.

The national forecast graphics are a useful chart for forecasters who want a quick summary of expected weather across the nation. They are often copied verbatim by broadcast media outlets to show the next day's weather. The graphics provide excellent information on the extent of precipitation areas and the expected transition lines between rain and snow.

The forecast tools tab provides a number of useful tools. Every forecaster should take some time to explore what's available here:

■ **Detailed national change maps** are available, showing change in many different parameters at 1, 3, 6, and 24 hour intervals

■ **Ensemble Situational Awareness Table** provides regional and local identification of anomalous temperature, wind, pressure, moisture, precipitable water, IVT, and other values extracted from the GEFS and European ensemble models

■ **Prototype Excessive Rainfall Maps** are quite detailed provide a closer look at forecast hazards within each NWS forecast area and are ideal for agencies that must monitor flooding risks.

■ **Weather In Context** is an interface identifies record-setting temperature and precipitation forecasts from the National Digital Forecast Database.

key details

WPC is a NCEP service center, which makes it a sister facility of the Storm Prediction Center, the National Hurricane Center, the Climate Prediction Center, and the Aviation Weather Center.

Most of the HPC prognosis charts are human-made and incorporate a blend of model guidance and forecaster experience. However check the product descriptions carefully before you begin using them as part of your forecast process.

Explore the WPC's text products as there are bulletins available which offer extensive forecast insight into continental and hemispheric patterns.

The United States 12/24/36/48 hour prognosis charts are "instantaneous" graphics, therefore the movement of weather systems and precipitation areas must be considered in between valid time periods.

HPC forecast charts were originally transmitted via DIFAX circuits. The charts adopted an Internet graphic format during the 1990s, and the fax version was discontinued on 15 January 2003.

The Weather Prediction Center is responsible for the Daily Weather Map (DWM) product, the official plot of surface conditions each morning that becomes part of the climatological record. This product has been generated since the late 19th century! The WPC holdings only go back six months, however the NOAA Central Library rescue project has maps for 1871 to 2004, some of which are highly detailed in the 1950s and 1960s. The WPC Daily Weather Map page has a link to this Central Library page.

real-time Internet sources

www.wpc.ncep.noaa.gov
www.wpc.ncep.noaa.gov/html/discuss.shtml

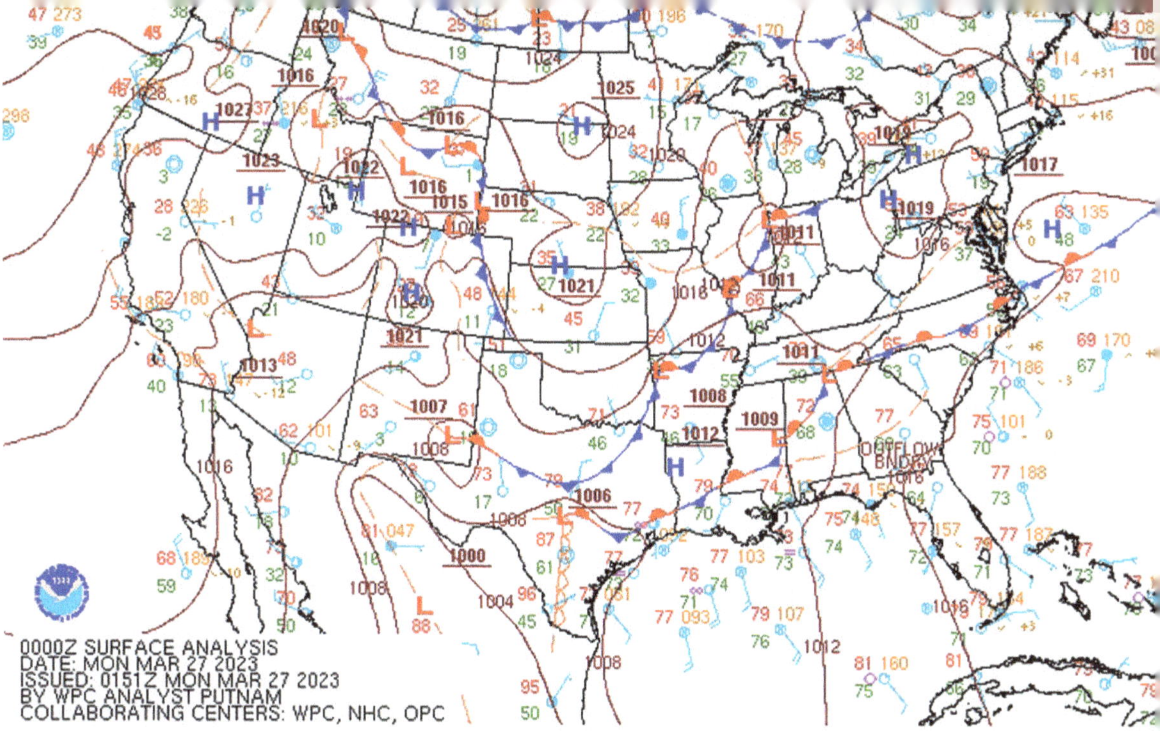

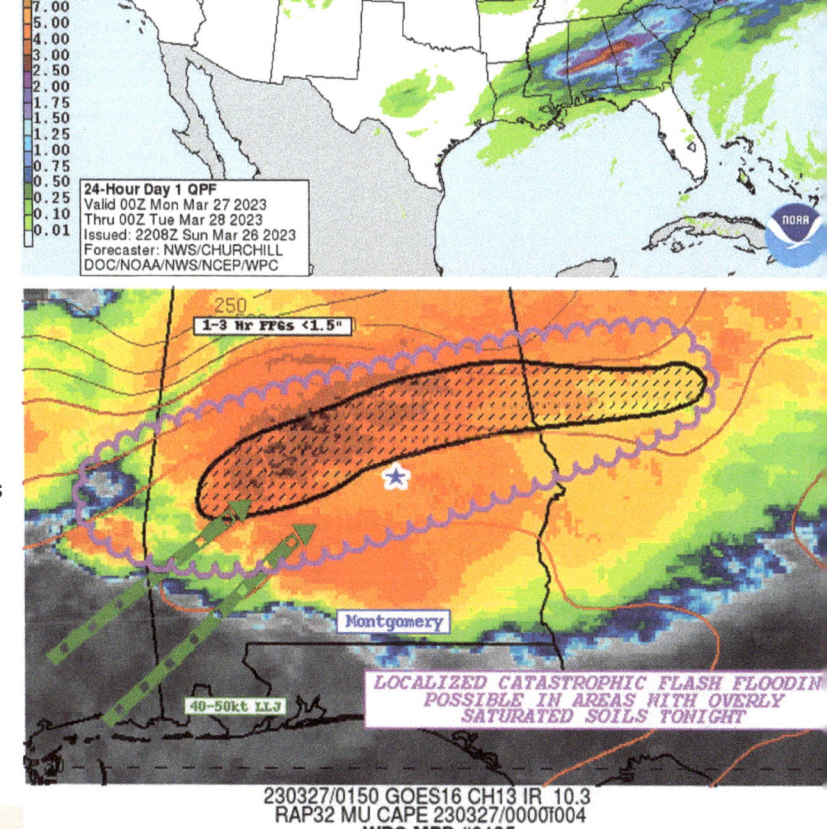

Top: The WPC surface analysis products are not as detailed as they were in the 1990s and earlier, but they still provide a good first-look at placement of fronts and other features. All fronts are still located manually by human forecasters; there is still no computer algorithm that is able to construct frontal placements from raw data, nor is one likely in the near future. *(NOAA/WPC)*

Right: WPC QPF (quantitative precipitation forecast) which offers some of the best estimates of potential rainfall at different time periods out to 7 days in the future. *(NOAA/WPC)*

Bottom: WPC flash flood guidance, with an accompanying discussion (not shown). WPC coordinates and tracks all hazards from flooding due to weather. *(NOAA/WPC)*

230327/0150 GOES16 CH13 IR 10.3
RAP32 MU CAPE 230327/0000f004
WPC MPD #0135

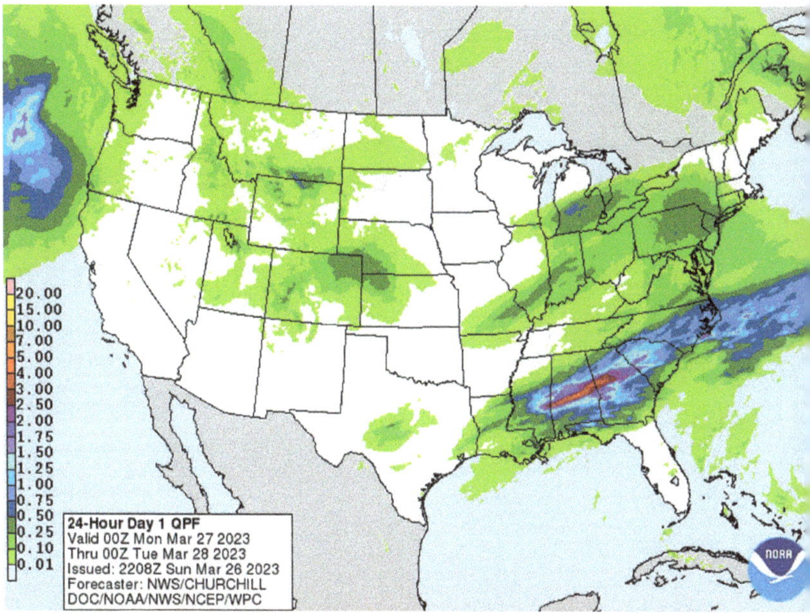

Area forecast discussion

There is probably no better insight into a National Weather Service forecaster's mind than the Area Forecast Discussion (AFD). These bulletins are composed before a forecast package is issued, typically twice a day with updates two more times per day. Different portions are now written by different forecast positions at each office.

The content of the AFD is governed by National Weather Service Instruction 10-503, and specifies that they are intended for federal agencies, weather-sensitive officials, and the media, and may be used for coordination between weather offices. Contents are expected to include the primary forecast problem, an indication of confidence, reasons behind advisory/warning issuance, reasons for deviating from model guidance, and other topics.

The AFD was known as an SFD (State Forecast Discussion) and fell under the FPUS3 header before a restructuring initiative took effect in 1999. The bulletins are now transmitted under the FXUS headers.

key details

The Area Forecast Discussion (AFD) is designed to allow the forecaster to explain their forecast thinking to federal, state, and local government officials and to the public. They are sometimes analyzed by researchers to understand the forecast process and improve the available tools.

Who wrote the bulletin? The text is usually followed either by the forecaster's last name or an internal code identifying the forecaster.

real-time Internet sources

weather.cod.edu/text
www.nws.noaa.gov/view/states.php
tgftp.nws.noaa.gov/data/raw/fx/ (raw bulletins)

FORECAST DISCUSSION CONTRACTIONS

ABNDT abundant	CDFNT cold front	FQT frequent	LFTG lifting	NGM nested grid model	QPF quantitative precipitation forecast
ABT about	CG cloud to ground	FRM form	LGRNG long range	NGT night	
ABV above	CHC chance	FROPA frontal passage	LGWV long wave	NIL none	RCV receive
AC convective outlook	CI cirrus	FRZN frozen	LI lifted index	NLY northerly	RGL regional model
ACLT accelerate	CIG ceiling	FT terminal forecast	LIS lifted indices	NNNN end of message	RGN region
ACPY accompany	CNTY county	FWD forward	LK lake	NRLY nearly	RNFL rainfall
ACTV active	COR correction	GEN general	LKLY likely	NTFY notify	ROT rotate
ADJ adjacent	CPBL capable	GND ground	LLJ low level jet	NVA negative vorticity advection	RPLC replace
ADL additional	CRNR corner	GRDL gradual	LMTD limited		RSG rising
ADVCT advect	CSDRBL considerable	GRT great	LN line	NXT next	SA surface observation
ADVN advance	CST coast	HAZ hazard	LRG large	OBND outbound	SCT scatter/scattered
ADVY advisory	DBL double	HGT height	LST local standard time	OBS observation	SEPN separation
ACFTG affecting	DCR decrease	HLF half	LTD limited	OBSC obscure	SGFNT significant
AFT after	DFNT definite	HLSTO hailstones	LTG lightning	OCLD occlude	SKC sky clear
AFTN afternoon	DISC discussion	HV have	LTL little	OCNL occasional	SLP slope or pressure
ALG along	DLA delay	HVY heavy	LYR layer	OCR occur	SPRD spread
ALQDS all quadrants	DLT delete	HWY highway	MAX maximum	OFC office	SR sunrise
AMD amend	DMG damage	IC ice	MDFY modify	OFP occluded frontal passage	SS sunset
AMS air mass	DPND deepened	IMDT immediate	MDL model		STFR stratus fractus
AMT amount	DSCNT descent	IMPL impulse	MDT moderate	OMTNS over mountains	SX stability index
ANAL analysis	DSIPT dissipate	INCL include	MED medium	OTLK outlook	TNTV tentative
AOA at or above	DURG during	INCR increase	MEGG merging	OTP on top	TROF trough
AOB at or below	DVLP develop	INDC indicate	MESO mesoscale	OTR other	TS thunderstorm
APRNT apparent	DVRG diverge	INDEF indefinitely	MET meteorological	OTRW otherwise	UPR upper
ARPT airport	DVRGG diverging	INSTBY instability	MID middle	OVC overcast	UVV upward motion
ATTM at this time	DW downward motion	INTS intense	MISG missing	OVTK overtake	VC vicinity
AWT awaiting	EBND eastbound	IOVC in overcast	MOV move	PBL boundary layer	VFR visual flight rules
BCM become	ELNGTD elongated	INVOF in vicinity of	MRGL marginal	PCPN precipitation	VRG veering
BD blowing dust	EMBDD embedded	IPV improve	MRNG morning	PLNS plains	VV vertical velocity
BDRY boundary	ERN eastern	ISOL isolate	MSG message	POS positive	WAA warm advection
BFR before	ERY easterly	ISOLD isolated	MSL mean sea level	PRST persist	WKN weaken
BLD build	ETA Eta model	KFRST killing frost	MST most	PSBL possible	WTR water
BLO below	EWD eastward	KT knots	MXD mixed	PTLY partly	WX weather
BYD beyond	FA aviation forecast	LAT latitude	NAB not above	PVA positive vorticity advection	YDA yesterday
CAA cold air advection	FLRY flurry	LCL local	NBND northbound		ZN zone
CCLDS clear of clouds	FLW follow	LCTMP little temp chg	NEG negative	PVL prevail	

```
FXUS63 KPAH 110030
AFDPAH

Area Forecast Discussion
National Weather Service Paducah KY
630 PM CST Fri Dec 10 2021

.UPDATE...
Issued at 630 PM CST Fri Dec 10 2021

Updated aviation discussion.

&&

.SHORT TERM...(Tonight through Saturday night)
Issued at 1250 PM CST Fri Dec 10 2021

Shortwave energy out over the south central Plains will continue
ENE toward the area through sunset, and provide the initial focus
for scattered convection, especially from 4 or 5 pm to 7 p.m. The
short range hi-res guidance continues to be more and more bullish
on scattered severe storms, likely Supercells in some cases.
Looking at what the models are depicting, this event seems to
compare favorably to 2/28/17 and 11/17/13 (Perryville, Crossville
and Brookport). So the concern for significant severe including
strong tornadoes is elevated. The shear and instability are more
than adequate, and in some of the data, slightly stronger.
Anomalously high dew point air is over the area, serving to lower
LCL's coupled with increasing SRH and MLCAPE values over the next
several hours.

The activity will continue tonight, coming in a couple of rounds.
Scattered activity initially into the evening hours, that may be
followed by slightly more organization in a linear fashion toward
midnight and beyond (especially west KY). Outside of the
convection, we may see near Wind Advisory level winds at times,
SSW gusts 30 to 40 mph tonight. Will hold for now on a headline
given it's borderline and with all that's going on.


000
TTAA00 KPIT 311831
STATE FORECAST DISCUSSION
NATIONAL WEATHER SERVICE PITTSBURGH, PA
216 PM EDT FRI MAY 31 1985

VERY WARM AND UNSTABLE AIRMASS AHEAD OF FRONT IN EASTERN OHIO
AND WESTERN PENNSYLVANIA. SEEMS THAT ONLY THING INHIBITING TSTM
DEVELOPMENT AT PRESENT TIME IS SUNSHINE TO HEAT THINGS UP.
NOW LOOKING FOR TSTMS TO START IN EASTERN OHIO BY LATE AFTERNOON.
FRONT NOW PAST TOL AND DTW AND ABOUT ON TIME FOR A LATE EVENING FROPA
IN W PA.

AS FOR REST OF FORECAST..LOOKS LIKE A QUICK SHOT OF COOLER AIR
SATURDAY..THEN WARMING UP ON SUNDAY. SHOULD BE SUFFICIENT WARM
ADVECTION TO AT LEAST PRODUCE SOME SIGNIFICANT MIDDLE AND HIGH CLOUDS
ON SUNDAY SO I PLAYED TEMPS A BIT LOWER THAN MOS AND HOPE THAT IT
DOESN'T  RAIN BEFORE EVENING.

WJD
```

Above: Forecast discussions for two major tornado outbreaks, one in 2021 (top) and another in 1985 (bottom). The gradual increase in detail and length over the years is typical of modern forecast discussions and reflects the much larger toolbox of today's forecaster, the increase in meteorological knowledge, and the wider outreach to media and emergency managers through this product.

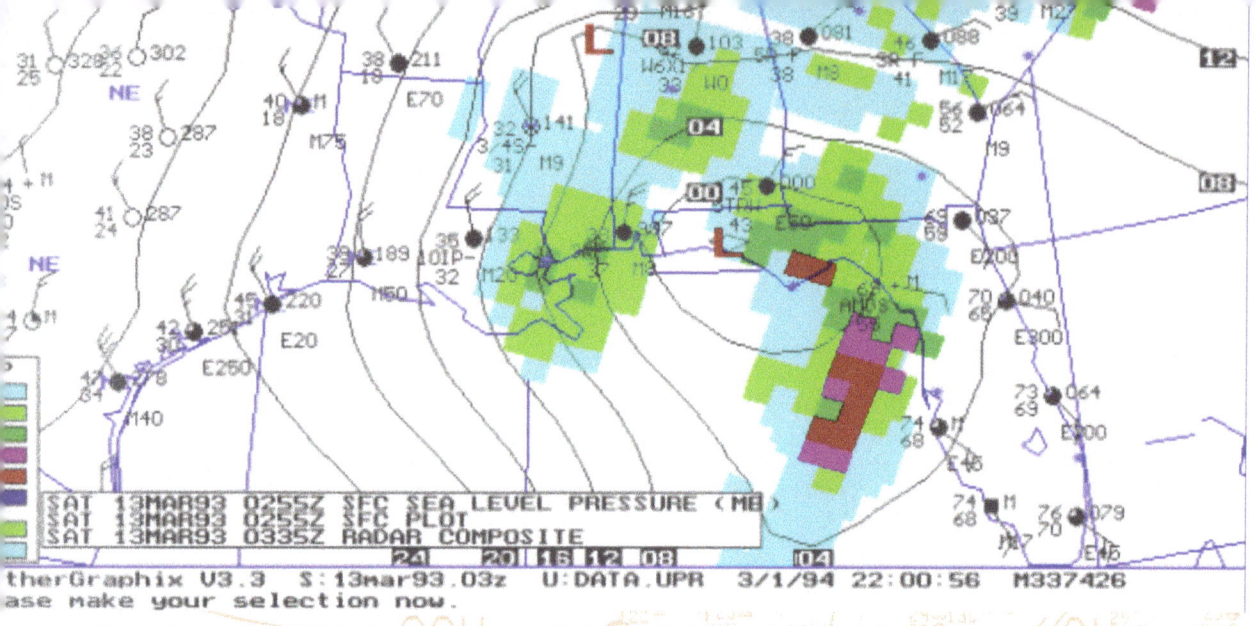

Above: A typical plot available to hobbyists on 13 March 1993, date of the so-called Storm of the Century, using WeatherGraphix (a tool written by the author). This uses the 40-km MDR (manually digitized radar) grid included with RAREP reports. MDR data was introduced in the 1970s, and the capability was added to the WSR-88D radar, but it was mostly discontinued during the early 2000s.

Below: The tools available to NWS forecasters on the AWIPS workstation 20 years later on 31 May 2013, date of the legendary El Reno, Oklahoma supercell. The displays are better but the basics have not changed.

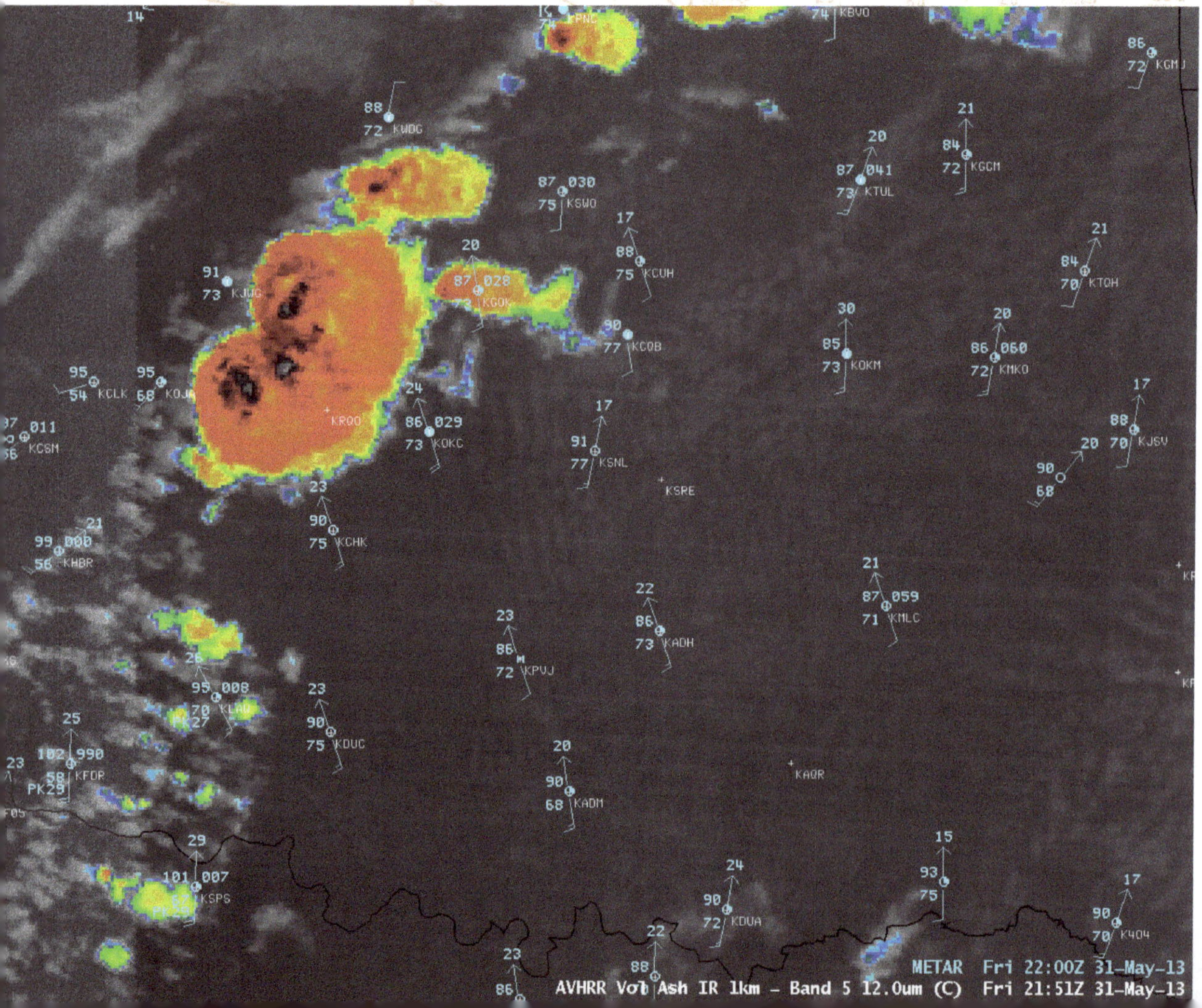

```
        jt (k) = max(1, min(4, int(3.0 + tem1) ))
        jt1(k) = max(1, min(4, int(3.0 + tem2) ))
!  ---  restrict extrapolation ranges by limiting abs(det t
        ft  = max(-0.5, min(1.5, tem1 - float(jt (k) - 3)
        ft1 = max(-0.5, min(1.5, tem2 - float(jt1(k) - 3)
!org    ft  = tem1 - float(jt (k) - 3)
!org    ft1 = tem2 - float(jt1(k) - 3)

!  ---  we have now isolated the layer ln pressure and temp
!       between two reference pressures and two reference t
!       (for each reference pressure).  we multiply the pre
!       fraction fp with the appropriate temperature fracti
!       the factors that will be needed for the interpolati
!       the optical depths (performed in routines taugbn fo

        fac10(k) = (1.0 - fp) * ft
        fac00(k) = (1.0 - fp) * (1.0 - ft)
        fac11(k) = fp * ft1
        fac01(k) = fp * (1.0 - ft1)

      enddo

!  ---  if the pressure is less than ~100mb, perform a diff
!       set of species interpolations.

      do k = 1, NLAY
        if (plog(k) > 4.56) then
          laytrop =  laytrop + 1

!  ---  for one band, the "switch" occurs at ~300 mb.
          if (plog(k) >= 5.76) layswtch = layswtch + 1
```

CHAPTER 5

NUMERICAL MODELS

Numerical Weather Prediction

Numerical weather prediction is the science of predicting the future state of the atmosphere using physical equations. In the 1904 paper *The Problem of Weather Forecasting from the Standpoint of Mechanics and Physics*, Vilhelm Bjerknes proposed that hydrodynamics and thermodynamics could be brought together as part of a numerical weather prediction system.

The application of numerical weather forecasting was expanded on in 1923 by Lewis Frye Richardson, a British meteorologist, in his book Weather Prediction by Numerical Process. He pictured a massive gallery where "a myriad [of human] computers are at work upon the weather of the part of the map where each sits, but each computer attends only to one equation or part of an equation. The work of each region is coordinated by an official of higher rank." The problem was seen as too complicated and expensive, and forecasting through dynamical means had not yet been proven.

With the breakthroughs in electronic computing technology during the 1950s, researchers had fantastic opportunities for laying down the craft of numerical forecasting. Universities and government institutions worked together to develop simple barotropic models, which became more sophisticated as scientific knowledge grew. The first operational predictions were available by 1960, and were followed by fantastic leaps and bounds in the decades ahead.

What is a numerical model?

While the physics and dynamics of models are far beyond the scope of this book, it is most important to point out that each model runs at a unique scale. Depending on the weather agency and purpose, a model is either run at a coarse resolution covering the entire world, or is run at a fine resolution covering a small area (usually the agency's home continent). The worldwide model is called a "global" model, while the fine model is called a "regional" model. Even finer models covering smaller areas, such as portions of a country, are called "mesoscale" models.

Detailed models are very accurate but are limited by their enormous processing requirements, which limits them to a specific area of the globe. As a result they suffer from boundary errors on their edges and their forecasts begin degrading just one or two days into the future. During the first day or two, forecasters are encouraged to use the regional models for their greater detail, but past this point should use global models, which lack boundary errors.

Forecasters must also be aware of the unique characteristics of each model, which are detailed section by section in this book, where applicable. These can range from eccentricities in the param-

---

Top: HPE Cray EX supercomputer. Used by NCEP, this model of supercomputer was the fastest supercomputer in the world in 2022.
Previous page: GFS model source code.

Meteorological Computing Power

NATIONAL CENTERS FOR ENVIRONMENTAL PREDICTION (NOAA)
Production Systems

Year	Model	Proc	System speed (Tflop)	Memory	Disk storage	Typical model
1956	IBM 701	1	0.000000001	1 kB	9 kB	(Research)
1958	IBM 704	1	0.000000008	20 kB	144 kB	Barotropic
1960	IBM 7090	1	0.000067	100 kB	50 MB	3-lvl QG model
1966	CDC 6600	1	0.000003	1 MB	75 MB	6-lvl PE, LFM
1974	IBM 360/195	1	0.00001	4 MB	300 MB	LFM, 7-lvl PE
1983	CDC Cyber 205 [1]	1	0.0001	32 MB	7.2 GB	OI, GSM, NGM, LFM
1987	CDC Cyber 205	2	0.0002	64 MB	14.4 GB	OI, GSM, NGM, LFM
1990	Cray Y/MP8 VII	1	0.0026	512 MB	2 GB	NGM, ETA
1994	Cray C90/16256 [2]	16	0.0153	2 GB	200 GB	NGM, ETA, RUC, LFM
1999	IBM RS/6000 SP	768	0.7	192 GB	4.6 TB	ETA, RUC2, etc
2000	IBM RS/6000 SP	2048	2.5	256 GB	7.5 TB	ETA, RUC2, etc
2003	IBM pSeries 690	44	7.3	1.4 TB	42 TB	ETA, RUC2, etc
2004	IBM pSeries 655	1152	12.46	6 TB	91 TB	GFS, NAM, etc
2006	IBM Power6/P575	2496	15.47	5 TB	160 TB	GFS, NAM, etc
2009	IBM Power6/P575	4992	69.7	19.7 TB	330 TB	GFS, NAM, etc
2013	IBM iDataPlex	10240	178.1	80.4 TB	2.59 PB	GFS, NAM, etc
2015	IBM iDataPlex	37312	705.9	80.4 TB	2.59 PB	Same (Tide/Gyre)
2015	Cray XC40	50176	2,890	Unknown	3.6 PB	Same (Luna/Surge)
2022	HPE Cray EX	327680	10,000	1.3 PB	26 PB	Same (Dogwood/Cactus)

CONSUMER DEVICE COMPARISON

Era	Model	Proc	System speed (Tflop)	Memory	Storage	Typical model
1981	Commodore 64	1	0.0000016	64 kB	144 kB	—
1992	Intel 486DX2/66	1	0.0000029	32 MB	340 MB	ETA, PE
1997	Intel Pentium II	1	0.000170	64 MB	2 GB	ETA, PE
2002	Intel Pentium 4 / 1.8	1	0.000860	256 MB	80 GB	ETA, PE
2010	Intel Core i7 860	4	0.0214	4 GB	2 TB	WRF
2014	Intel Core i7 4790K	4	0.043	8 GB	3 TB	WRF
2017	NVIDIA Titan V GPU	6	110	12 GB	—	—
2022	iPhone 14 Pro	2	2.0	6 GB	128 GB RAM	—
2023	Intel Core i9 12900KF	16	1.13	32 GB	4 TB	WRF

kB = 10^3 bytes; MB = 10^6 bytes; GB = 10^9 bytes; TB = 10^{12} bytes; PB = 10^{15} bytes
[1] Acquisition date given as August 1983 (Dey, 1989, *Wea. and Forecasting*).
[2] Caught fire 27 Sep 1999 and was destroyed. Models ran on backup Cray C5 until replaced by IBM RS/6000 SP on 18 Nov

eterization of radiation and convection, which show up in unusual ways, to shortfalls in vertical and horizontal resolution which give the model weaknesses in certain weather regimes. All of these are referred to as model biases.

When the hemispheric wave number, the number of long waves around a hemisphere, undergoes a change, it is referred to as a wave number transition. A common example is when a low wave number suddenly increases from 2 or 3 to 5 or 6 over one to two days. One signal of a possible pending transition is when a very strong, long, broad polar jet becomes established in the North Pacific Ocean. Transition events signal a major shift in weather regimes, and are notorious for causing errors and inconsistencies in numerical weather forecasts.

Getting model data

There are three primary sources of model data graphics on the Internet: government, academic, and private sector. The exception is "link sites", which contain no original content and simply refer to one of the sources below.

Though they may be handy for navigating to the right product, such sites are not included in this book as they are often incomplete, outdated, or abandoned.

- **Government agencies**. Common sense says that getting it from the source is preferred. However most products placed online at originating agencies are not as useful as those from other sources. This is understandable as display of the information is not their primary mission.
- **Universities and institutions**. Academic sources often use meteorological graphics as a means to facilitate discussion, encourage technical proficiency in their students, and often as a means to attract new talent. On the other hand there are numerous universities with excellent meteorology programs but with no inhouse weather graphics online.
- **Private sector**. A range of private companies and dedicated hobbyists offer access to various models. Some suggestions are provided in the inset at right. Internet weather graphics have often been generated using GEMPAK, Unisys WXP, or GRADS software, but other tools have entered the picture.

See the inset at right for a list of various web sites.

Other global models

Finally it should be mentioned that this is not the complete inventory of global models which are available. Dozens of nations have operational global models, including Germany, Russia, China, and India.

However they are generally not included in this book either because of translation barriers or because the host agency will not share their graphics publicly. Furthermore, for model data that does exist publicly, most non-American models have neither detailed fields nor gridded output data available for further inspection.

Since the mid-20th century and continuing to the present, the United States has embraced a truly free interchange of model data and modelling software, a policy which accelerated around 2000 with the modernization of the government Internet infrastructure. Literally all data is available online. The implications of this cannot be understated as it represents a true commitment to scientific progress. Almost any real time digital product can be found on the <*tgftp.nws.noaa.gov*> web site, where many terabytes of live data are stored. The notable exception to this rule is the system of private weather station data that is fed to federal agencies as part of the MADIS program.

Data from global sources has been much more difficult to obtain, and it is common for non-U.S. weather agencies to withhold their operational data. WMO Resolution 40, imposed in 1995, has long been at the root of this problem, which provided exemptions allowing various countries to withhold observational data for monetization purposes. National and regional systems have had similar problems. Full-resolution radar data, for example, is withheld from free distribution in much of Europe, and the UK-MET model is likewise restricted from free public access. This is part of the reason that tools like GRLevelX do not exist outside the United States. The ECMWF moved to remove restrictions on its data in 2020, which represented a small step back toward the idea of scientific data exchange.

Abbreviated list of model graphics sources

GOVERNMENT SITES
mag.ncep.noaa.gov (official NCEP chart collection)
spc.noaa.gov/exper

ACADEMIC SITES
weather.cod.edu
weather.rap.ucar.edu
wx.erau.edu/teaching/milsyn
www.meteo.psu.edu/ewall
arctic.som.ou.edu/tburg/models
cams.nssl.noaa.gov
inside.nssl.noaa.gov/tgalarneau/real-time-qg-diagnostics
www.atmos.albany.edu/student/abentley/realtime

COMMERCIAL SITES
www.weathernerds.org
www.tropicaltidbits.com/analysis/models
pivotalweather.com
wxcharts.com (international)
www.weatheronline.co.uk (international)
www.wright-weather.com
www.twisterdata.com
maps.weatherbell.com ($)

Updated March 2023. Subject to change. Contact the author (see front of the book) to suggest other sites with original content for a future edition.

Historical Perspective

an excerpt from

Weather Prediction by Numerical Process

L. F. Richardson, 1922

It took me the best part of six weeks to draw up the computing forms and to work out the new distribution in two vertical columns for the first time. My office was a heap of hay in a cold rest billet. With practice the work of an average computer might go perhaps ten times faster. If the time-step were 3 hours, then 32 individuals could just compute two points so as to keep pace with the weather, if we allow nothing for the very great gain in speed which is invariably noticed when a complicated operation is divided up into simpler parts, upon which individuals specialize. If the co-ordinate chequer were 200 km square in plan, there would be 3200 columns on the complete map of the globe. In the tropics the weather is often foreknown, so that we may say 2000 active columns. So that 32 x 2000 = 64,000 computers would be needed to race the weather for the whole globe. That is a staggering figure. Perhaps in some years' time it may be possible to report a simplification of the process. But in any case, the organization indicated is a central forecast-factory for the whole globe, or for portions extending to boundaries where the weather is steady, with individual computers specializing on the separate equations. Let us hope for their sakes that they are moved on from time to time to new operations.

After so much hard reasoning, may one play with a fantasy? Imagine a large hall like a theatre, except that the circles and galleries go right round through the space usually occupied by the stage. The walls of this chamber are painted to form a map of the globe. The ceiling represents the north polar regions, England is in the gallery, the tropics in the upper circle, Australia on the dress circle and the antarctic in the pit. A myriad computers are at work upon the weather of the part of the map where each sits, but each computer attends only to one equation or part of an equation. The work of each region is coordinated by an official of higher rank. Numerous little "night signs" display the instantaneous values so that neighbouring computers can read them. Each number is thus displayed in three adjacent zones so as to maintain communication to the North and South on the map. From the floor of the pit a tall pillar rises to half the height of the hall. It carries a large pulpit on its top. In this sits the man in charge of the whole theatre; he is surrounded by several assistants and messengers. One of his duties is to maintain a uniform speed of progress in all parts of the globe. In this respect he is like the conductor of an orchestra in which the instruments are slide-rules and calculating machines. But instead of waving a baton he turns a beam of rosy light upon any region that is running ahead of the rest, and a beam of blue light upon those who are behindhand.

Four senior clerks in the central pulpit are collecting the future weather as fast as it is being computed, and despatching it by pneumatic carrier to a quiet room. There it will be coded and telephoned to the radio transmitting station.

Messengers carry piles of used computing forms down to a storehouse in the cellar.

In a neighbouring building there is a research department, where they invent improvements. But these is much experimenting on a small scale before any change is made in the complex routine of the computing theatre. In a basement an enthusiast is observing eddies in the liquid lining of a huge spinning bowl, but so far the arithmetic proves the better way. In another building are all the usual financial, correspondence and administrative offices. Outside are playing fields, houses, mountains and lakes, for it was thought that those who compute the weather should breathe of it freely.

GFS model

In the 1960s the National Meteorological Center (NMC, the forerunner to NCEP) began operating a model covering the northern hemisphere. The first model spanning both hemispheres was the Global Spectral Model (GSM), implemented on 18 March 1981. It was a 30-wave (465 km resolution) 12-level model that used a spectral configuration in which the atmosphere was represented as waves rather than gridpoint values. For this reason it was often referred to as the "spectral model".

In March 1985, the GSM became the Medium Range Forecasting (MRF) model, which took advantage of the larger data structures offered by the new CYBER 205 computer. A separate version of the MRF known as the Aviation (AVN) run was started early in the forecast cycle to get products out to airlines and dispatchers as soon as possible, then the quality-controlled MRF itself followed several hours later. On 23 April 2002, the MRF and AVN were merged into a single run known as the Global Forecast System (GFS). The 1990s through the 2010s, with the vast leaps in computing power, brought a series of upgrades that tremendously enhanced the model's resolution. By 2015 it was a 1534-wave (13 km) 64 layer model.

In June 2019 NCEP switched to a gridpoint model configuration. This was known as the FV3 (Finite Volume). The horizontal resolution at press time was 13 km, equivalent to the previous spectral version. The vertical resolution was increased to 127 layers in 2020.

Model biases

Biases in the GFS FV3 model has not been fully established, but it does appear to continue a cold bias that often becomes noticeable around the 200-hour point, especially in northerly latitudes. The previous spectral GFS model was already known for being too aggressive with amplification of the pattern and being bullish with cold air outbreaks.

The GFS is known for having a fast bias, moving systems faster than actually forecast in the extended period. The ECMWF is often consulted for information on timing. Likewise, the GFS is known for being too aggressive in bringing Gulf moisture northward ahead of a developing Southwest system. The previous spectral GFS model was known for phasing waves in split flow patterns resulting in large long wave troughs and ridges, but it is unknown whether this bias continues in the FV3.

key details

The GFS is a 13 km 127-layer gridpoint model. This makes the model mesoscale in resolution, although it is not a convection-allowing model (does not explicitly model convection). It is being integrated into the Unified Forecast System (UFS), an initiative to simplify model architecture at NCEP.

Beginning June 2022 the operational GFS ran on supercomputers named Dogwood and Cactus, leased from General Dynamics Information Technologies and operated in Virginia and Arizona respectively. At the time of commissioning they were the 49th and 50th fastest computers in the world.

The GFS is in widespread use for worldwide forecasting, even in Europe and Asia, since it is one of the only full-scale planetary weather models whose output is disseminated in its entirety without restriction. Only limited portions of other global models are available for public consumption, though ECMWF data restrictions were rolled back in 2020. It should be noted that due to poor resolution of upper air data in data-sparse regions like Africa, the Indian Ocean, and southern Asia, significant errors occasionally exist in spite of the model's tremendous resolution.

The GFS is run four times a day (00Z, 06Z, 12Z, and 18Z) out to 384 hours. The model is usually started at H+2:47, and the 120 hour products are normally completed by H+4:10, as of 2023.

The current GFS version at press time (early 2023) was FV3-GFS V3.0 (GFSv17).

real-time Internet sources

www.nco.ncep.noaa.gov/pmb/nwprod/analysis
www.emc.ncep.noaa.gov/modelinfo/
wxweb.meteostar.com
meteocentre.com/models/

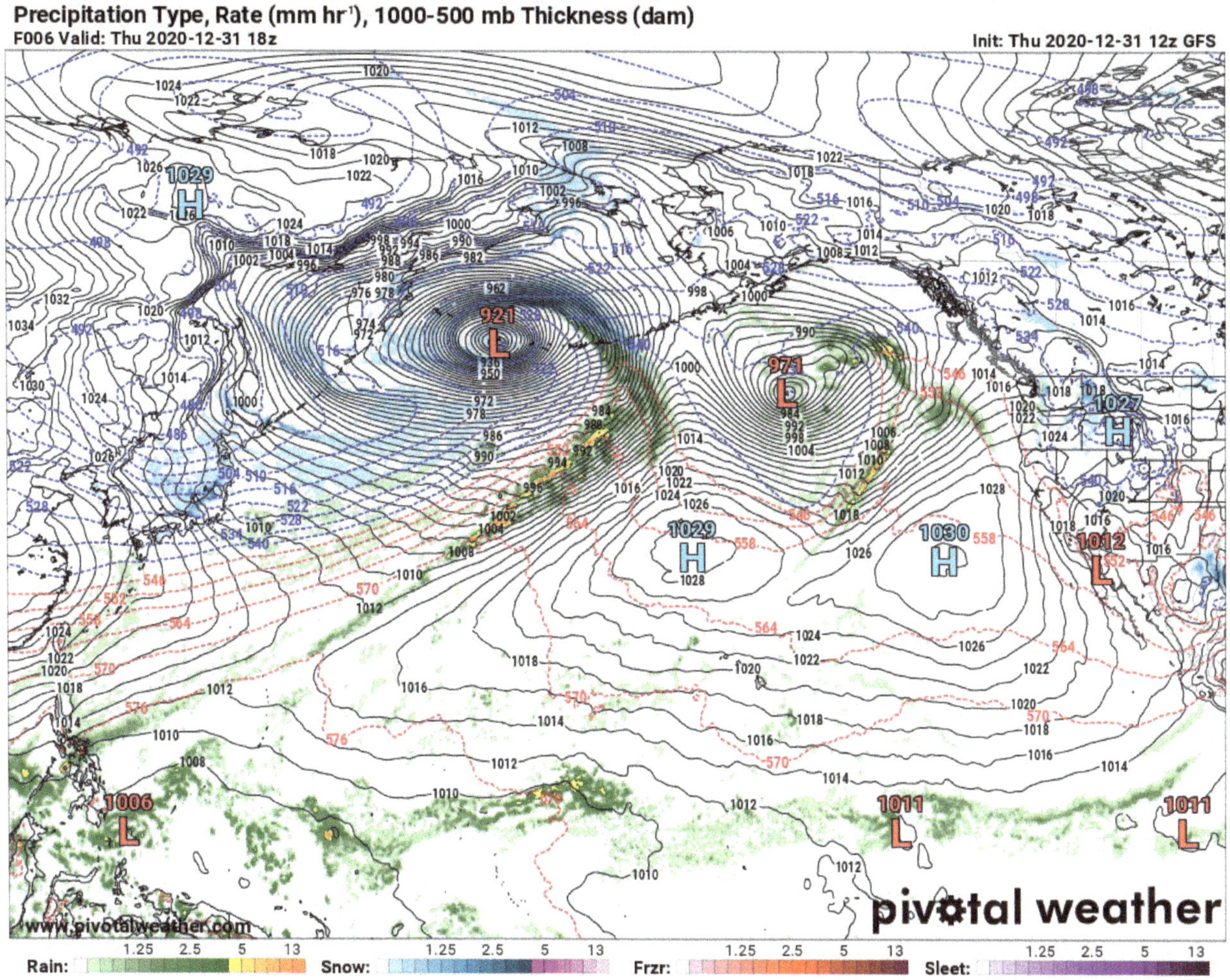

Precipitation Type, Rate (mm hr⁻¹), 1000-500 mb Thickness (dam)
F006 Valid: Thu 2020-12-31 18z Init: Thu 2020-12-31 12z GFS

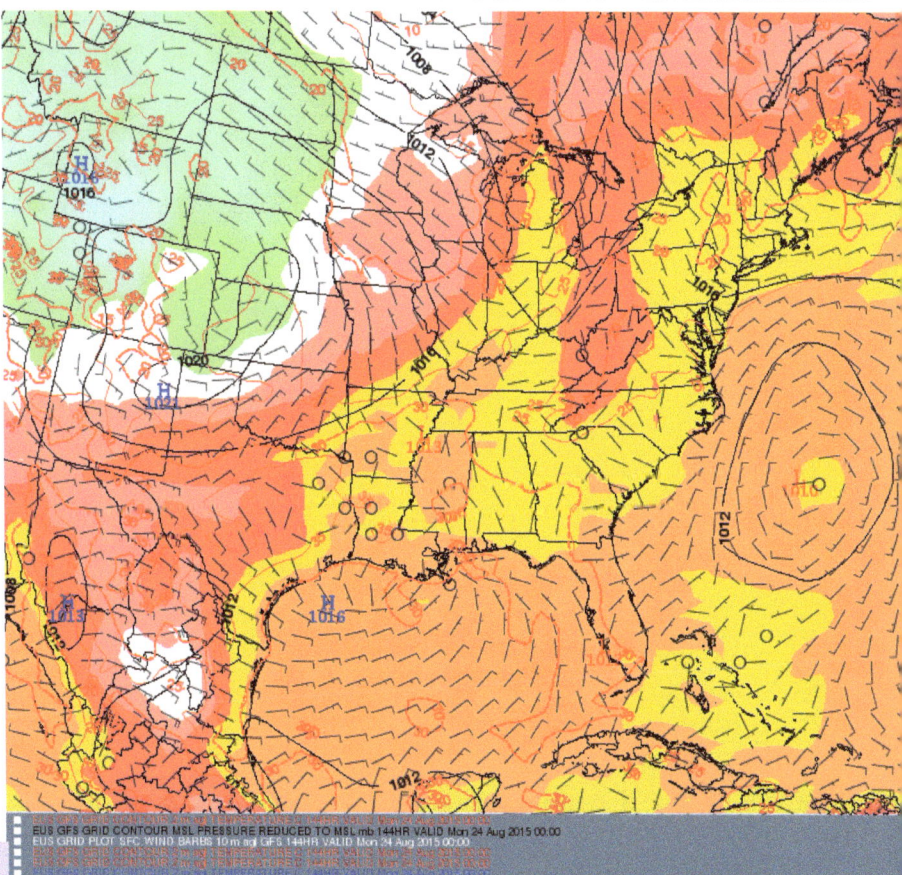

Above: The GFS model has long been known for its worldwide coverage. This 6-hour GFS forecast for the Pacific Ocean showed a record-setting Aleutian low with an indicated pressure of 921 mb (27.20"). It is unknown whether this pressure was achieved, but if it did it would set an all-time extratropical record for the Northern Hemisphere. *(Pivotal Weather)*

Right: Traditional GFS output of pressure, temperature, and dew point. *(Meteostar)*

NAM model

The NAM model (North American Mesoscale) model is NCEP's regional model for the United States. It is a replacement for the older Eta, NGM, and LFM models. The NAM has used the Weather Research and Forecasting (WRF) model since 2006.

The WRF model was completed under a partnership of U.S. government and academic users who aspired to emulate Europe's development of a robust "community" numerical model. The new model had to be flexible, portable, and scalable. A prototype was fielded in 2000. The model was officially implemented by National Centers for Environmental Prediction (NCEP) in June 2006, replacing the Eta in its NAM (North American Mesoscale) suite.

The NAM is run at a resolution of 12 km, with 60 vertical layers. A nested window on the continental U.S. provides a resolution of 4 km. The nest will be replaced in by 2024 with the Rapid Refresh Forecast System (RRFS).

The WRF is public domain. Binary and source code is freely available for download. It offers users a choice of algorithms from different contributors. The source code uses a layered software architecture that encapsulates low-level processing within the model to make custom routines flexible, independent, and efficient.

There are two primary versions of the WRF: the **ARW** (advanced research WRF; formerly called "EM" for Eulerian mass) maintained by the National Center for Atmospheric Research, and the **NMM** (non-hydrostatic mesoscale model) maintained by the National Centers for Environmental Prediction (NCEP). Each has a different set of equations at its core. Preliminary studies found little skill difference between the ARW and NMM.

Model biases

Early on, the NAM WRF was recognized for being aggressive with moisture, the opposite of the problem that used to plague the Eta model. The effect was particularly evident in broad forested regions such as the eastern United States. The parameterizations used for evapotranspiration were adjusted in 2007, alleviating this problem.

A 2008 study of turbulence found that the NAM tends to overforecast wind speeds in the upper troposphere, especially when associated with a subtropical jet with anticyclonic curvature.

The NAM is favored during the cold season because of its superior handling of shallow polar air masses. The GFS and other global models tend to be too aggressive at eroding these types of cold air outbreaks.

key details

The NAM is based on the WRF, which is a finite difference (gridpoint) model. It may use either eta or zeta vertical coordinate systems. The horizontal and vertical resolution is completely dependent on user specifications, and is often less than 10 km. In the NAM, the CONUS grid is 4 km.

The first release of the WRF was on November 30, 2000. Real-time model runs began at NCAR in March 2001 and from NCEP in June 2006. The WRF allows unlimited options for dynamics and physics packages, and can be tailored to a substantial degree. This allows the end user to select the ideal configuration for a given purpose. The WRF, taking on dual roles as a mesoscale and regional model, is expected to eventually replace all NCEP model runs. Full information about the WRF model can be found at <www.wrf-model.org>

The high-resolution nested components of the NAM model will be replaced with the Rapid Refresh Forecast System (RRFS) in late 2023 or 2024. This is part of an initiative to migrate models to a unified architecture. The RRFS will also replace the HRRR and RAP.

The NAM is normally started at H+1:15 past the cycle hours of 00, 06, 12, and 18Z, and the 84 hour panels are completed by H+2:45, as of 2023.

real-time Internet sources

mag.ncep.noaa.gov
wxweb.meteostar.com/models
meteocentre.com/models/
pivotalweather.com

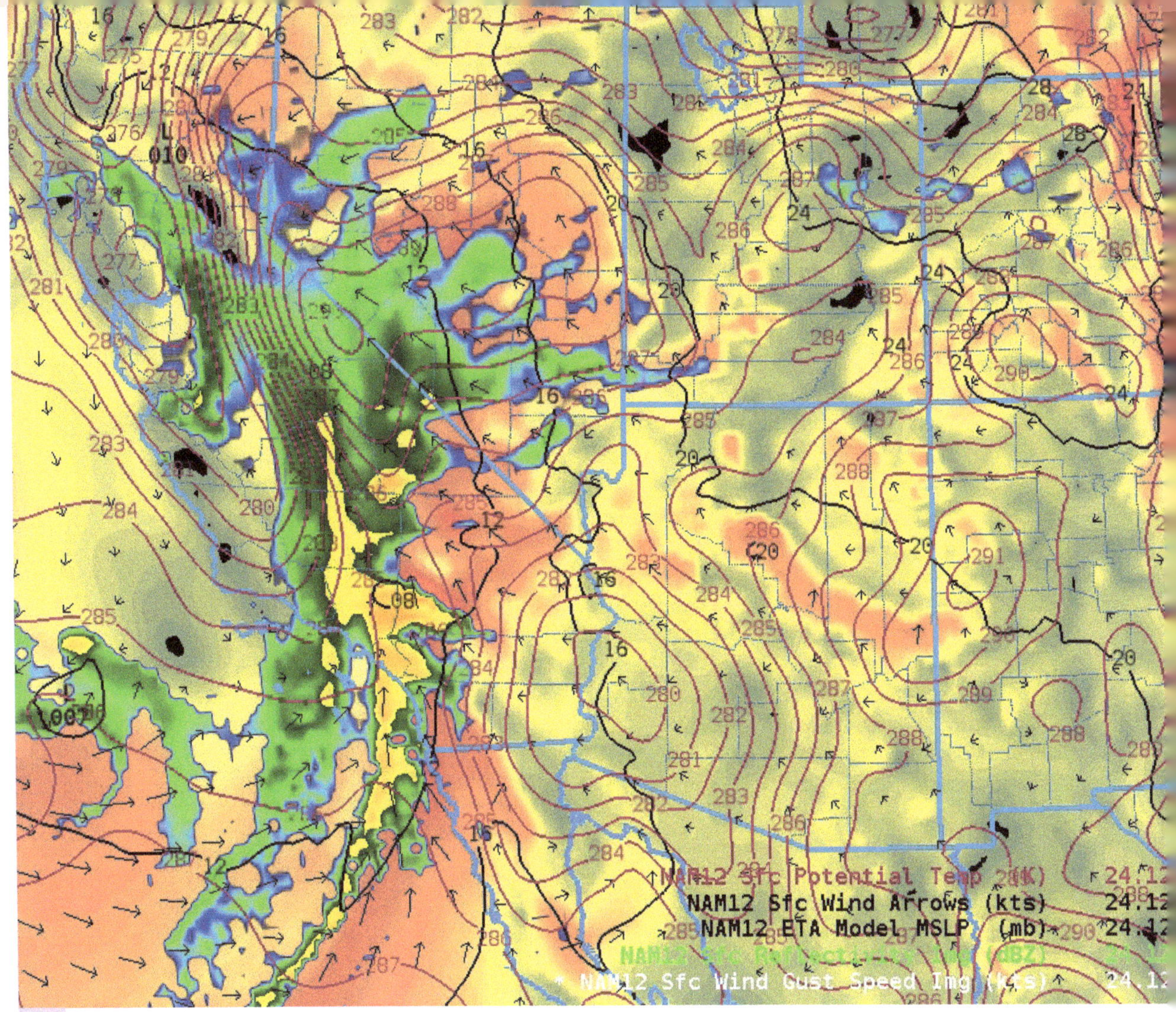

Top: NAM forecast of reflectivity (bright green and enclosed yellow shading), wind gust speed (dull green, yellow, and red shading), potential temperature (red lines), and wind (vectors) with a strong Pacific system moving inland on 25 February 2023. It brought blizzard conditions to the mountains north of Los Angeles resulting in a shutdown of Interstate 5 at Tejon Pass. *(AWIPS-II)*

Right: NAM domain with nested windows on the U.S. Alaska, Hawaii, and Puerto Rico. The small window shown in the desert southwest is a movable 1.33 km resolution window that can be placed over areas of interest. *(NCEP)*

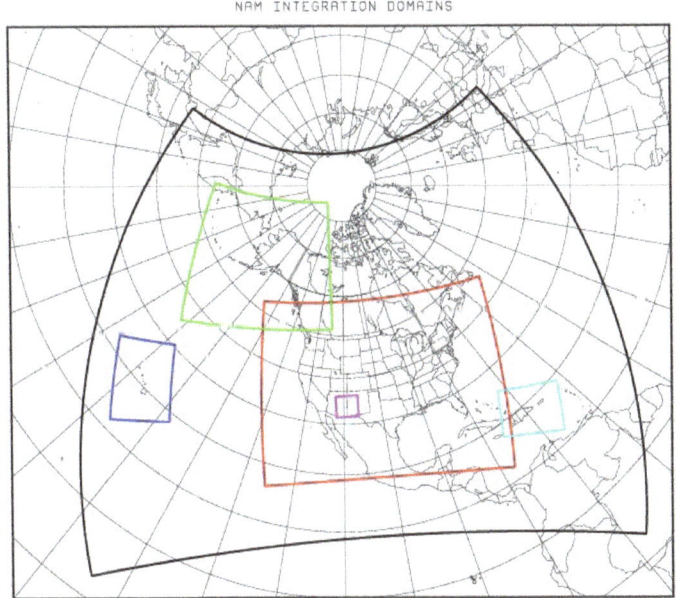

RAP & HRRR models

In the 1980s, NMC (NCEP) began offered "early look" models to serve the needs of customers like airlines, the military, and emergency management agencies. Initially the old LFM model run on faster computers provided this function, and by 1994 the modern Rapid Update Cycle (RUC) model filled this role.

In 2012 the RUC was replaced with new technology, consisting of the RAP and HRRR (detailed below). In late 2023 or 2024 these models are in turn expected to be replaced with the Rapid Refresh Forecast System (RRFS), as part of the move to a unified modeling architecture. This will also replace the high-resolution NAM nests.

RAP model

Although the RUC model was highly capable, a successor was in devleopment based on the robust, modular Weather Research & Forecasting (WRF) model. It emerged as the Rapid Refresh (RAP) model, implemented on 20 March 2012, using the WRF-ARW model. The RUC model was immediately discontinued.

The model uses a gridpoint array of 758x567 points, offering 13 km resolution. There are 50 sigma layers, and the model is initialized from the global GFS model.

The RAP model covers a much larger area than the RUC, a domain that includes Alaska, northern Canada, and much of the Caribbean. Sea surface temperature is processed at a resolution of 4 km, versus 24 km with the RUC. Comparison of the RAP with its forerunner, the ROC, shows that it offers significant improvements in winds, temperature, and dewpoint.

HRRR model

The High Resolution Rapid Refresh (HRRR) model is a subset of the Rapid Refresh model, which offers a resolution of 3 km instead of 13 km. This makes the HRRR a storm-scale model instead of a mesoscale model. Since it is impossible to sample storm-scale processes in real thunderstorms, the HRRR must actually spin up its own showers and thunderstorms based on input from the RAP fields. If errors occur in the RAP wind, temperature, and moisture fields, this has enormous impacts on the HRRR radar forecasts.

The HRRR is better able to take advantage of terrain interactions, which means in the western United States it has the potential to more accurately forecast trends in wind, ceiling, and visibility.

key details

The RAP is a WAF-ARW gridpoint model with a resolution of 13 km. This is coarser than the NAM's CONUS resolution of 4 km. There are 50 vertical levels, similar to the NAM's 60 layers.

The High Resolution Rapid Refresh (HRRR) model is a subset of the RAP model that offers a base resolution of 3 km.

The RAP model is based on the WRF-ARW model. ARW stands for Advanced Research WRF, and was developed by National Center for Atmospheric Research (NCAR). The ARW core is considered slightly superior for low-level temperature and humidity, allowing better accuracy with ceiling and visibility. Preliminary testing suggests it is slightly less accurate with precipitation coverage. The ARW and NMM cores, in spite of their significant differences, overall have very similar accuracy.

The lowest level of the RAP model is 8 meters (26 feet) above ground level, while the highest level is 50 hPa (about 68,000 ft).

The RAP products are generated hourly, normally starting at H+0:26 and completing at H+0:55 (for the HRRR, H+1:30) as of 2023. There are delays around 00Z and 12Z due to supercomputer time being allocated for other essential model runs.

The RAP and HRRR are slated to be replaced with the Rapid Refresh Forecast System (RRFS) in late 2023 or 2024. This will also replace the NAM nest.

real-time Internet sources

rapidrefresh.noaa.gov
rapidrefresh.noaa.gov/HRRR
meteocentre.com/models/

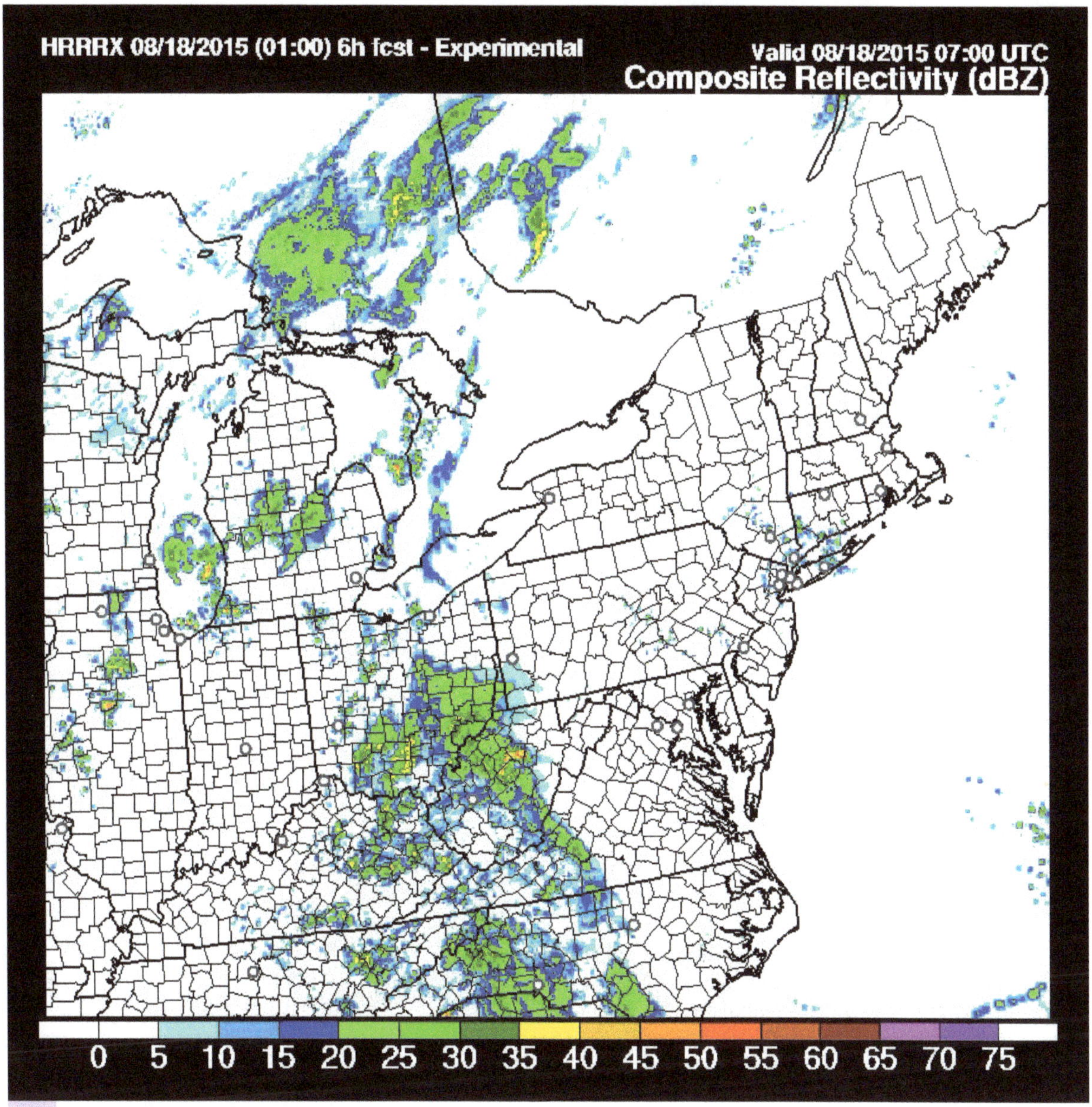

Top: High Resolution Rapid Refresh (HRRR) depiction of composite reflectivity, that is, the highest radar reflectivity found in a vertical column at each point. Though this product might seem like a holy grail of forecasting, significant errors occasionally occur. Careful verification is needed when this product is used in a forecast. *(NCEP)*

Right: Forecast domain for the Rapid Refresh model (striped) and the old RUC model (the box zoomed closely on the conterminous United States). *(NCEP)*

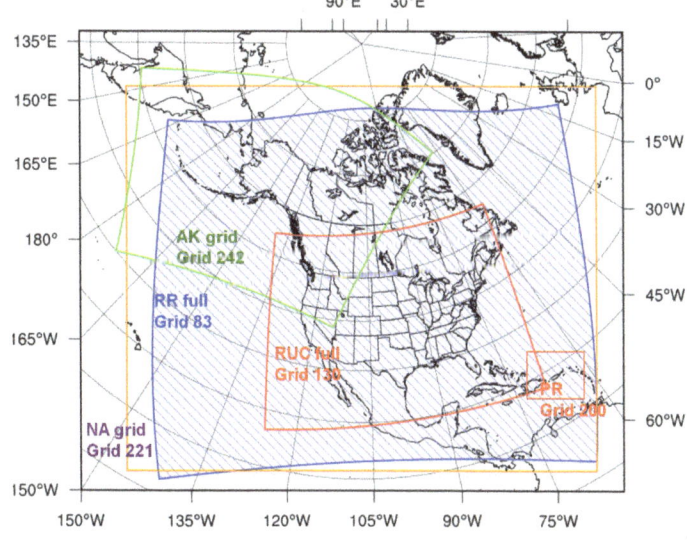

ECMWF model

The "ECMWF" is the traditional name for a model specifically known as the Integrated Forecast System (IFS) High Resolution (HRES) model. It is known as the ECMWF model because it is operated by the European Centre for Medium-Range Weather Forecasts (ECMWF). This is a multinational forecast center that was proposed in 1967 and became operational in 1973.

The first operational ECMWF model forecast was produced on 1 August 1979 using a Cray 1A supercomputer. It was a 15-level gridpoint model with a resolution of 200 km. In April 1983, ECMWF adopted a spectral model with 63 waves and 16 layers. In 1992 an ensemble forecasting system was added. In April 1995 parameterization of clouds was added. In 1996 the Optimum Interpolation analysis scheme was replaced by the 3DVAR method. The WAM ocean wave model was introduced in 1992, followed by integration with the atmospheric model in 1998.

Model biases

Due to frequent upgrades in the ECMWF, the biases are not well known. A lot of what is provided here is based on historical biases that have already been established.

The ECMWF has been historically considered to be superior at forecasting upper-level heights during the cold season across North America, particularly with respect to wave number transitions and the onset of +PNA circulation episodes (west Canada ridge with polar air affecting the central and eastern U.S.). It has excelled at timing the outbreak of cold fronts into the central US.

The ECMWF is notorious for overdeveloping or overpopulating cutoff lows, particularly in the southwestern U.S. However this yields somewhat better skill than other models in spring when cutoff lows are most common. The model is also too slow or even retrogressive with cutoff lows, which are sometimes even erroneously shunted westward underneath the Pacific subtropical high.

The model has historically shown a meridional bias, making upper air patterns unusually amplified and surface systems more intense and more slow. Therefore a progressive pattern being forecast by the ECMWF is significant, as this is associated with high hemispheric wave numbers and fast zonal flow.

key details

The ECMWF model is a 1279-wave hydrostatic spectral model (about 16 km resolution) with triangular truncation. It uses 137 hybrid sigma-pressure levels, consisting of sigma (terrain-following) surfaces near the ground and pressure surfaces aloft, and moves forward in 10-minute time steps.

The ECMWF model as of 2023 was being run on a Atos BullSequana XH2000 cluster which uses over 1 million AMD Epyc Rome cores. The published performance is 30,000 teraflops. This is about 100 times faster than the Cray XC30 system which was in operation almost ten years previously.

ECMWF moved to an open data policy in 2020. Previously, only a very limited subset of weather data was releasable to the general public. Many chart sets are made available at the new ECMWF graphics portal at charts.ecmwf.int. Not all of the standard forecast charts are available, so it is necessary to visit third party services like pivotalweather.com to get a wider range of graphics with more available regions. The author has also noted that the ECMWF chart viewer is erratic and sometimes fails to load charts, requiring the graphic to be reloaded.

The ECMWF uses a 4DVAR data assimilation system that ingests a wide variety of surface observations, radiosonde, aircraft ACARS, and geostationary satellite wind data.

This model is supported by 28 European member states, and is generally considered to be the most sophisticated and computationally expensive numerical prediction model used in global forecasting.

real-time Internet sources

charts.ecmwf.int
www.ecmwf.int/en/forecasts/charts
meteocentre.com
weather.cod.edu/forecast
www.wetterzentrale.de

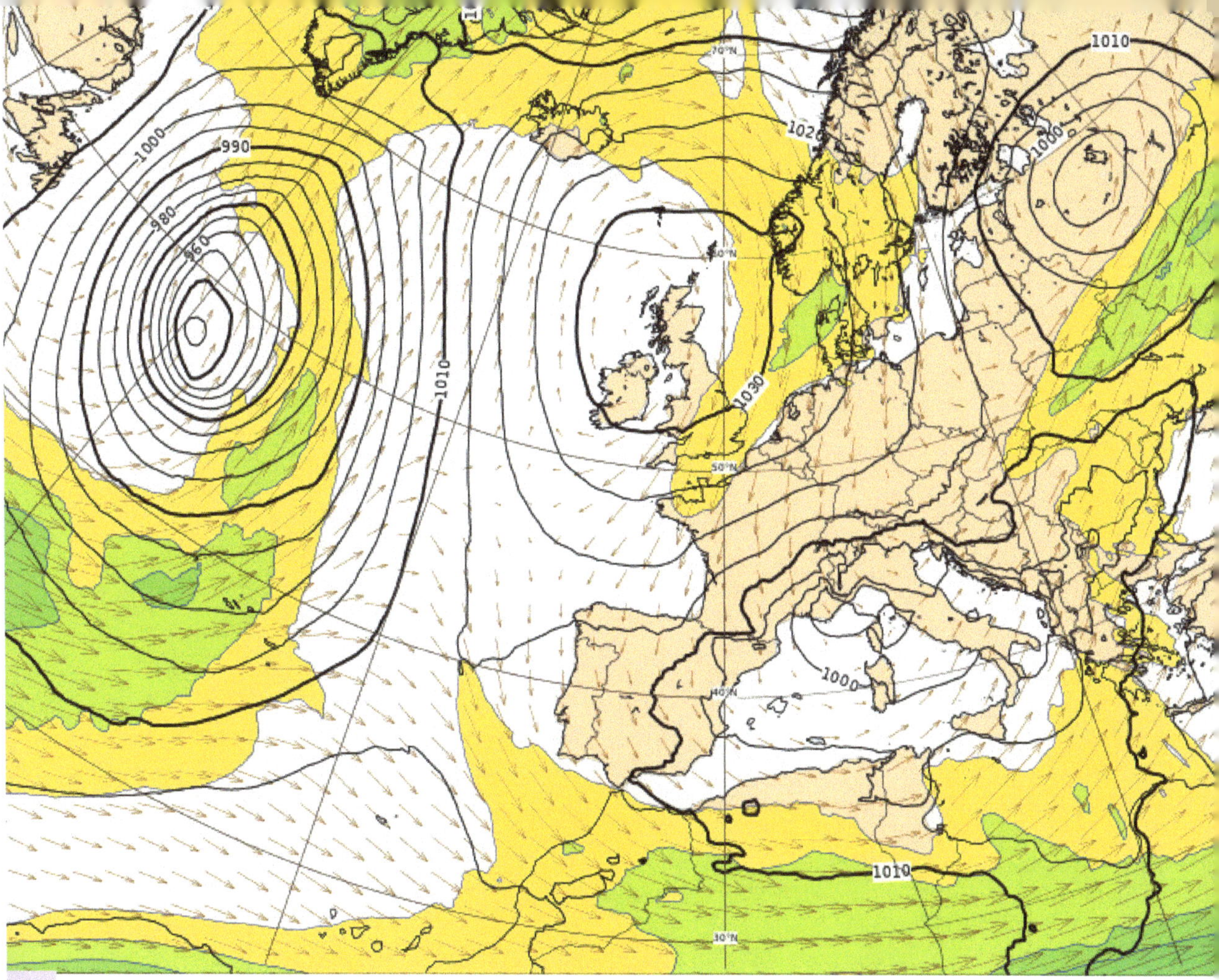

Above: ECMWF IFS HRES plot of sea level pressure and 200 mb wind on 26 Feb 2023 at 1200 UTC, as obtained from the charts.ecmwf.int website. This also includes 200 mb wind vectors (brown arrows) and isotachs (color bands). *(ECMWF)*

Right: Sites like Pivotal-Weather.com are recommended for detailed use of the ECMWF model. Here we see a strong Pacific system moving across the Lower Mississippi River valley, bringing a threat of severe thunderstorms on 3 March 2023. *(Pivotal Weather)*

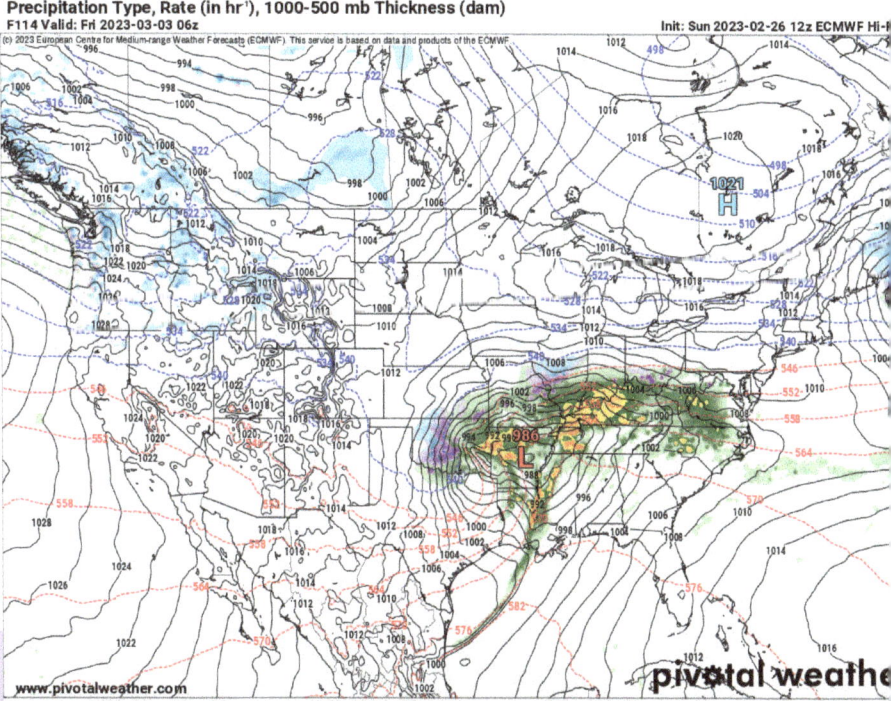

Other important models

GEM model (Canada)

The Canadian Global Environmental Multi-scale (GEM) model is a single model made up of a global run (GDPS, formerly known as the GLB or SEF) and a regional subset (RDPS, formerly known as the regional fine-elements model: REG, RFE, and EFR). There is also a very-high resolution regional subset known as the HRDPS. The GEM is a global spectral system run at the CMC computing facility in Dorval, Quebec.

The GEM/GDPS global model, comparable to the US GFS model, forecasts out to 72 hours (12Z base time), 240 hours (00Z base time), and 360 hours (Saturdays). Gridpoints were originally based on latitude-longitude pairs, but in 2015 the GDPS adopted a Yin-Yang grid, where forecasts are computed on two "interlocking" 270-degree shells of the Earth oriented perpendicular to one another. The global model uses a resolution of 15 km, with 84 levels in the vertical (beginning 2019). Assimilation used the 4DEnVar method.

The GEM/RDPS regional model, comparable to the US NAM, is a variable-gridpoint model that produces output from a base time of 00Z and 12Z, forecasting out to 48 hours. It operates with 10 km resolution and 84 hybrid sigma-Z levels (beginning 2019), using the Kain-Fritsch deep convection scheme, and the Kuo transient scheme for shallow convection. The high-resolution HRDRPS operates with a 2.5 km resolution and 62 levels (beginning 2019).

Canada's technological developments began in 1950 with the introduction of an experimental barotropic model running on an IBM 650 at McGill University. Operational charts did not begin until July 1963 when a barotropic model began producing output for 500 mb. On 18 February 1976 the first global spectral model was implemented. It featured 20 hemispheric waves with rhomboidal truncation and five vertical levels. In 1987 CMC acquired a Cray X/MP 28 for its model suite.

On 22 April 1986, the first regional model, analogous to the NGM and Eta, was introduced. It was called the RFE (Regional Finite Element), also known as EFR (Éléments Finis Régionaux) and featured 15 layers with 190 km resolution, and was placed on the Cray X/MP in 1987 with better resolution.

On 24 February 1997 the RFE became the GEM Regional model, and on 14 October 1998 the SEF was decommissioned. It was replaced by the GEM Global model, a 0.9-deg global uniform grid model with 28 eta levels.

The GEM has proven to be an excellent, accurate model. Since it uses sigma coordinates, it does have difficulty handling lee-mountain effects such as lee cyclogenesis and cold air damming. The GEM model has also been faulted in rejecting too much upper air data which appear to be valid but don't fit the first-guess fields.

The regional model is known to have a cold bias at the surface during wintertime nocturnal cooling episodes owing to the model's deficit in downward long-wave radiation. The model also tends to overforecast precipitation on the windward side of mountains, especially during the spinup cycle (roughly the first 6 hours).

UKMET (Great Britain)

The term UKMET is a shorthand phrase describing the Unified Model (UM) system, a sophisticated set of numerical forecast models operated by the UK Met Office. It consists of the Global (UKMET-G) and the UK mesoscale window (UKMET/UKV). These models are noteworthy for continuously operating as gridpoint models instead of the spectral models that were run by NCEP and ECMWF until the mid-2010s. As of 2023, the model was being run with a global resolution of 10 km with a 1.5 km window on the UK. There are 70 vertical levels.

Great Britain's expertise with numerical models goes back to 1959, when the University of Manchester successfully ran a numerical prediction on a Ferranti Mark 1 computer. Almost immediately the British Met Office put a Ferranti Mercury computer to work, producing experimental 36-hour forecasts for the eastern Atlantic and western Europe. It took about 6 hours to produce each complete forecast.

With a new computer purchase in 1965, the

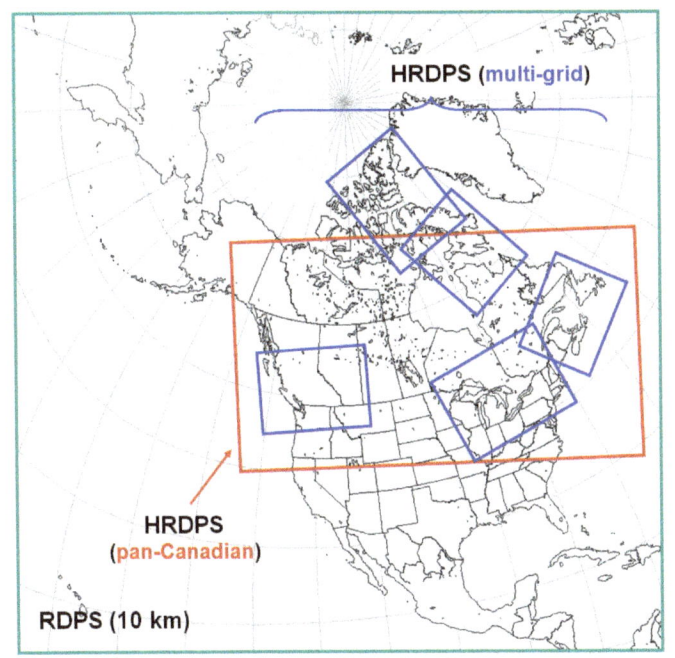

Left: Canadian RDPS (regional) and HRDPS (high-resolution subsets. The RDPS is made up of the entire window including Siberia, Mexico, and Hawaii, while the HRDPS consists of smaller provincial subsets followed by a pan-Canadian national sector that was introduced in 2014. *(CMC)*

Met Office introduced its first operational model that covered 30 hours and up to 72 hours experimentally. This was improved further with an IBM 360 acquisition in 1972 and introduction of a nested grid scheme. The model has undergone upgrades since then, the latest being the addition of a new physics package nicknamed "New Dynamics" on 7 August 2002. The model was moved to new NEC SX-6 systems in March 2004.

The UM system has been run for decades at the Bracknell facility west of London, but during the summer of 2003 the facility was moved to the Met Office's new headquarters at Exeter in Devon.

In 2013 the UK Met Office approved a £97 million ($156 million) contract on a new Cray XC40 supercomputer. This system came online in December 2016. It provided a 15,000 Tflop capability and replaced the older Power 775 systems, made up of 18,432 Power7 8C cores which delivered 476 Tflop of power.

This system was expected to be replaced in late 2023 by a Microsoft Azure system made up of HPE Cray EX supercomputers. This would bring processing speed up to 60,000 Tflop.

NAVGEM/NOGAPS (US Navy)

The NAVGEM model replaced the NOGAPS (Navy Operational Global Atmospheric Prediction System) in February 2013. This is a global spectral model which as of 2015 had a horizontal resolution of 425 waves (37 km) and a vertical resolution of 50 hybrid-sigma levels. The 2013 transition from NOGAPS to NAVGEM added advanced physics packages, such as the SL/SI (Semi-Langrangian/Semi-Implicit) core which solves the trajectory of parcels in reverse fashion to find which one at the previous time step ended up at the forecast location.

The Navy had been experimenting with numerical models as early as 1959. In the mid-1970s there was a push for a global spectral model to be developed, and the Naval Research Laboratory produced the first version of NOGAPS in 1982. It was a finite-difference (gridpoint) model. During long-term testing in the 1980s it was found that NOGAPS was a poor contender to the AVN/MRF and ECMWF models. NOGAPS was rebuilt as a spectral model, yielding a much more robust and accurate system that began forecasts in January 1988. It transitioned to the NAVGEM model in February 2013.

Data from the NAVGEM model is available at www.nrlmry.navy.mil/metoc/nogaps/ .

Ensemble predictions

Ensemble predictions are collections of two or more numerical weather prediction solutions, valid for the same location and forecast time. Combined with each other, each solution, called a member, gives a range of possible outcomes. Much of the work done with ensembles in the 2000s and 2010s also focused on developing methods of visual display that are useful to forecasters.

Ensemble predictions may be constructed by combining data across many different model systems, by using different sets of physics packages and initialized conditions in a single dynamical model, or by using prior model runs. All three methods may also be used together.

A crude form of ensemble forecasting has been done by hand since the 1960s, in which forecasters manually compare one model system to another. For example, the 500 mb GFS chart might be compared against the 500 mb ECMWF panel. This is sometimes referred to as the "poor man's ensemble". This technique is still available in many ensemble model displays.

Ensemble output

Ensemble data can be displayed to operational forecasters in a number of ways:

- **Ensemble mean (MN)**. This chart is a mathematical average of all available model solutions for a given point in time. Often appears smoothed.
- **Standard deviation (SD)**. Usually paired with mean (MN) to show consistency between members. Low values indicate good agreement.
- **Spaghetti diagram (SP)**. Selects one or two isopleths from each ensemble member combined together on one map. The more spread out the isopleths are, the weaker the agreement between ensemble members. Generally used only for contoured fields such as height or pressure.
- **Stamps (ST)**. Multiple "postage stamps" of plotted data are shown, one stamp per member. This can be done with contoured fields but works best for outlines or shapes, such as those of storm cells.
- **Paintball (PB)**. Shapes are combined on one map and color coded according to each member. It is often difficult to evaluate closely clustered areas which overlap.
- **Probability (PR)**. A single map color-colored to indicate the percentage of members matching a given criteria.
- **Conditional probability (CPR)**. Same as probability except showing matches that are constrained to a specific range of values. Often appears as a colored "band" where there is the greatest disagreement between members.

key details

The ensemble approach is based on the idea that a near-infinite number of possible forecast solutions exist for an initial atmospheric state. Some solutions will be more accurate than others. By evaluating each solution individually or by using statistical methods, forecast success is more likely than by using any single model solution.

Most dynamical models have an ensemble component. The most widely-used ones (with the deterministic component in parentheses) are GEFS (GFS), EPS (ECMWF/IFS), GEPS (GDPS), REPS (RDPS), SREF (NCEP regional models), and HREF (NCEP and UCAR community convective-allowing models).

Ensembles may alert forecasters to patterns shown by only a few ensemble members which might have great significance.

As models are being reconfigured at an increasingly rapid rate, making it difficult to establish a collective knowledge of model biases, ensembles have taken an increasingly important role in weather forecasting.

Spaghetti diagrams are excellent tools for showing contoured data, along with a limited visual depiction of the probability distribution. However the spaghetti diagram does not show the full field from each ensemble member and it may be necessary to view spaghetti charts for different values or inspect the actual ensemble members.

Much of ensemble forecasting is rooted in the work of MIT meteorologist Edward Lorenz and his study of deterministic chaos.

real-time Internet sources

www.emc.ncep.noaa.gov/mmb/SREF/SREF.html
www.esrl.noaa.gov/psd/map/images/ens/ens.html
www.cpc.ncep.noaa.gov/products/predictions
www.spc.noaa.gov/exper/href/
www.spc.noaa.gov/exper/sref/
weatheroffice.ec.gc.ca/ensemble

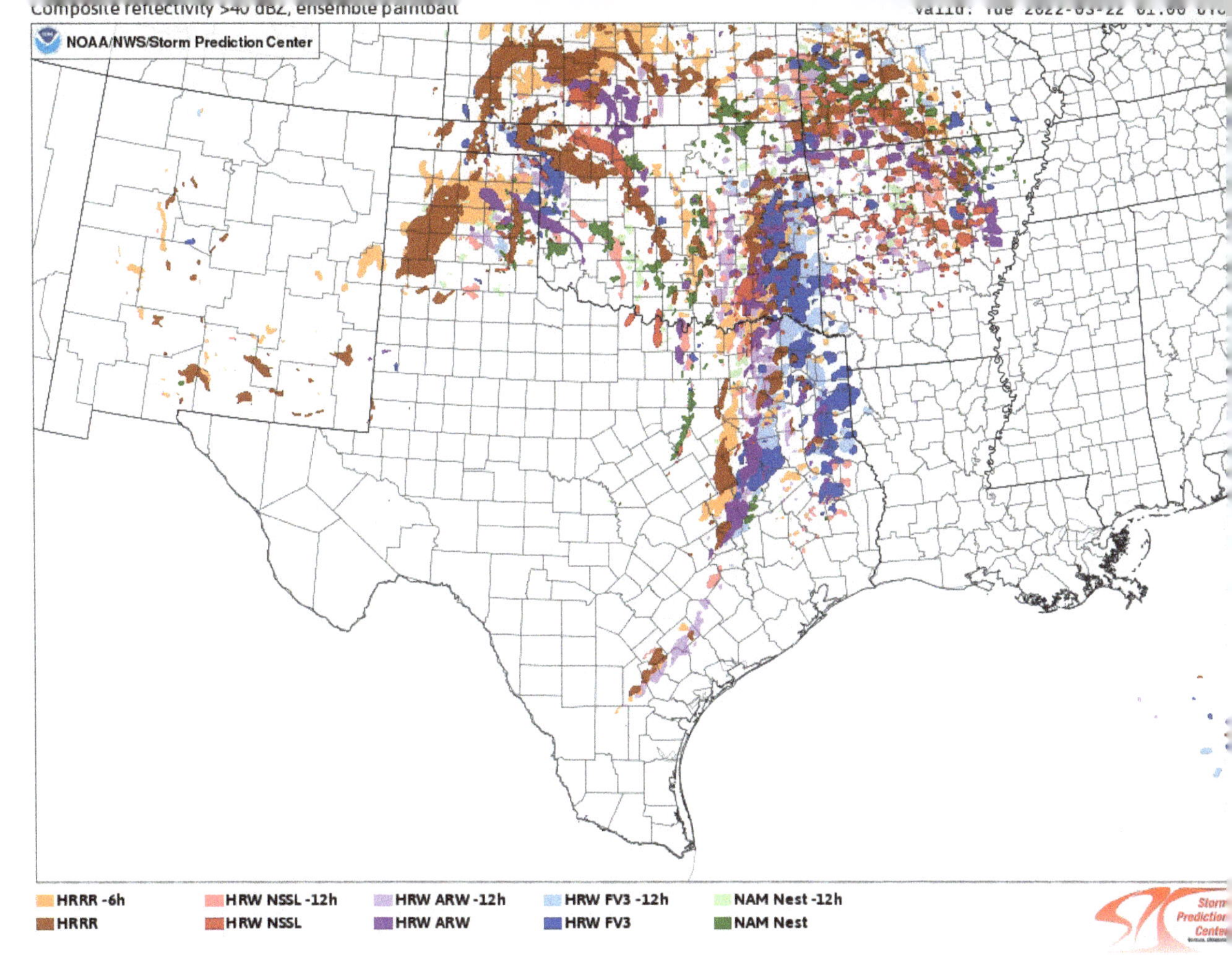

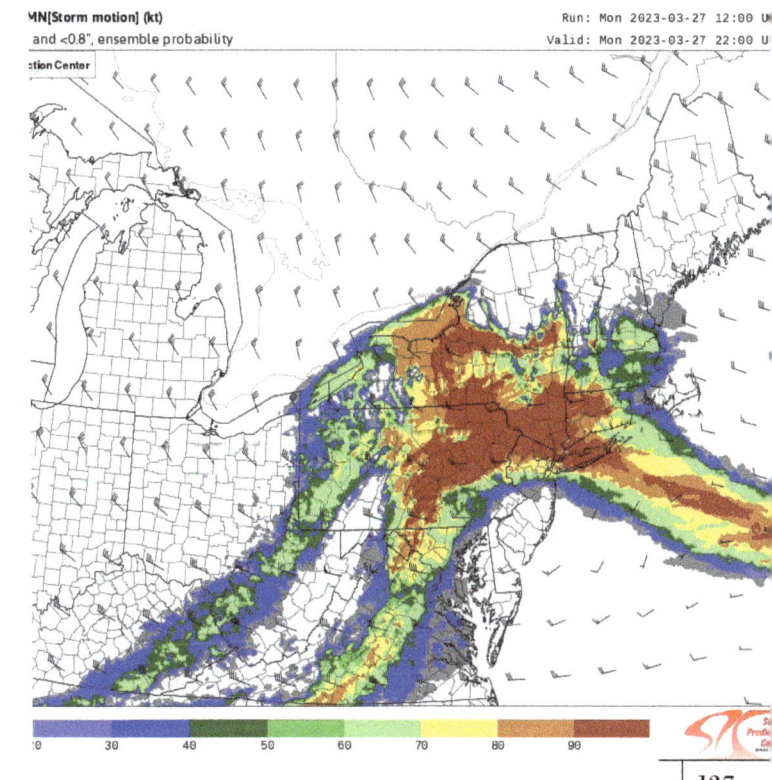

Top: 13-hour forecast HREF ensemble paintball plot of moderate or higher reflectivity (exceeding 40 dBZ) during the severe weather outbreak of 21 March 2022. Some immediate indications are the FV3 HRW suggesting earlier and more energetic development (a bias often noted in that model), with the HRRR being a slow outlier.

Right: HREF probabilistic ensemble from the Storm Prediction Center. Red and orange areas indicates a strong consensus between all ensemble members that precipitable water will equal 0.5 to 0.8 inches. This chart also shows very strong agreement that southern New Jersey will exceed a PW of 0.8 inches. *(SPC)*

Model soundings

Model soundings allow forecasters to view dynamical model output in the form of a sounding chart such as a Skew-T. In the 1980s and 1990s when these tools were first introduced, model output was too coarse to provide much useful information, and the physics packages of the day were rather primitive.

During the 2000s there was enormous improvement in model resolution, especially in the vertical, and better handling of subsynoptic detail. Data assimilation systems also became very robust. As a result, highly accurate forecast soundings were commonplace by the 2010s, and many model display web sites began allowing these diagrams to be easily constructed. Model soundings are now accurate enough where they can serve as important components of the forecast process, providing data is crosschecked where possible.

A groundbreaking framework widely used for viewing model soundings is the SHARPpy plug-in, developed collaboratively in 2015 by P. Marsh (SPC), K. Halbert (UW Madison), G. Blumberg (NASA GSFC), and T. Supinie (OU). Written in Python, this tool was developed to provide a universal, modular solution for sounding display. SHARPpy has been integrated into a number of operational sounding display systems such as those at the Storm Prediction Center, as well as on private vendor sites such as Pivotal Weather. This tool is explained in great detail in the author's book *Instability, Skew-T & Hodograph Handbook*.

There are a multitude of possible uses of model forecast soundings. Air masses can be evaluated in the vertical, and frontal transition zones can be identified and characterized. This helps provide a detailed picture of winter precipitation type. In severe weather situations, the depth and relative strength of the elevated mixed layer (EML, above the cap) is apparent. Heights of inversions, instability and shear parameters, and much more can easily be determined.

It's essential to sample model data at a number of different points to eliminate small-scale circulations that are spun up by the model. Some web sites allow users to draw a box to construct an averaged sounding, which helps filter out small scale artifacts. Forecasters should also view the precipitation fields and make sure that soundings are not being constructed in a downdraft area or close to a storm, as these areas may not be representative of the local environment.

key details

Model forecast soundings are an extraction of data from a dynamical model in a vertical column above a given point. This may be obtained for various valid times in the model run. Area-averaged soundings can also be constructed to filter small-scale noise.

Model forecast soundings can be obtained from any numerical weather prediction model: NAM, GFS, ECMWF, and so forth. Crude model soundings can even be constructed by hand using forecast data from constant-pressure charts. This was often done by forecasters in the late 20th century, at a time when no computer software was available to postprocess model data in forecast offices.

Severe weather forecasting is enhanced by the use of model soundings, especially from high-resolution convection-allowing models such as the HRRR. The forecaster should evaluate as many points as possible and look over the integrity of the sounding to ensure it provides a representative estimate of the storm environment.

The model forecast sounding is most prone to errors in the boundary layer, where there are a greater number of factors that can affect the sounding. These include radiation processes, topographical and vegetation effects.

Model soundings are heavily influenced by convection generated within the model. Forecasters should use caution when selecting a model sounding location. Moisture and temperature effects from these precipitation areas will also advect to other areas over time and influence soundings at those locations. These effects may or may not be representative and require additional interpretation by the forecaster.

real-time Internet sources

www.pivotalweather.com/model.php
weather.cod.edu/forecast/
www.ready.noaa.gov/READYcmet.php
rucsoundings.noaa.gov
www.wright-weather.com ($)

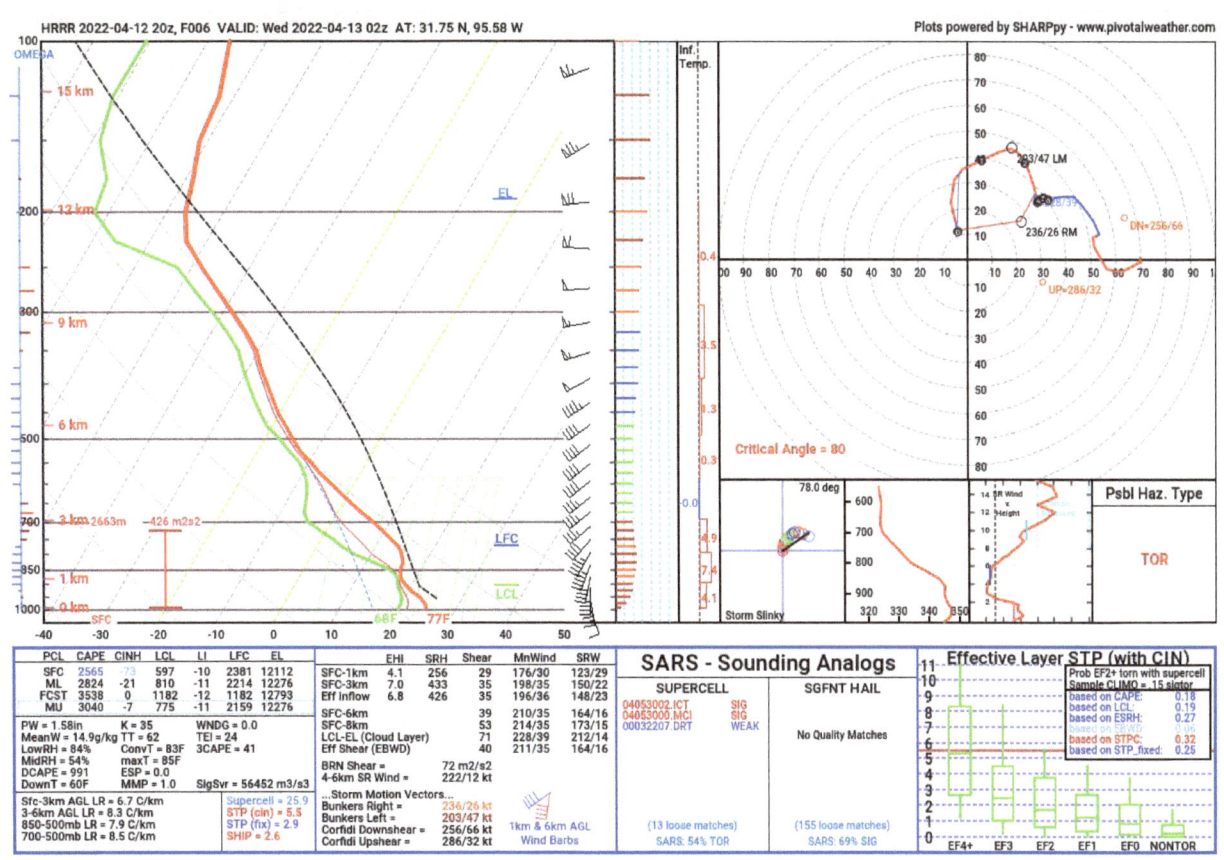

Above: HRRR forecast sounding from Pivotal Weather showing conditions on a severe weather day in Texas. *(PivotalWeather)*

Below: The SHARPpy program provides a highly detailed breakdown of stability quantities and severe weather parameters (left), and kinematic values (right). The various values of CAPE are useful for assessing the potential for strong convective updrafts. CINH (convective inhibition) indicates the degree of mid-tropospheric capping, or suppression, that resists growth of deep convection. Very low CINH is associated with upscale growth into mesoscale convective systems (MCSs). The LCL (lifted condensation level) is given in meters and is strongly tied to relative humidity; high LCLs are detrimental to tornado development but favor downburst winds and hail if instability values are high enough. The quantity LI is lifted index, which was widely used in the 1960s through the 1980s to evaluate instability. The parameter LFC is lifted condensation level (m), which measures the altitude above ground level at which a buoyant parcel will become buoyant and undergo deep convection. High LFCs are not supportive of tornadoes. EL is equilibrium level, which gives a rough idea of anvil cloud top height (not the overshoot). Other parameters are described in various references online as well as in the author's *Instability, Skew-T & Hodograph Handbook*.

Historical models

Today's models are best understood in the context of models that dominated meteorology during the past few decades. Furthermore, as meteorologists often refer to journal papers which actually used some of these older models, this reference section is useful for understanding their qualities, configuration. and performance.

This section will focus on United States operational models, as these are most familiar to the readers and have been universally available to the public in many different formats.

Global forecast suite

The earliest fully automated model was the 3-level quasigeostrophic model, which entered service in October 1955 and covered the United States.

The problem of mapping the entire world and keeping track of new systems coming into the forecast window from the Pacific was partially solved in April 1958 with the implementation of a Northern Hemisphere barotropic model. A three-layer baroclinic model was put into operation in March 1962, offering significant improvement in the ability to forecast extratropical cyclones.

A truly global model was not fielded until September 1974 with the 6LPE (6-layer primitive equation model). This as upgraded to 7LPE in 1978. In 1980 it was converted into a 12-layer spectral model. By spectral, this means that the fields are treated as a series of waves instead of measurements between gridpoints. This model had a very coarse 381 km resolution between gridpoints.

On 17 April 1985, NCEP introduced the Medium Range Forecast (MRF) model as part of its global suite. This was a special development of the spectral model that took advantage of the new CYBER 205 computer's ability to handle long vectors. The MRF model has been extensively revised since its 1985 implementation and now forms the backbone of today's GFS system.

Limited-area Fine Mesh (LFM)

The Limited-area Fine Mesh (LFM) model is completely gone from operational forecasting, but it left a deep impact on North American forecasting from 1971 to 1993. It was the first regional (continental-sized) model to enter operational use and was the result of a cooperative effort between the U.S. Air Force and the U.S. Weather Bureau.

The LFM made its first operational forecast on 29 September 1971. It was initially a 6-layer model, with a resolution of 190.5 km (contrast this with 50 layers and 13 km with the current RUC!). The model was initially run only to 24 hours, but by 1975 and 1976 it was extended out to 36 then 48 hours.

In 1979, a major upgrade to the LFM resulted in the LFM-II, which featured 7 layers and 127 km resolution, however in 1981 the earlier 190.5-km resolution was restored but added in a fourth-order finite difference approximation to compensate for the downgrade. Computation of the LFM-II on NMC's IBM 360 systems took about 20 minutes for the full run, decreasing to 75 seconds on the CDC Cyber 205 computers introduced in 1983.

During the reign of the LFM in the 1970s and 1980s, it served as the basis for NMC to develop numerous sets of model output statistics (MOS), which allowed computers to automatically produce accurate forecasts for specific cities, taking into account LFM forecast fields, predictors, and empirical rules. The results were disseminated in the form of tables every 12 hours over teletype circuits.

By 1985, the end was in sight for the LFM as the cutting-edge NGM came online with almost twice the resolution in each dimension and much better accuracy. However the LFM excelled in its simplicity, so NMC assigned it a slot called the ERL (early) run, giving forecasters preliminary maps and charts just hours after radiosonde release time until the NGM package was available. The Cyber 205 systems took only 75 seconds to produce the full 48-hour LFM run.

The final nail in the coffin for the LFM came several years later as an even more sophisticated model, the Eta, made its introduction. The LFM was finally removed from the production suite in June 1993, replaced by the brand new Eta, and was completely phased out of all NCEP activities on 29 February 1996.

Nested Grid Model (NGM)

The NGM (Nested Grid Model) was developed as the "next generation" improvement to the old LFM (Limited-area Fine Mesh model) that predominated United States forecasting in the 1970's. It was created by research meteorologist Norman A. Phillips in 1978.

Operational use of the NGM commenced on 27 March 1985 on the new Cyber 205 supercomputer. An improved physics package was added in August 1986, followed by a fine-mesh grid expansion in February 1987. The model was scaled down to two grids in August 1991, and that concluded work on it with the bigger and better Eta expected to come online. The model was scheduled to be axed at NCEP by 1998, however this date was extended until March 3, 2009.

As implied by its name, the NGM uses multiple nested grids: a large, coarse one to handle distant systems in Asia, Europe, and elsewhere, and a smaller, denser one focused on North America where the highly detailed computations are made. This nesting is a concept embraced by the Canadian GEM Regional and UKMET models.

Overall, the NGM's physics were changed very little, and this allowed many model biases to be documented. For example, the NGM worked best in the warm season. It tended to forecast surface lows and highs as being too strong over land and too weak over the ocean, and had a poleward bias, in which weather systems erred slightly to the north in the Northern Hemisphere. Digging upper-level troughs were not portrayed strong enough, and plunging cold air outbreaks were not moved fast enough and were allowed to stagnate over southern areas too long when the air mass was actually eroding by that time.

At the time of its implementation, the NGM was recognized as the official NOAA regional model and was part of a suite called the RAFS (Regional Analysis and Forecasting System). RAFS is the predecessor to the NAM of today.

Eta model (ETA)

The Eta model formed the basis of most operational forecasting in the United States beginning with its implementation in June 1993 and lasting until its discontinuation in 2006.

The Eta model is named after the greek letter "eta", thus is pronounced EHH-ta. It is not an acronym, even though it is often capitalized in old references. Unlike most early models, the Eta model used terrain-following layers which drastically simplified computations over rough terrain. The model was developed in 1978 at the University of Belgrade in Yugoslavia by Zavisa Janjic and Fedor Mesinger. Janjic brought the model to NMC in the mid-1980s. Its exceptional skill with terrain, which gave it some of the qualities of a mesoscale model, led to its official implementation in 1993.

The model was initially run at 00Z and 12Z with a resolution of 80 km at 38 layers. Output fields were generated out to 48 hours. The resolution was boosted to 48 km on 12 October 1995 with major physics improvements, then eventu-

Above: Vertical resolution of the primary NOAA/NMC (NCEP) models of the 1980s. The GFS's predecessor used 12 levels, but in 2023 it used 127.

ally changed to 12 km.

Naturally, the Eta was superior to the LFM and NGM, especially in capturing mountain interactions such as cold air damming in the Carolinas. It was also considerably less susceptible to convective feedback ("blowup") from storm activity. However early on it had a drying bias in the late spring, where forecasts were too aggressive with the dryline, and by the late 1990s this flip-flopped to a moist bias. The Eta model accuracy also deteriorated rapidly past the 36 to 48 hour point.

The Eta was discontinued in June 2006 after the WRF/NMM was brought online to drive the NAM product suite. However it is still an important part of forecasting history and went on to be used as an ensemble member at NCEP.

Rapid Update Cycle (RUC)

During the late 1980s, ACARS aircraft observations, wind profiler data, and an increasingly dense surface network were all rapidly coming online. It became quite obvious that a new model was needed to synthesize all of this mesoscale data at a rapid rate. There was also a need for a model that would end the twice-daily wait at weather offices for new model runs and allow forecasters to respond to changes more quickly. Thus, the RUC (Rapid Update Cycle) model was born.

The model was initially developed by the NOAA Forecast Systems Laboratory, where it was known as MAPS (Mesoscale Analysis and Prediction System). It underwent about three years of testing, and finally the completed model was implemented at the National Meteorological Center (NMC, now NCEP) in September 1994.

In April 1998 the RUC-2 model was introduced, replacing the original RUC. It added considerable improvements to the physics, parameterizations, and model resolution. GOES precipitable water data was introduced with this upgrade.

In April 2002, the RUC20 model was introduced, replacing the RUC-2. It doubled the resolution and added 10 new vertical layers. There were also vast improvements to integration of surface data, improved diagnostic routines, better cloud parameterization, and inputs every 6 hours from the Eta run. Analysis was done through old-fashioned optimum interpolation (OI) up until sometime in 2003 when the 3DVAR method was integrated.

In June 2005 the resolution was increased to 13 km operationally. With this change it has begun using more input data, such as METAR cloud/visibility data and GOES cloud top data.

The model was retired on May 1, 2012 and replaced with the Rapid Refresh (RAP) model.

MM5 model

The MM5 was never part of the official NOAA operational forecast suite, but deserves special mention as it was one of the first mesoscale models and was a predecessor to the WRF which is in widespread use nowadays.

The MM5 is not the name for a suite of products but rather for a "brand" of model, the Mesoscale Model Fifth Generation developed by the University Corporation for Atmospheric Research (UCAR) in collaboration with Pennsylvania State University (PSU). It was designed to solve a pressing requirement in the research community for a portable, small-scale model that was not locked to a large, regional domain.

The model was first developed as early as 1972. As it came to the forefront of mesoscale meteorology with the arrival of powerful computing technology, the MM5 was released to the weather community in early 1994. The first generation of the model was optimized for Cray systems. The second version, released the same year, supported more operating systems and included improved physics packages.

During the 2000s, the WRF replaced the MM5 as the community model of choice due to its modernized design and flexible architecture. The U.S. Air Force was one of the biggest supporters of the MM5 during the late 1990s but replaced all of its MM5 systems with WRF-ARW models in late 2007.

Since it is adapted for high-resolution terrain and uses non-hydrostatic calculations, the MM5 is well suited for modelling extremely small-scale processes. It can be used to examine mesoscale convective systems, fronts, tertiary circulations, and urban meteorology effects. Furthermore its design for portability allows it to be run on a broad range of Unix systems, with vast flexibility on the resolution, the domain size, and other configuration options. The source code is in FORTRAN and can be easily modified to customize the model.

The AFWA configuration of the MM5 was noted in 2000 by NCEP to have problems with overforecast rain totals in the southeast and eastern U.S., particularly in regions of synoptic-scale forcing. The objective analysis scheme was suspected as a culprit.

Unfortunately it is impossible to give a complete description of MM5 model biases since it is run by different users in a variety of configurations and with varying data sources. However the consensus is that the MM5 does an excellent job with mesoscale weather systems, true to its design.

Below: 24 hour forecast from the Eta during its final year, showing the advance of Hurricane Rita directly toward Houston in September 2005. Forecast solutions showing this path contributed to what ended up being a chaotic mass evacuation from the city. Landfall was actually in Louisiana slightly east of the state line.

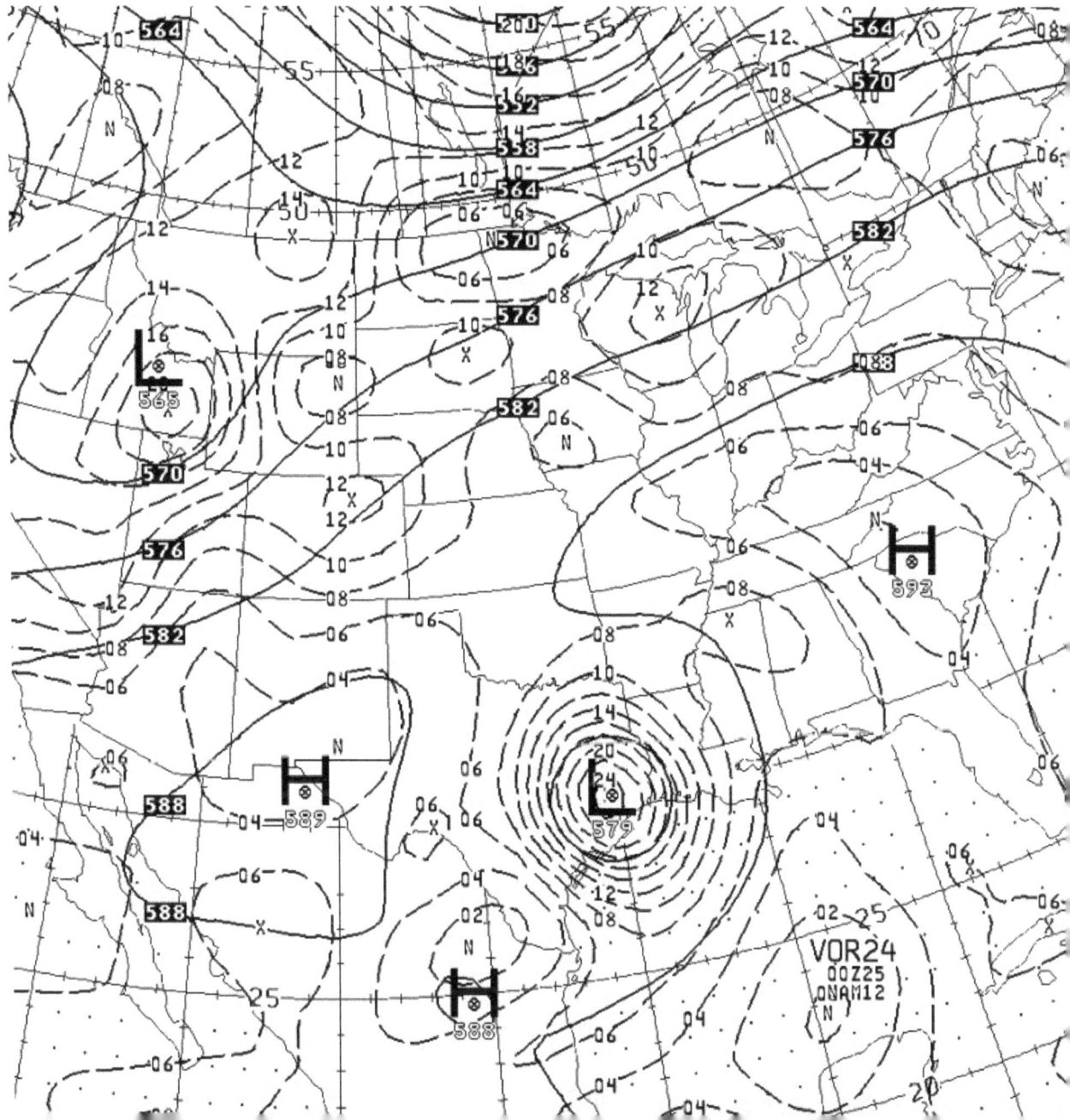

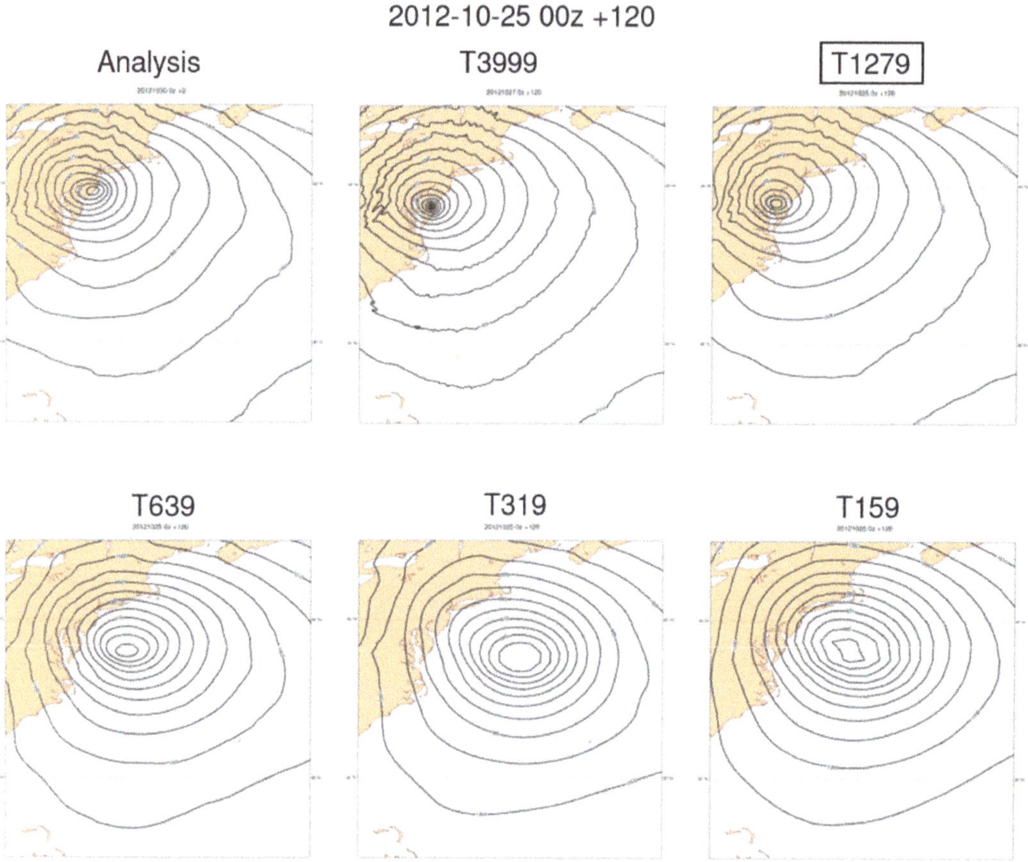

2012-10-25 00z +120

Analysis	T3999	T1279
T639	T319	T159

Above: ECMWF 5-day forecast for Hurricane Sandy's landfall, based on different resolutions. The T1279 (16 km) example shows the 1279-wave configuration introduced by ECMWF in 2010 and serving as the operational model of the day, which accurately brought the storm into coastal New Jersey, dramatically increasing the storm surge risk on the right half of the storm. The T639 (32 km) model is representative of the ECMWF model resolution in 2004. The T319 (64 km) model is what would have existed in 1998. The T159 (125 km) is representative of the model as it existed in 1989. The T3999 (5 km) model is experimental and may be achievable operationally around 2025-2030; note this shows a slight eastward correction to the coast that is consistent with the analysis data. *(ECMWF)*

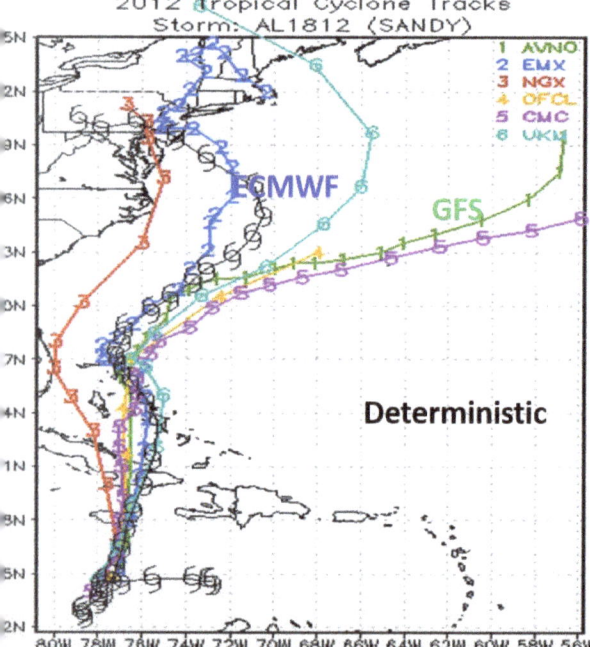

Left: Comparison of 5-day forecast of ECMWF and GFS for Hurricane Sandy. At this time the GFS was running in T574 (23 km) configuration while the ECMWF was in T1279 (16 km) configuration. This was probably a contributing factor to the GFS's weak response. The GFS did begin converging on the New York City and New Jersey area about 3 days in advance.

```
SAUS70 KWBC 210000
METAR
K47A 202345Z AUTO 32007KT 10SM CLR 02/M13 A3021 RMK AO2=

ETX50H
950
SAUS70 KWBC 210000
METAR
KBKS 202345Z AUTO 02008KT 10SM SCT011 SCT016 13/12 A3024 RMK
     AO1=
KBWD 202345Z AUTO 36008KT 10SM OVC035 06/02 A3034 RMK=
KCPT 202345Z AUTO 29003KT 1 1/2SM BR BKN004 OVC012 04/04 A3038
     AO2=
KDKB 202345Z AUTO 25004KT 5SM HZ CLR M08/M11 A3023 RMK AO2=
KDKR 202345Z AUTO 05003KT 10SM FEW048 07/03 A3038 RMK AO2=
KDUX 202345Z AUTO 26003KT 10SM CLR 04/M03 A3028 RMK AO2=
KE38 202345Z AUTO 08007KT 10SM CLR 05/00 A3030 RMK AO2=
KECU 202345Z AUTO 34008KT 10SM CLR 11/03 A3032 RMK AO2=
KERV 202345Z AUTO 02006KT 10SM BKN049 OVC055 13/02 A3027 RMK
     AO2=
KETB 202345Z AUTO 29003KT 10SM BKN022 OVC035 M08/M11 A3020 RMK
     AO2=
KF05 202346Z AUTO 16005KT 7SM CLR 01/00 A3034 RMK AO2=
KFEP 202345Z AUTO 25003KT 10SM CLR M10/M11 A3022 RMK AO2=
KFWS 202345Z AUTO 00000KT 7SM BKN011 OVC016 05/02 A3037 RMK
     AO2=
KFOA 202346Z AUTO 20005KT 10SM SCT045 M02/M07 A3034 RMK AO2=
KFWC 202345Z AUTO 25003KT 10SM SCT050 M02/M08 A3036 RMK AO2=
KGBG 202345Z AUTO 00000KT 7SM CLR M08/M09 A3028 RMK AO2=
KGDJ 202345Z AUTO 28003KT 5SM BR OVC014 05/03 A3039 RMK AO2=
KGNC 202345Z AUTO 00000KT 10SM CLR 06/02 A3033 RMK AO2=
KGVT 202345Z AUTO 00000KT 10SM OVC008 04/03 A3037 RMK AO2=
KGYI 202345Z AUTO 03003KT 7SM OVC016 04/02 A3037 RMK AO2=
KHBV 202345Z AUTO 04007KT 10SM OVC014 13/10 A3025 RMK AO2=
KHHF 202345Z AUTO 00000KT 10SM CLR 04/M05 A3029 RMK AO2=
KHSB 202345Z AUTO 22004KT 10SM CLR M01/M07 A3036 RMK AO2=
KLOT 202346Z AUTO 24006KT 7SM CLR M09/M12 A3024 RMK AO2=
```

CHAPTER 6
RAW DATA

SYNOP surface data

```
AAXX 12124
72594 35566 81004 10083 20056 30238 40266 56002 91153 333 10139
20072 555 91212
72791 17448 81708 10083 20072 30172 40176 56017 60051 761// 91155
333 10133 20083 70079 96010 555 91212
72384 35964 00000 10106 21017 30068 40249 53010 91154 333 10211
20100 555 91212
72677 11366 80105 10072 20061 30903 40145 51002 69951 761// 91153
333 10161 20072 70095 555 91212
72683 36866 82004 10033 21033 38772 40209 50007 91153 333 10144
20017 555 91212
72681 32966 01605 10017 21011 39183 40199 51011 91153 333 10161
20017 555 91212
72750 35966 02310 10061 21128 38063 40068 53012 91153 333 10244
```

The SYNOP format is given in WMO Pub. 306, "Manual on Codes" in section FM-12. It specifies the format that must be used worldwide for all non-aviation weather observations (aviation reports are coded under METAR format).

The format is as follows:

AAXX hhmmw
CCCCC rxhvv NDDFF 1sttt 2SDDD 3pppp 4PPPP 5appp 6RRRT 7wwpp 8NLMH 9hhmm

If any particular group is missing, it is assumed the category is not applicable. For example, if the sky is clear, the 8NLMH cloud code group will be omitted. Solidii ("/") usually indicate missing data.

Header (AAXX hhmmw). The header usually appears at the top of a group of reports collected by a single office. It always starts with "AAXX" ("BBXX" indicates a ship report with more elaborate coding standards that are beyond the scope of this book). The time follows in hours (hh) and minutes (mm). If the wind indicator (w) is 0 or 1, winds are in m/s; if 2 or 3 winds are in knots; an even value is estimated and odd is measured.

Station location (CCCCC). This represents the five-digit WMO identifier where the weather was observed.

Miscellaneous and visibility group (rxhvv). The precipitation indicator (r) tells whether a supplementary block is used. Station type (x) is 1-3 if manned, 4-7 if automated. Lowest cloud height (h) is a coded value. Visibility (vv) is also a coded value; when below 50 it generally indicates the visibility in tenths of kilometers.

Wind group (NDDFF). This group begins with the total cloud cover (N) in eighths; if it is 9 the sky is obscured and if a solidus it is not known. The wind direction hundreds and tens digits follow in degrees relative to true north (dd) and speed (ff). The units are given in the header (see above).

Temperature group (1sttt). Exact temperature (ttt) in tenths of degrees Celsius. If the sign value (s) is "1", the temperature value is negative.

Dewpoint group (2SDDD). Exact dewpoint temperature (DDD) in tenths of degrees Celsius. If the sign value (S) is "1", the dewpoint value is negative.

Station pressure (3pppp). The station pressure (pppp) is in tenths of millibars.

Sea-level pressure (4PPPP). The sea-level pressure (pppp) is in tenths of millibars. If the first digit is 1, 2, 5, 7, or 8 this indicates that this is not a sea-level pressure and is instead a geopotential height value.

Sea-level pressure (5appp). Pressure tendency (a) with 0-3 risen, 4 steady, and 5-8 fallen, and change (ppp) in tenths of millibars.

Precipitation (6RRRT). Liquid precipitation amount (RRR) in whole mm. "990" is a trace, and with higher values the last digit is the value in tenths of a mm. The duration of the period (T) is "4" if 24 hours, "2" if 12 hours, and "1" if 6 hours.

Weather (7wwpp). The current weather (ww) and past weather (pp) is expressed as a two-digit code (see code table). In general the higher the number the more significant the phenomenon.

Cloud group (8NLMH). The amount of low or middle clouds (N) is in eighths. The code for any low (L), middle (M), or high (H) clouds is given.

Time group (9hhmm). The observation time is given in hours (hh) and minutes (mm) UTC.

Below: Plot of extreme cold in Siberia on 13 January 2023, plotted with the Digital Atmosphere software analysis program, written by the author. The chart is constructed almost entirely from SYNOP reports.

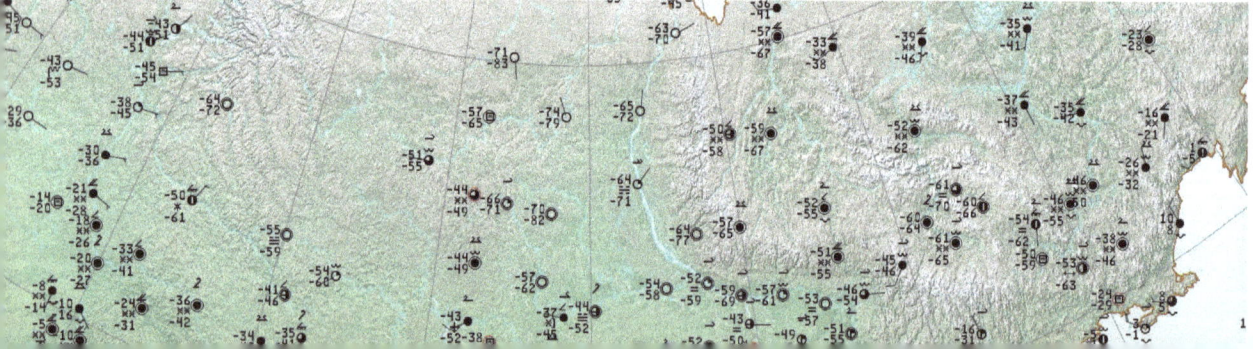

409
SMRS15 RUMS 241800
AAXX 24181
22282 32598 71803 10134 20113 40097 56002 85214   333 10163=
22292 32473 61303 10196 20166 40081 56001 85930   333 10241=
22365 NIL=
22438 32571 71802 10182 20165 40105 53003 82274   333 10226=
22446 NIL=
22563 32882 72102 10220 20173 40113 53008 86083   333 10278=
22641 11684 32101 10220 20175 40121 53002 69900 70281 81242   333 10265=
22695 NIL=
22778 11472 79901 10193 20182 40152 52007 69940 70298 82461
333 10274=
22798 12984 71802 10208 20174 40158 54000 69900 85051   333 10275=
22854 11884 31702 10216 20156 40148 57002 60010 70198 81031   333
10236=
22867 11971 21801 10216 20187 40160 57001 69900 70281 80001   333
10242=
22939 32697 21503 10212 20148 40158 53001 81501 333 10272=
22996 12598 60201 10210 20159 40176 53004 69900 83905 333 10280=
27008 32582 72302 10235 20161 40156 57001 87900 333 10284
85920=
27051 32960 82301 10212 20126 40175 53001 80001 333 10268=
27225 33568 22003 10208 20133 40186 53001 82101   333 10262=
27252 NIL=
27369 32581 72702 10222 20158 40191 52004 87300   333 10274 87925=
27393 32598 20000 10224 20144 40189 53001 81401   333 10279=
27479 32599 32401 10225 20153 40193 53006 83200   333 10276=
27532 32997 30000 10198 20149 40193 56001 82031   333 10252=
27648 32997 20000 10192 20155 40203 53002 82030   333 10268=
27679 32980 22302 10232 20100 40198 57006 82040   333 10270=
28214 32996 20000 10203 20141 40192 53003 80008   333 10277=

SMCI07 BABJ 241800 RRA
AAXX 24181
51716 31958 23605 10244 20105 38809 49994 52028 70600 80002
333 00556 10344=
51765 32980 70000 10224 20123 39071 49987 52008 82032 333
00253 10342=
51777 32968 00703 10276 20056 39027 49971 52011 333 00300
10358=
53149 32680 12702 10175 20166 38675 49986 54000 81500 333
00055 10279=
53231 32980 02204 10177 20092 38402 52002 333 00357 10257=
53336 32980 00000 10206 20105 38623 49992 52010 333 00000
10276=

real-time Internet sources

weather.cod.edu/digatmos/syn/
www.ogimet.com/usynops.phtml.en

METAR surface data

METAR stands for Meteorological Airport Report, and is the worldwide standard for transmitting weather reports from airfields. It is the backbone of weather reports in the United States, Europe, and the Pacific Rim.

Familiarity with METAR format is important for a forecaster to be able to pick up on minor trends that might occur at a weather station. The format is:

CCCC ddhhmmZ (AUTO) dddff VV ww CCCHHH tt/dd P (RMK)

Station location (CCCC). This represents the four-letter ICAO identifier where the weather was observed.

Observation time (ddhhmmZ). The UTC time the observation was taken: calendar day (dd); hour (hh); and minute (mm). The "Z" ending is a reminder that the time zone is Zulu (UTC) time.

Auto flag (AUTO). If this flag is present, it indicates the observation was taken by a machine.

Wind (dddff). The wind direction in degrees relative to true north (ddd) and speed (ff). If winds are gusting, the group takes the form dddffGgg, where gg is the gust speed. The group is always appended with units: KT (knots), MPS (meters per second) or KMH (km/h). If the wind direction will be variable, ddd is encoded as VRB. It is also permissible to encode the group as dddVddd to indicate a range of wind directions exceeding 60 degrees.

Prevailing visibility (VV). A number that may be whole or a fraction. Always ends with SM (statute miles) or nothing (meters).

Weather (ww). A two-letter standard abbreviation for any weather that will occur, with appropriate modifiers. The term CAVOK may be used if all clouds are above 5000 ft, visibility is above 10 km, and no significant precipitation is occurring. The United States does not use CAVOK.

Cloud condition (CCCHHH). Assigned for each cloud layer and may repeat. Consists of cloud cover (CCC) and height in hundreds of feet (HHH). Cloud cover may be clear, few (FEW, 1 to 2 eighths coverage), scattered (SCT, 3 to 4 eighths), broken (BKN, 5 to 7 eighths), or overcast (OVC). When the sky is obscured, CCC will be encoded as VV for vertical visibility and the HHH value will indicate the visibility into the obscuration.

Temperature/dewpoint (tt/dd). The temperature (tt) and dewpoint (dd) in whole degrees Celsius. If any value is negative, it is preceded by an "M".

Pressure (P). If this value starts with "A", it indicates altimeter setting with the value in hundredths of inches (e.g. A2997 corresponds to 29.97 inches). If the value starts with "Q" it indicates sea-level pressure with the value in whole millibars.

Remarks (RMK). If the word "RMK" appears, it indicates that supplementary information follows.

Below: Chart constructed from METAR reports, from the Aviation Weather Center website. *(NOAA/AWC)*

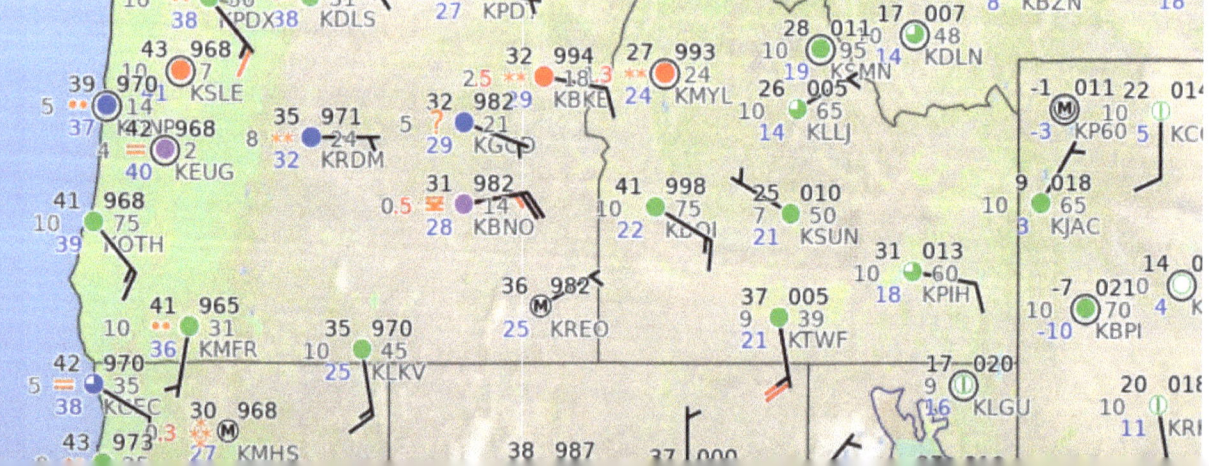

```
METAR KBUF 231054Z 13005KT 8SM -RA OVC012 03/03 A2903 RMK AO2 SLP836 P0001
T00330028=
METAR KBUF 231154Z 19004KT 3SM RA BR FEW003 BKN007 OVC015 03/03 A2899 RMK
AO2 SLP818 4/001 P0006 60010 70035 T00280028 10044 20028 $=
METAR KBUF 231254Z 23023G36KT 1 1/2SM -SN BR BKN005 BKN013 OVC022 01/00
A2898 RMK AO2 PK WND 21036/1249 RAE54SNB1157E23B42 SLP814 CIG 003V008 P0005
T00110000 $=
METAR KBUF 231354Z 23032G56KT 1/8SM R23/1400V2200FT +SN BLSN VV004 M02/M03
A2902 RMK AO2 PK WND 23059/1328 SLP834 P0000 T10221033=
METAR KBUF 231454Z 24039G57KT 1/8SM R23/1600V2400FT +SN BLSN VV007 M05/
M07 A2909 RMK AO2 PK WND 25063/1440 SLP859 SNINCR 1/2 P0000 60005 T10501067
53038=
METAR KBUF 231554Z 24034G58KT 1/16SM R23/0800V1600FT +SN BLSN VV005 M06/M08
A2912 RMK AO2 PK WND 24058/1548 SLP868 P0004 T10611078=
METAR KBUF 231654Z 24035G58KT 1/16SM R23/0400V0500FT +SN BLSN VV002 M09/M10
A2914 RMK AO2 PK WND 24059/1605 SLP877 SNINCR 1/3 P0000 T10891100=
METAR KBUF 231754Z 22037G60KT 1/16SM +SN BLSN VV002 M11/M13 A2918 RMK AO2 PK
WND 21060/1751 SLP889 SNINCR 2/5 4/005 933005 P0007 60016 T11111128 10033
21111 53019 RVRNO $=
METAR KBUF 231854Z 22029G52KT 0SM +SN BLSN VV001 M12/M13 A2917 RMK AO2 PK
WND 23056/1758 SLP889 SNINCR 2/7 P0007 T11221133 RVRNO $=
```

Above: METARs at Buffalo, New York, during the great blizzard of December 23-24, 2022.

real-time Internet sources

www.aviationweather.gov/adds/metar/
weather.cod.edu/digatmos/sao/
www.ogimet.com/metars.phtml.en (choose type: SA)

METAR remarks (U.S.)

Most METAR reports around the world are brief, but at larger airports the United States includes some of the key groups from synoptic reports in its remarks. There is also other data that is added at most domestic stations. Here are some of the remarks that are frequently encountered.

AO1 or AO2. Automated station. If appended with "A", the observation is human-augmented.
SLPppp. Sea-level pressure, where ppp is the tens, units, and tenths value in millibars.
Tatttbddd. Exact temperature (tttt) and dewpoint (dddd) in tenths of degrees Celsius. The elements a and b are sign flags: when it is "1" the value that follows it is negative.
1xxxx. Six-hour max temperature (xxxx) in tenths of degrees Celsius.
2nnnn. Six-hour minimum temperature (nnnn) in tenths of degrees Celsius.
4/sss. Snow depth (sss) in whole inches.
4axxxbnnn. Twenty-four hour maximum (xxx) and minimum (nnn) temperature in tenths of degrees Celsius. The elements a and b are sign flags: when it is "1" the value that follows it is negative.
5tppp. Pressure tendency (t) with 0-3 risen, 4 steady, and 5-8 fallen, and change (ppp) in tenths of millibars.
6pppp. Six-hour precipitation (pppp) in hundredths of inches.
7pppp. Twenty-four hour precipitation (pppp) in hundredths of inches.
8lmh. Cloud type codes. See Appendix 5 for decoding tables.
PCPN pppp or **P pppp**. One-hour precipitation (pppp) in hundredths of inches.

TAF forecast

```
                    FM281300 27008KT P6SM SCT250
                    FM282000 24012KT P6SM FEW250
          KMKE 280300Z 2803/2906 VRB03KT P6SM FEW250
                    FM281400 33005KT P6SM BKN250
                    FM281900 25007KT P6SM FEW090
                    FM290000 23010KT P6SM FEW250
          KORD 280300Z 2803/2906 VRB03KT P6SM SCT200
                    FM281500 28007KT P6SM SCT060 BKN150
          KDSM 272328Z 2800/2824 02007KT P6SM FEW250 BKN300
                    FM280600 01007KT P6SM OVC130
                    FM280800 36007KT P6SM SCT030 OVC130
```

The Terminal Aerodrome Forecast (TAF) is the worldwide standard for encoding standardized forecasts for any airport. It is based on the METAR observation format. For many decades the United States used domestic FT (terminal forecast) style, an extension of their SAO airways observation format. Both the FT and SAO formats were discontinued after a 1993-1995 transition period, and are now historical relics.

A TAF forecast can be particularly useful to meteorologists to ascertain what is expected at another location. The general format is as follows:

CCCC ddhhmm DDHHEE dddff VV ww CCCHHH

Station location (CCCC). This represents the four-letter ICAO identifier where the forecasted weather will occur.

Issuance time (ddhhmm). The UTC time the forecast was issued. The calendar day (dd); hour (hh); and minute (mm). A "Z" may be suffixed to the end as a reminder it is Zulu (UTC) time.

Forecast period (DDHHEE). The UTC time of the forecast period, with the starting day (DD) and hour (HH), and the ending hour (EE) (usually on the next day).

Wind (dddff). The wind direction in degrees relative to true north (ddd) and speed (ff). If winds are gusting, the group takes the form dddffGgg, where gg is the gust speed. The group is always appended with units: KT (knots), MPS (meters per second) or KMH (km/h). If the wind direction will be variable, ddd is encoded as VRB. It is also permissible to encode the group as dddVddd to indicate a range of wind directions exceeding 60 degrees.

Prevailing visibility (VV). A number that may be whole or a fraction. Always ends with SM (statute miles) or nothing (meters).

Weather (ww). A two-letter standard abbreviation for any weather that will occur, with appropriate modifiers. The term CAVOK may be used if all clouds are above 5000 ft, visibility is above 10 km, and no significant precipitation is occurring; the United States does not use CAVOK.

Cloud code group (CCCHHH). Assigned for each cloud layer and may repeat. Consists of cloud cover (CCC) and height in hundreds of feet (HHH). Cloud cover may be clear, few (FEW, 1 to 2 eighths coverage), scattered (SCT, 3 to 4 eighths), broken (BKN, 5 to 7 eighths), or overcast (OVC). When the sky is obscured, CCC will be encoded as VV for vertical visibility and the HHH value will indicate the visibility into the obscuration.

Wind shear group (WShhh/dddff). Sometimes, particularly in the United States, a low-level wind shear alert will be encoded. The value hhh specifies the maximum height above the surface in hundreds of feet, and ddd and ff specify the wind direction and speed above that height.

Transition identifier. These introduce new groups of weather conditions that will occur.

- FM hhmm indicates a significant change will take place at hour hh and minute mm.

- TEMPO hhee indicates a temporary condition lasting a total of less than half the time period will occur between hour hh and hour ee.

- BECMG hhee indicates a transition period that will begin at hour hh and end at hour ee, and after this time the new condition will become predominant.

- PROBpp hhee indicates a temporary condition with a probability value. The probability in percent is pp, and the duration of the expected weather ranges from hour hh to hour ee. This is used primarily in the United States. Only 30 or 40 is used; if there is a higher probability, then TEMPO is used.

Other groups. The U.S. military tends to use two groups:

- QNHppppINS, where pppp is the lowest expected altimeter setting in hundredths of inches. This is used by U.S. military stations.

- Ttt/hhZ is a maximum and minimum temperature group for the forecast period (two groups are used), where tt is the temperature and hh is the hour of occurrence. The group may also appear as TNtt/hhZ TXtt/hhZ.

```
FTUS41 KBUF 231248 AAA
TAFBUF
TAF AMD
KBUF 231248Z 2313/2412 22022G38KT 2SM -RASN BR BKN005 OVC020
     FM231500 24028G48KT 1SM -SN BR OVC012
     FM231600 24032G54KT 3/4SM -SN BLSN OVC015
     FM231800 22034G53KT 1/2SM SN BLSN OVC020
     FM231900 23035G54KT 1/4SM +SN BLSN OVC025=

349
FTUS41 KBUF 232331
TAFBUF
TAF
KBUF 232331Z 2400/2424 23035G56KT 1/4SM +SN BLSN OVC001
     FM241100 23031G50KT 1/4SM +SN BLSN OVC001=

555
FTUS41 KBUF 241129
TAFBUF
TAF
KBUF 241129Z 2412/2512 22028G48KT 1/4SM SN BLSN OVC002
     FM250300 24025G40KT 1/2SM SN BLSN OVC015
     FM250500 25025G38KT 2SM -SN BLSN OVC015
     FM251100 26022G35KT P6SM BKN045=
```

Above: TAFs at Buffalo, New York, during the great blizzard of December 23-24, 2022.

Below: TAF for February 28, 1984, showing the old international TAF format (top), only issued for international airports, and the version for United States domestic use (bottom). Most of the abbreviations on the international format will be familiar. Sky condition preceding cloud type is given in oktas. The number preceding each of the weather phenomena groups indicates the numerical position in the "ww" Present Weather table published in WMO Pub. 306. Needless to say, this old format offered more descriptiveness.

```
000
FTUS31 KCHI 281600
TAF
KORD 1818 03022/32 9999 7SC020 7AC080 INTER 1804 3200 71SN 38BLSN
9//012 PROB20 1200 17SN 38BLSN 9//006 GRADU 0304 34018/28 4800 71SN
38BLSN 8SC020 PROB40 1200 71SN 38BLSN 9//007=

000
FTUS1 KCHI 281606
ORD FT AMD 1 281615 1605Z C20 BKN 80 BKN 0322G32 OCNL C12 X 2S-BS
     CHC C6 X 3/4S-BS.  04Z C20 OVC 3S-BS 3418G28 CHC C7 X 3/4S-BS.
     09Z MVFR CIG S BS WND..
```

real-time Internet sources

www.aviationweather.gov/adds/tafs
www.ogimet.com/metars.phtml.en (choose type: FT)

```
                                    MANNKX

                                    72293 TTAA  62121 72293
                                    99004 10058 02003 00172 11660 08001 92827 12672 36018
                                    85533 09076 35510 70132 04484 31014 50578 14387 26531
                                    40743 28557 27540 30943 43368 26051 25063 50569 26567
                                    20206 58166 25577 15387 56779 25579 10640 62775 24546
                                    00197 58567 25500 88121 61775 25059
                                    77175 26007 41209

                                    72293 TTBB  62128 72293
```

TEMP radiosonde data

Nearly all radiosonde data is transmitted in the TEMP format prescribed by the WMO in Publication 306, Section FM-35. It is broken up into three major blocks, TTAA (significant level), TTBB (mandatory level), and PPBB (winds aloft) data. Other blocks such as TTCC and TTDD pertain to data in the stratosphere and is not generally used by forecasters.

Mandatory level block (TTAA)

This block shows wind, temperature, and dewpoint at predesignated levels, such as 200 and 500 mb.

The block usually begins with:
TTAA ddhhi ccccc

The TTAA is a flag that shows this is the mandatory level block. Then follows the calendar day (dd) and hour (hh). A value of 50 is added to the day if the wind units are in knots, otherwise the wind units are in m/s. The highest wind data (i) is a coded figure which is roughly in hundreds of millibars. Finally the WMO station identifier (ccccc) is indicated.

Following this is a series of repeating blocks in the format:
pphhh TTTDD dddff

The level (pp) is expressed in tens of millibars, e.g. "85" indicates 850 mb. The exception is "99", which always is the first block and indicates the ground, and "92", which is 925 mb, and "88" and "77" are special use (see below). Following this is the height (hhh) in different expressions of meters, except for level "99" (ground) in which hhh is the surface pressure in tens, units, and tenths of a millibar. The hhh expression is in whole meters from 1000 to 700 mb (it is 1hhh meters at 850 mb and 2hhh or 3hhh meters at 700 mb, whichever brings it closer to 3000 m). From 500 to 400 mb hhh is expressed in decameters. From 300 to 100 mb hhh is 1hhh decameters. Following this is the temperature block TTTDD, with temperature (TTT) in tenths of degrees Celsius and dewpoint depression (DD) in units and tenths of degrees Celsius if at or below "50" and in whole degrees Celsius if above "50" (subtract 50 before using). Finally the wind is presented as direction (ddd) relative to true north and speed (ff). Direction always ends with "0" or "5", and "1" is added to it for each hundred units of wind speed (e.g. a dddff of 25604 indicates a wind from 255° at a speed of 104).

Tropopause information is encoded as 88ppp TTTDD dddff which indicates conditions at the tropopause: most importantly its pressure. Maximum winds are encoded as 77ppp dddff.

Significant level block (TTBB)

The significant level block is designed to show temperature and dewpoint only at levels bounded by strong changes. The header and format is much the same as the mandatory level (TTAA) block, except that the repeating data block is in the format:
nnppp TTTDD

where nn occurs in a repeating sequence (00 for the surface, followed by 11, 22, 33, 44, 55, 66, 77, 88, 99, 11, 22, etc). The rest of the block is identical to the TTAA block with the omission of wind data, and no tropopause or maximum wind data.

Winds aloft block (PPBB)

This block contains wind data only, and it is graduated in feet rather than millibars. Again, the header and format are similar to TTAA and TTBB format, except that repeating data is in the format:
9habc aaaAA bbbBB cccCC

The "9" is a marker that makes it easy to pick out the group elements. The rest of the group 9habc indicates heights of the block, followed by wind data aaaAA bbbBB cccCC (encoded the same way as in the TTAA/TTBB sections) at three levels. The ten-thousands place for height is indicated by h and the thousands place for each of the three groups by a, b, and c. The height value ha000 ft applies to wind group aaaAA, hb000 ft applies to wind group bbbBB, and hc000 ft applies to wind group cccCC. Not all three groups need to be encoded; if one or two are omitted, a, b, or c will contain a solidus.

```
520
USUS50 KWBC 241200 RRC

TTAA 74121 72318 99944 13606 26005 00145 ///// ///// 92814 14421
28513 85525 12657 30015 70126 02256 25526 50577 13563 24026 40743
23777 23056 30948 36173 22089 25072 44569 22101 20218 55167 21602
15401 55770 22556 10657 60771 21021 88183 56767 22083 77262 22105
41606 51515 10164 00004 10194 30015 27018=

647
UMUS41 KRNK 241217
SGLRNK

72318 TTBB  74120 72318 00944 13606 11931 14822 22877 11624
33869 13057 44850 12657 55700 02256 66631 01763 77620 02760
88578 05563 99509 12561 11470 17165 22430 19981 33196 55966
44112 56772 55100 60771 31313 45202 81106 41414 50961=

PPBB  74120 72318 90034 26005 30518 31517 90678 28515 28016
25517 909// 25522 91124 25523 25020 23519 916// 24023 92058
24525 23059 22582 93045 22587 22105 22102 9404/ 21602 22068
95024 22536 23528 21521=

445
USRE01 FMEE 241200

TTAA 74111 61976 99017 26260 12019 00157 22857 12021 92830 17025
12528 85548 13014 13032 70166 11091 14020 50590 04384 11515 40761
16179 10512 30972 30974 10509 25098 415// 04504 20246 537// 00514
15424 677// 02065 10662 807// 24508 88999 77149 02066 31313 47408
81059=

519
UKRE01 FMEE 241200

TTBB 74118 61976 00017 26260 11009 23458 22968 20034 33947 18229
44788 09205 55774 12459 66758 12091 77725 11091 88672 11291 99588
03086 11548 00885 22443 11380 33339 23776 44257 40171 55164 643//
66147 685// 77137 669// 88102 799// 21212 00017 12019 11867 12534
22803 14028 33766 11523 44702 14520 55641 12508 66555 13517 77443
09013 88328 13011 99216 00000 11149 02066 22142 01556 33/// /////
44118 30006 55104 18508 31313 47408 81059 41414 48501 51515 92830
17025 12528 77318 12258 12024 60439 04286 13012=
```

real-time Internet sources

weather.cod.edu/digatmos/upa/
weather.uwyo.edu/upperair/sounding.html

```
CDS SP 1809 -X E9 OVC 1/12BD 2435G41/957/D6
CDS SA 1751 -X 120 SCT 3BD 990/83/21/2431G47/958/D1 PK WND 2447/43/ 812 64
% CLC    No report available.
CLL SA 1754 E19 BKN 45 OVC 7 081/78/71/1710/978/ 810 70
COT SA 1751 E20 OVC 10 83/69/1505/970/ 714 73
CRP SA 1750 M15 OVC 10 073/80/73/1318/974/ 807 15// 73
DAL SA 1750 M29 BKN 41 OVC 10 79/65/174G20/971/ //// 71
DFW SA 1750 M28 BKN 35 OVC 15 046/77/66/1817G25/969/ 814 15// 70
DHT SA 1756 -X 10 SCT 3BD 958/58/22/2340G50/951/BD5 PK WND 2351/05/ 812 49
DLF SA 1755 40 SCT 10 044/84/68/1504/972/ 705 1100=
DRT SP 1808 AUTOB CLR BLO 60 BV8 86/65/1506/973 PK WND 10 000
DRT SA 1748 AUTOB CLR BLO 60 BV8 85/65/1405/974 PK WND 08 000
DWH SA 1750 20 SCT E50 BKN 15 78/69/1908/982
DYS SA RTD 1755 25 SCT E60 BKN 250 BKN 15 016/86/53/E2518G26/966/CB
    15E-13SE MOVG NE/ 814 1903=
EFD SA 1755 20 SCT M30 BKN 8 097/80/69/E1512G20/982/CIG RGD/ 805 1800
=
ELP SA 1753 CLR 30 066/64/28/2421G31/989/BD NE/ 814 51
_____

GGG 1725 AREA 4TRWU/NEW 303/200 275/195 20W C2430=
GLS 1732 AREA 1RW-/NEW 347/80 15/45 299/50 MT 170 AT 311/41=
HDO 1725 PPINE=
MAF 1725 AREA 1RW-/NC 43/240 85/130 90W C2715 MT 250 AT 80/89
 MSTLY TRWU
 ^ES9 FS9 HU9 IU9 JT99 KR1 LQ1=
SEP 1725 SPL AREA 3TRW++/- 354/200 294/25 100W A2720 C3430 MT 410 AT
 356/73 TOP 400 AT 338/45 TROP 402
 ^HL21 IL31 JK141 KK141 LK341 MJ122=
OKC 1730 SPL  AREA 2TRWXX/+ 336/220 147/50 196/250  MT 490 AT 227/102
 OUN TROP 398 PTLY TRWU
 ^JK21 KK11 LL66 ML66 NJ6066 OJ4031 PJ6 QJ601 RL11 SL54 TL99 UL99=
LCH 1725 AREA 1RW+/NEW 8/55 42/70 66/80 93/75 113/35 356/40 MT 180
 AT 96/50
 ^KM222 LP2 MO32=
SIL 1725 PPINA UFN ALTN BTR=
LZK 1735 PPINA UFN ALTN FSM=
MCI 1725 3TRW++/NEW 123/100 144/100 20W MT 440 AT 137/95
 ^OP4 PP1=
STL 1725 CELL RWU/NEW 250/140 D5=
UMN 1725 AREA 4TRW++/+ 20/112 71/55 60W C2225 MT 340 AT 19/80
 CVR INCRG
 ^HM21 IM4 JM32 KN3 LN2=
ICT 1735 AREA 1RW++6R- 283/110 187/125 65W C2046 MT 460 AT 191/120
 AUTO
 ^MG231 NG3542 OH4421 PI2221 QJ2366 RJ2231 SK55=
LIC 1732 COR AREA TRW++/+ 316/130 202/130 80W MT 350 AT 225/86 MOVMT
 N PTN C1830 S PTN C0505
 AREA 4RW-/NC 21/80 125/85 45W C1945 MT 240 AT 68/64
 ^IJ1 JI1 JN1 KI11100111 LI20100111 MI3 MO11 NI132 OJ2401 PI1431 QK13=
ZAB 1730
 SCIN5 1730 ///PPINA@
 FMMN5 1730 ///PPINE@
 ELAT2 1730 ///PPINE@
 MRIN5 1730 ///PPIDE@
 AQQN5 1730 IT1 KR1 MN1101 NR1 QR1 PR1/QN2420C//SW MU@
 PNXA3 1730 ///PPINE=
```

Above: Two highly common data formats that were used every day by forecasters between the 1950s and the mid-1990s. The top example shows hourly airways code, the predecessor to METAR in the United States and Canada. The code below shows a radar reporting code known as RAREP. Thse were always coded by a human radar operator, and describes the echos in terms of azimuth and distance, along with other descriptive code. The digital code formats following the caret symbol consists of digital radar code, a primitive type of raster display based on 41 x 41 km boxes aligned with the LFM model grid. It was possible to plot all of this data by hand.

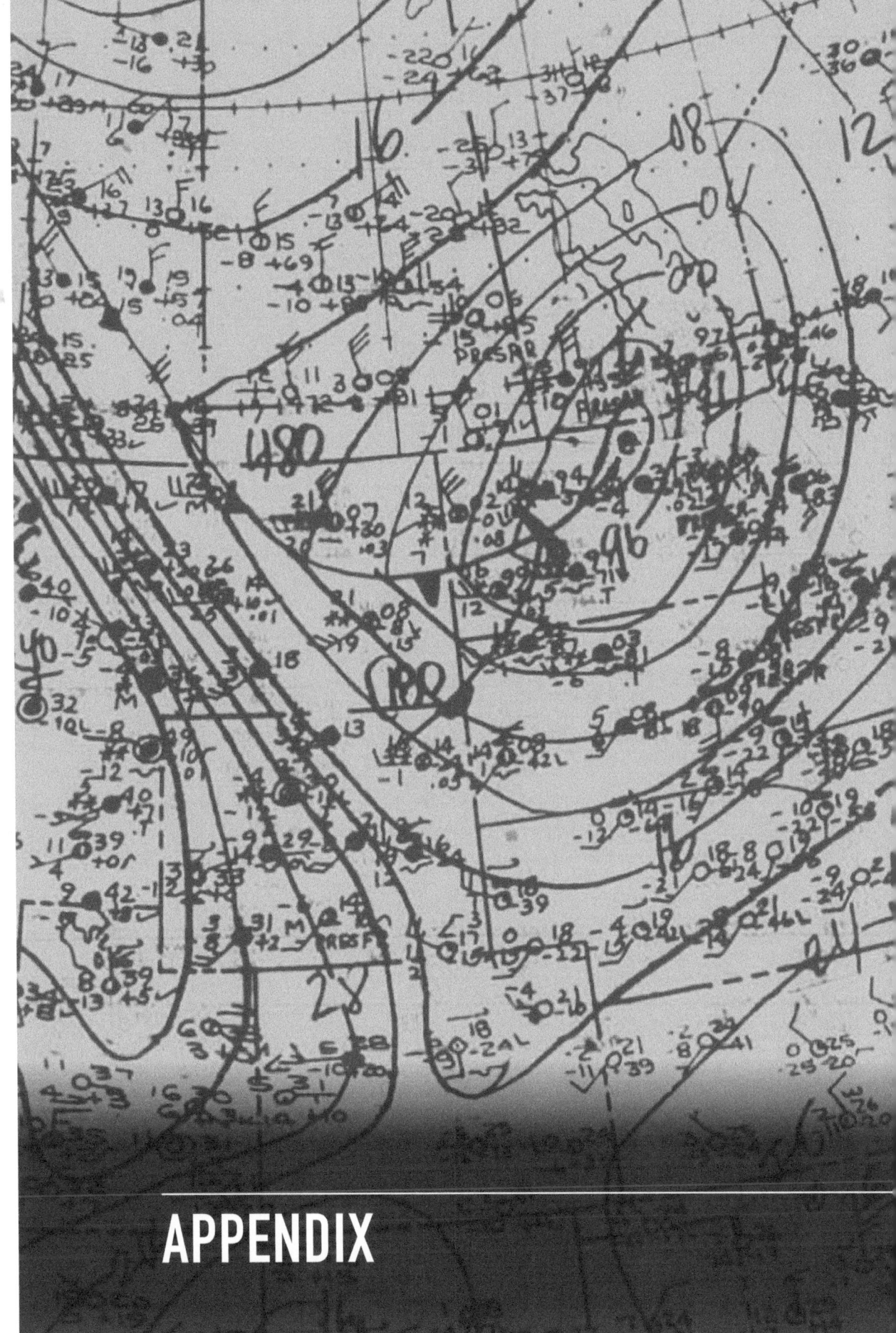

APPENDIX

Appendix 1A. Surface Plot Schematic

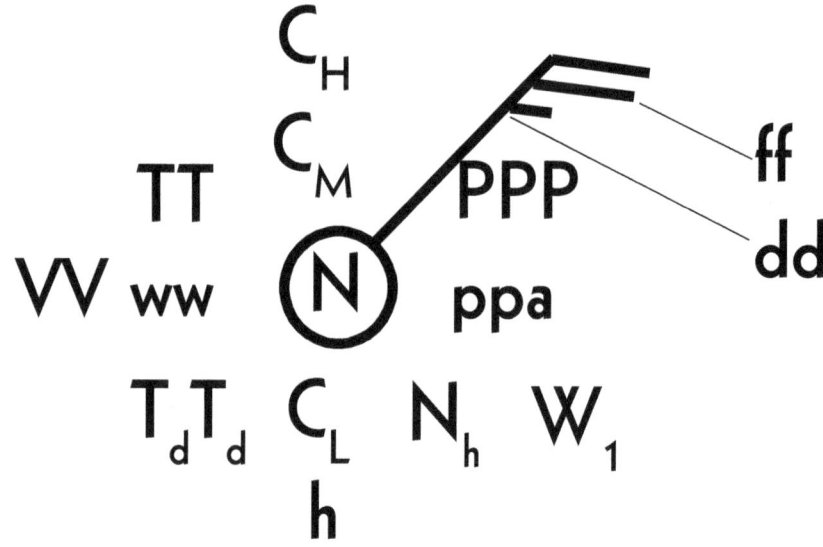

Presented here is the universally-adopted surface plot form.

TT — Temperature in degrees Celsius or Fahrenheit. Usually in whole degrees but may be expressed in tenths.

VV — Visibility in statute miles or meters. Mile values will appear as either whole numbers or fractional numbers. Meter values frequently appear as four digits, e.g. 0700.

ww — Symbol for weather type (see Table XXXX).

T_dT_d — Dewpoint temperature in degrees Celsius or Fahrenheit. Usually in whole degrees but may be expressed in tenths.

C_H — High cloud symbol.

C_M — Middle cloud symbol.

N — Total amount of cloud cover in oktas (eighths). The amount of the circle filled in is proportional to the amount of cloud cover. ○=Clear; ◔=1 okta; ◒=2 oktas; ◓=3 oktas; ◑=4 oktas; ◕=5 oktas; ◕=6 oktas; ◕=7 oktas; ●=8 oktas; ⊗=Sky obscured. When an automated station produced the observation, the symbol will be plotted as a square instead of a circle. Coloring may optionally be used: blue indicates MVFR flying conditions (ceiling 1000-3000 and/or visibility 3-5 sm); and red for IFR flying conditions (ceiling less than 1000 ft and/or visibility less than 3 miles).

C_L — Low cloud symbol.

h — Height of lowest low cloud layer, or if not present, lowest middle cloud layer. This is a coded single-digit value. 0=0-50 m; 1=50-100 m; 2=100-200 m; 3=200-300 m; 4=300-600m; 5=600-1000 m; 6=1000-1500 m; 7=1500-200 m; 8=2000-2500 m; 9=2500+ m; /=unknown.

PPP — Pressure in tens, units, and tenths of a millibar. Sometimes shows units, tenths, and hundredths of an inch of mercury.

pp — 3-hour pressure change in units and tenths of a millibar.

a — A two-segmented line representing pressure change during the past three-hours.

N_h — Amount of lowest low cloud layer, or if not present, lowest middle cloud layer. Expressed in oktas (eighths); if 9 the sky is obscured.

W_1 — Symbol for type of recent weather (See Table XXXX).

dd — Wind direction. Shaft points into the wind.

ff — Wind speed. A thick flag represents 50 kt, a long barb represents 10 kt, and each short barb represents 5 kt. The example shows 25 kt. If the wind is calm, the shaft is omitted and a circle is drawn around the station plot.

Appendix 1B. Upper Air Plot Schematic

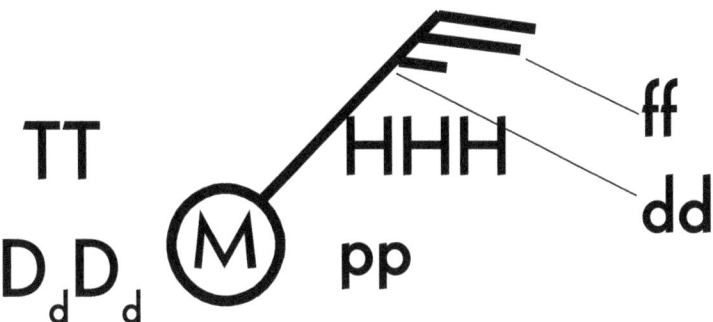

Presented here is the universally-adopted upper air plot form.

TT — Temperature in degrees Celsius. Usually in whole degrees but may be expressed in tenths.

$D_d D_d$ — Dewpoint depression in Celsius degrees. Usually in whole degrees but may be expressed in tenths.

M — The plot circle is filled whenever the dewpoint depression is 5 Celsius degrees or less. This signifies the possible presence of cloud material and perhaps the threat for icing. The element will appear as a square when the observation was made by an aircraft (such as ACARS data) or a dropsonde. It will appear as an asterisk when the observation was satellite-based.

HHH — Geopotential height. It is either thousands, hundreds, and tens of meters or hundreds, tens, and units of meters according to the level involved. See the table at right.

pp — 12-hour height change in meters.

dd — Wind direction. Shaft points into the wind.

ff — Wind speed. A thick flag represents 50 kt, a long barb represents 10 kt, and each short barb represents 5 kt. The example shows 25 kt. If the wind is calm, the shaft is omitted and a circle is drawn around the station plot.

HEIGHT DECODING RULES

1000 mb: Geopotential height is a whole number in hundreds, tens, and units of meters. If the figure exceeds 500, subtract it from 500 to get the correct negative height.

925 mb: Geopotential height is a whole number in hundreds, tens, and units of meters.

850 mb: Geopotential height is in hundreds, tens, and units of meters. The thousands place is always "1".

700 mb: Geopotential height is in hundreds, tens, and units of meters. The thousands place is always "2" or "3", whichever brings the entire figure closest to 3,000.

500 mb, 400 mb: Geopotential height is in thousands, hundreds, and tens of meters. In other words, it is an expression in whole decameters.

300 mb, 250 mb: Geopotential height is in thousands, hundreds, and tens of meters. The ten-thousands place is always "0" or "1", whichever brings the entire figure closest to 10,000.

200 mb, 150 mb, 100 mb: Geopotential height is in thousands, hundreds, and tens of meters. The ten-thousands place is "1".

Appendix 2A. ICAO Regions

An ICAO identifier tells much more than you might expect, even when you don't know where the station is. The first digit always indicates the geographic region of the station. The second digit, in some cases, indicates the country or sub-region of the station. (Source: ICAO Document 7910, Location Indicators)

Loc	Region	Examples	
Axxx	Antarctica, New Guinea, Solomon Islands	AYPY	Port Moresby, Papua New Guinea
Bxxx	Greenland and Iceland	BIRK	Reykjavik, Iceland
Cxxx	Canada	CYYZ	Toronto, Ontario
Dxxx	Northwest Africa	DNMM	Lagos, Nigeria
Exxx	Northern Europe	EHAM	Amsterdam, Netherlands
Fxxx	Southern and Central Africa	FACT	Capetown, South Africa
Gxxx	West Africa & East Atlantic	GMTT	Tangier, Morocco
Hxxx	East Africa	HECA	Cairo, Egypt
Kxxx	United States	KJFK	New York City, New York
Lxxx	Southern Europe	LIRF	Rome, Italy
Mxxx	Central America and West Caribbean	MMMX	Mexico City, Mexico
Nxxx	South Pacific	NZAA	Auckland, New Zealand
Oxxx	Middle East	OIII	Teheran, Iran
Pxxx	North Pacific, Alaska, and Hawaii	PANC	Anchorage, Alaska
Rxxx	Western Pacific	RJAA	Tokyo, Japan
Sxxx	South America	SAEZ	Buenos Aires, Argentina
Txxx	Atlantic and East Caribbean	TJSJ	San Juan, Puerto Rico
Uxxx	Former Soviet Republics	UUEE	Moscow
Vxxx	India and Indochina	VIDP	Delhi, India
Wxxx	Indonesia, Malaysia, and Singapore	WSSS	Singapore
Yxxx	Australia	YSSY	Sydney, Australia
Zxxx	China, Mongolia, and North Korea	ZBAA	Beijing, China

Appendix 2B. WMO Regions

WMO station numbers are used primarily in SYNOP surface reports and in TEMP upper air observations. As with ICAO identifers, a WMO identifier can reveal some information about its location, as the first and second digits relate the station to a geographic area. (Source: WMO Pub. 9, Vol A - Observing Stations)

Loc	Region	Examples
0xxxx	Northwest Europe	07150 - Paris, France
1xxxx	Southeast Europe	16240 - Rome, Italy
2xxxx	Northern former USSR	27515 - Moscow, Russia
3xxxx	Southern former USSR	38457 - Tashkent, Uzbekistan
4xxxx	Middle East and Pacific Rim	47662 - Tokyo, Japan
5xxxx	China	54511 - Beijing, China
6xxxx	Africa	61641 - Dakar, Senegal
7xxxx	North America	72530 - Chicago, United States
8xxxx	South America	83378 - Brasilia, Brazil
9xxxx	Australasia	94767 - Sydney, Australia

Appendix 3. Descriptors

These are the standardized weather and obscuration to vision types used in METAR and TAF forecasts. Only the SYNOP code format, containing up to 99 weather types, goes into further detail. The basic construction is in the order **intensity-proximity-precipitation-(space)-obscuration-miscellaneous**. Therefore freezing rain with fog is encoded as FZRA FG. All precipitation is assumed to be moderate unless a different intensity modifier (+ or -) is used. (Source: WMO Pub 306, Manual on Codes: Code Table 4678)

Abbv	Meaning	Type of item	Abbv	Meaning	Type of item
-	Light	Intensity	MI	Shallow	Descriptor
+	Heavy	Intensity	PE	Ice pellets (sleet)	*DISCONTINUED*
BC	Patches	Descriptor	PL	Ice pellets (sleet)	Precipitation
BL	Blowing	Descriptor	PO	Dust devils	Miscellaneous
BR	Mist	Obscuration	PR	Partial	Descriptor
DR	Drifting	Descriptor	PY	Spray	Obscuration
DS	Dust storm	Miscellaneous	RA	Rain	Precipitation
DU	Widespread dust	Obscuration	SA	Sand	Obscuration
DZ	Drizzle	Precipitation	SG	Snow Grains	Precipitation
FC	Funnel cloud	Miscellaneous	SH	Shower	Descriptor
FG	Fog	Obscuration	SN	Snow	Precipitation
FU	Smoke	Obscuration	SQ	Wind squalls	Miscellaneous
FZ	Freezing	Descriptor	SS	Sandstorm	Miscellaneous
GR	Hail	Precipitation	TS	Thunder	Descriptor
GS	Small hail	Precipitation	UP	Unknown precip	Precipitation
HZ	Haze	Obscuration	VA	Volcanic ash	Obscuration
IC	Ice crystals	Precipitation	VC	In vicinity	Proximity

Appendix 4. Present Weather

These two-digit numerical codes indicate the type of present weather that exists. They are used in SYNOP reports. The format is laid down in WMO Pub. 306, "Manual on Codes", table 4677.

Code	Sym	Meaning
00		Clear skies
01		Clouds dissolving
02		State of the sky unchanged
03		Clouds developing
04	⌒	Smoke
05	∞	Haze
06	S	Widespread dust not raised by wind
07	$	Dust or sand raised by wind
08	⨖	Dust devils
09	(S)	Duststorm or sandstorm not at station
10	=	Mist
11	==	Patches of shallow fog
12	==	Continuous shallow fog
13	<	Lightning visible, no thunder heard
14	⌣	Virga
15	)·(	Distant precipitation
16	(·)	Nearby precipitation
17	(R)	Thunderstorm with no precipitation
18	V	Wind squall
19	)(	Funnel cloud, waterspout, or tornado
20	]	Drizzle during past hour
21	]	Rain during past hour
22	]	Snow during past hour
23	]	Rain and snow during past hour
24	~]	Freezing rain during past hour
25	]	Rain showers during past hour
26	]	Snow showers during past hour
27	]	Hail showers during past hour
28	≡]	Fog during past hour
29	R]	Thunderstorm during past hour
30	S⊢	Slight-moderate duststorm, decreasing
31	S⊢	Slight-moderate duststorm, steady
32	⊢S	Slight-moderate duststorm, increasing
33	S⊢	Severe duststorm, decreasing
34	S⊢	Severe duststorm, steady
35	⊢S	Severe duststorm, increasing
36	+	Slight-moderate drifting snow
37	⇸	Heavy drifting snow
38	+	Slight-moderate blowing snow
39	⇸	Heavy blowing snow
40	(≡)	Fog at a distance
41	≡≡	Patches of fog
42	≡|	Fog, sky visible, thinning
43	≡|	Fog, sky not visible, thinning
44	≡	Fog, sky visible, no change
45	≡	Fog, sky not visible, no change
46	|≡	Fog, sky visible, becoming thicker
47	|≡	Fog, sky not visible, becoming thicker
48	⊻	Fog, depositing rime, sky visible
49	⊻	Fog, depositing rime, sky not visible

Code	Sym	Meaning
50	,	Drizzle, light, intermittent
51	,,	Drizzle, light, continuous
52	⁏	Drizzle, moderate, intermittent
53	⁏,	Drizzle, moderate, continuous
54	⁝	Drizzle, heavy, intermittent
55	⁝,	Drizzle, heavy, continuous
56	∿	Freezing drizzle, light
57	∿	Freezing drizzle, moderate or heavy
58	;	Drizzle and rain, light
59	⁝	Drizzle and rain, moderate or heavy
60	•	Rain, light, intermittent
61	••	Rain, light, continuous
62	⁞	Rain, moderate, intermittent
63	∴	Rain, moderate, continuous
64	⁞	Rain, heavy, intermittent
65	⁞	Rain, heavy, continuous
66	∿	Freezing rain, light
67	∿	Freezing rain, moderate or heavy
68	⁎	Rain and snow, light
69	⁎	Rain and snow, moderate or heavy
70	✱	Snow, light, intermittent
71	✱✱	Snow, light, continuous
72	✱	Snow, moderate, intermittent
73	✱✱	Snow, moderate, continuous
74	✱	Snow, heavy, intermittent
75	✱	Snow, heavy, continuous
76	↔	Diamond dust (ice crystals)
77	⇁△	Snow grains
78	─✱─	Snow crystals
79	△	Ice pellets
80	▽̇	Rain showers, light
81	▽̇	Rain showers, moderate to heavy
82	▽̇	Rain showers, violent
83	▽̇	Snow and rain showers, light
84	▽̇	Snow and rain showers, moderate to heavy
85	▽̇	Snow showers, light
86	▽̇	Snow showers, moderate to heavy
87	▽̂	Snow and ice pellet showers, light
88	▽̂	Snow and ice pellet showers, mod. to heavy
89	▽̂	Hail showers, light
90	▽̂	Hail showers, moderate to heavy
91	R]•	Recent thunderstorm, light rain
92	R]:	Recent thunderstorm, mod. to heavy rain
93	R]✱	Recent thunderstorm, light snow or mix
94	R]✱	Recent thunderstorm, mod-heavy snow/mix
95	R	Thunderstorm, light to moderate
96	R	Thunderstorm, light to moderate w/ hail
97	R	Thunderstorm, heavy
98	R	Thunderstorm, heavy, with duststorm
99	R	Thunderstorm, heavy, with hail

Appendix 5. Cloud code groups

These are the code forms used to represent low, middle, and high clouds. They are commonly encoded in both **SYNOP** and **METAR** reports. (Source: WMO Pub 306, Manual on Codes: Code Table 509, 513, and 515)

Low cloud types
1   CUMULUS, fair weather, no vertical development
2   CUMULUS, moderate vertical development
3   CUMULONIMBUS, no anvil
4   STRATOCUMULUS formed by spreading of cumulus
5   STRATOCUMULUS
6   STRATUS, of fair weather
7   STRATUS, of bad weather (scud)
8   CUMULUS AND STRATOCUMULUS with bases at different levels
9   CUMULONIMBUS with anvil cloud
0   No low clouds
/   Low clouds not visible due to darkness or obscuration

Middle cloud types
1   ALTOSTRATUS, mostly transparent
2   ALTOSTRATUS, opaque, or NIMBOSTRATUS
3   ALTOCUMULUS, mostly transparent
4   ALTOCUMULUS, patches
5   ALTOCUMULUS, invading the sky
6   ALTOCUMULUS, formed by spreading of cumulus
7   ALTOCUMULUS, at different layers
8   ALTOCUMULUS, castellanus (cumuliform)
9   ALTOCUMULUS, of a chaotic sky at random levels
0   No middle clouds
/   Middle clouds not visible due to darkness or obscuration

High cloud types
1   CIRRUS, fibrous
2   CIRRUS, in dense patches
3   CIRRUS, from cumulonimbus anvil
4   CIRRUS, progressively invading the sky
5   CIRRUS OR CIRROSTRATUS, invading sky, less than 45 deg above horizon
6   CIRRUS OR CIRROSTRATUS, invading sky, more than 45 deg above horizon
7   CIRROSTRATUS, covering the entire sky
8   CIRROSTRATUS, not covering the entire sky, not invading
9   CIRROCUMULUS
0   No high clouds
/   High clouds not visible due to darkness or obscuration

Appendix 6. Isopleths

What is a line called when it represents a certain quantity? This table will explain the technical name. (Source: AMS Glossary of Meteorology <amsglossary.allenpress.com/glossary>).

A line of equal . . .	Term
Temperature	isotherm
Potential temperature (theta)	isentrope
Dew point	isodrosotherm
Humidity	isohume
Wind speed	isotach, isovel
Wind direction	isogon
Shear	isoshear
Pressure	isobar
Density	isopycnal, isopycnic
Height	isoheight, contour, isohypse
Cloud cover	isoneph
Time	isochrone
Thunderstorm phase	isobront
Thunderstorm frequency or intensity	isoceraunic
Radar Doppler velocity	isodop
Precipitation	isohyet
Seasonal precipitation	isomer
Snowfall or snow depth	isonival, isochion
Sunlight	isohel
Aurora frequency	isochasm
Radar echo intensity	isoecho

Appendix 7. Chart analysis symbology

Shown here are standard markings used by Air Force Global Weather Central during the 1950s and 1960s, as published by Col. Robert C. Miller in Notes on Analysis and Severe-Storm Forecasting Procedures of the Air Force Global Weather Central (1972). Miller was a driving force in severe weather forecasting during the 1950s and 1960s and established many of the techniques used in Air Force forecasting during that era. He also created one of the few sets of meteorological symbology ever developed. While some styles have been adopted, many have become technically obsolete, fallen into disuse, or substituted with generic styles. Regardless of the state of modern-day techniques, it can be said that there has been no comparable set of conventions released since Miller's 1972 paper, and they serve as a fascinating reference.

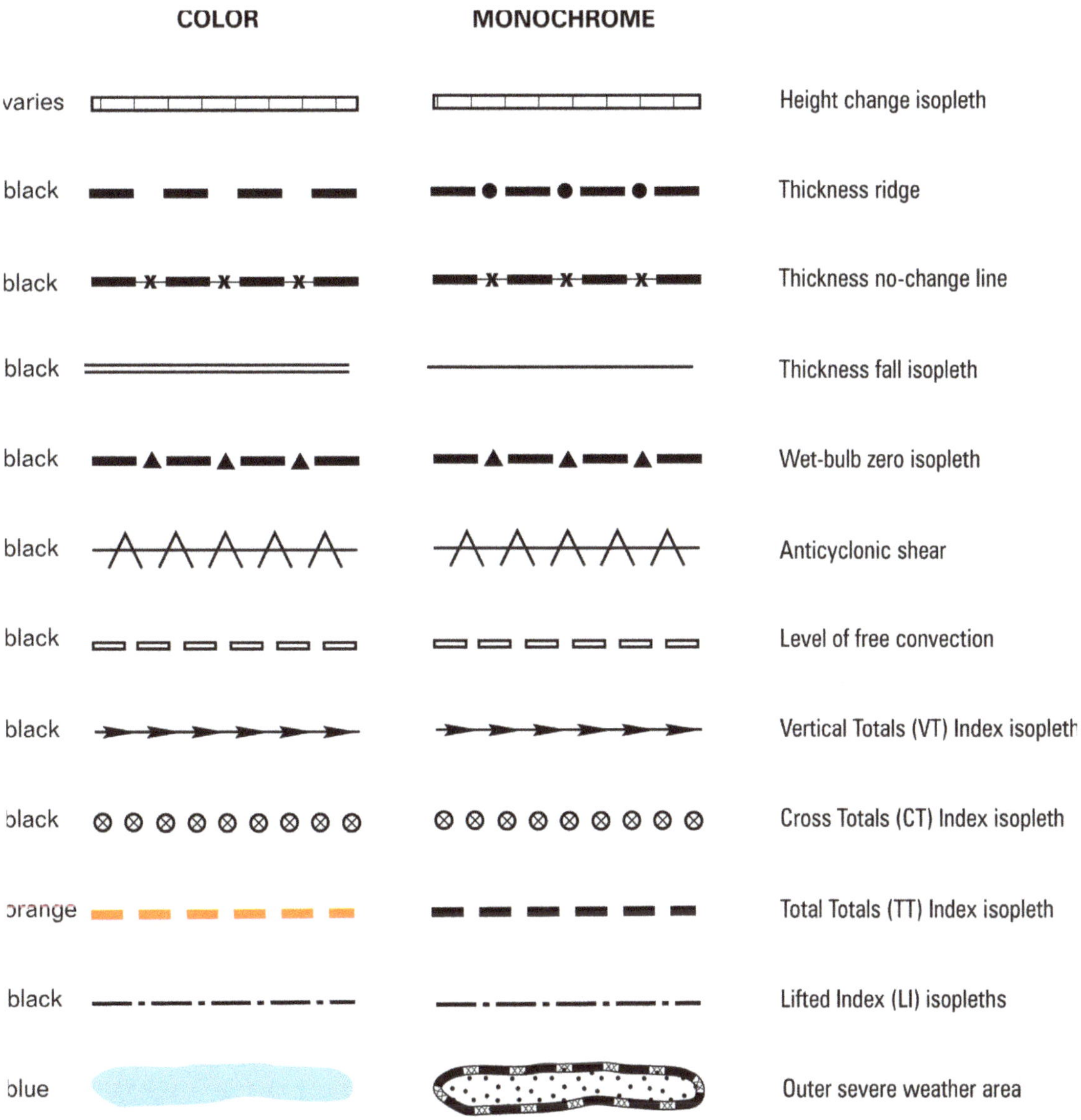

Color			Description
varies			Height change isopleth
black			Thickness ridge
black			Thickness no-change line
black			Thickness fall isopleth
black			Wet-bulb zero isopleth
black			Anticyclonic shear
black			Level of free convection
black			Vertical Totals (VT) Index isopleth
black			Cross Totals (CT) Index isopleth
orange			Total Totals (TT) Index isopleth
black			Lifted Index (LI) isopleths
blue			Outer severe weather area

APPENDIX 153

COLOR	MONOCHROME	
green		850 mb isodrosotherm/isohume
green		850 mb moisture axis
green (35 KT)	35 KT	850 mb jet axis
red		850 mb dryline / dry prod / dry intrusion edge
red		850 mb temperature ridge
red		850 mb axis of cold air advection
red (25 KT / 35 KT)	25 KT / 35 KT	850 mb shear
brown		700 dry intrusion edge
brown		700 mb moisture
brown	x—x—x	700 mb 12-hr no change (T or hgt)
brown	•x•x•x•x	700 mb temperature ridge
brown	△△△△△	700 mb thermal trough
brown		700 mb convergence zone
brown		700 mb axis of cold air advection
brown (55 KT)	55 KT	700 mb jet axis (dry)
brown		700 mb diffluence
brown		700 mb significant height falls
brown		700 mb significant temperature falls

APPENDIX

COLOR	MONOCHROME	
blue		500 mb isotherms
blue		500 mb critical isotherm
blue		500 mb thermal trough
blue		500 mb significant height falls
blue		500 mb significant temperature falls
green		500 mb moisture
yellow		500 mb PVA zone
blue	70 KT	500 mb jet axis
blue		500 mb shear
blue		500 mb diffluence
purple	90 KT	300-200 mb jet axis
purple		300-200 mb jet max
purple		300-200 mb diffluence

APPENDIX 155

Appendix 8. Stability Indices

The long, hard road of understanding the thunderstorm is demonstrated by the plethora of stability indices. Many of them exist as simple rules of thumb, created in a time where calculators and slide rules precluded lengthy calculations for multiple forecast points. Since the 1990s, the explosion of computing power has made computations of integrated stability (CAPE) and shear a relatively trivial process. It must be remembered that these are all simplifications of very complex processes, and in no way do they replace a meaningful understanding of the sounding. The equations used here are presented in as simple a format as possible. Terms used are T=temperature (deg C), T_d=dew point (deg C), D=dewpoint depression (C deg), FF=wind speed (kt); DD=wind direction (deg C). Other terms are explained where they occur. Indices are listed in a very rough descending order of frequency in operational forecasting and in journals with some grouping (e.g. LI with MLI, and VT with CT and TT).

■ Convective Available Potential Energy (CAPE)

CAPE is currently the most widely used predictor for both thunderstorm potential and severe weather risk. It was defined in 1976 by Moncrieff and Miller. This form of the equation yields integrated instability in joules per kilogram.

$$\text{CAPE} = \left(\sum_{LFC}^{EL} \left[\frac{(T_{op} - T_e)}{T_e} \vec{g} \right] \right) \Delta z$$

<300: Mostly stable, little or no convection
300-1000: Marginally unstable; weak thunderstorm activity
1000-2500: Moderately unstable; possible severe thunderstorms
2500-3500: Very unstable; severe thunderstorms; possible tornadoes
3500+: Extremely unstable; severe thunderstorms; tornadoes likely

■ Convective Inhibition (CINH)

Convective Inhibition is calculated in the same manner as CAPE except for areas along the parcel lift where the parcel is colder than the surrounding air. In effect, it figures areas that are negatively buoyant. It was developed in 1984 by Frank Colby.
<0: No cap
0 to 20: Weak capping
21-50: Moderate capping
51-99: Strong capping
100+: Intense cap; storms not likely

■ BRN Shear

BRN shear is simply a measure of the vector difference in the winds through the vertical. The greater the BRN shear, the more likely that a thunderstorm downdraft and precipitation cascade will be separated from the updraft.
BRN Shear = 0.5 $(U_{AVG})^2$
 where UAVG is the vector difference between the 0 to 6 km AGL winds and the winds in the lowest 0.5 km of the atmosphere
40-80 $m^2 s^{-2}$: Supercells possible
80-100 $m^2 s^{-2}$: Favors long-lived supercells

■ Energy-Helicity Index (EHI)

EHI is a product of CAPE and 0-6 km shear, thus it is high when either parameter is high. It was developed in 1991 by John Hart and Josh Korotky.
EHI = (CAPE x SRH) / 160,000
1.0-2.0: Heightened threat of tornadoes
2.0-2.4: Tornadoes possible but unlikely strong
2.5-2.9: Tornadoes likely
3.0-3.9: Strong tornadoes possible
4.0+: Violent tornadoes possible

■ Storm Relative Helicity (SRH)

EHI is a product of CAPE and 0-6 km shear, thus it is high when either parameter is high. The SRH is not used widely because it has a high dependence on the correct storm motion vector and it is extremely sensitive to the wind field. The SRH was defined in 1990 by Robert Davies-Jones, Don Burgess, and Mike Foster.
SRH = w · (v - c) Δz
 where w = k x dV/dz from 0 to 6 km, v is the wind vector, and c is the storm vector
150-299: Weak tornado potential
300-449: Moderate tornado potential
450+: Strong tornado potential

■ Bulk Richardson Number (BRN)

The Bulk Richardson Number is a ratio between instability and 0-6 km vertical shear (contrasting with EHI which is a product). It is a dis-

criminator of storm type, not a predictor. High values indicate unstable and/or weakly sheared environments, while low values indicate weak instability and/or strong shear. It was defined in 1986 by Morris Weisman and Joseph Klemp.
BRN = CAPE / [0.5 * U^2]
  where U is the difference in wind speed in meters per second between 0 and 6 km
<10: Severe weather unlikely
10-45: Associated with supercell development
>50: Weak multicell storms

■ Lifted Index (LI)
The Lifted Index was widely favored in 1980s before integrated stability measures became widespread. It lifts a parcel from the surface to 500 mb and compares the parcel temperature to the environmental temperature. The parcel's starting dewpoint should originate from an average mixing ratio of the lowest 100 mb and from the forecast afternoon temperature. The historical origin of the Lifted Index is attributed to Joseph Galway in 1956.
LI = T_{ENVIR}(500) - T_{PARCEL}(500)
  Parcel lifted from surface
>2 - No significant activity
2 to 0: Showers/thunderstorms possible with other source of lift
0 to -2: Thunderstorms possible
-2 to -4: Thunderstorms probable, only a few severe
<-4: Severe thunderstorms possible

■ Modified Lifted Index (MLI)
The Modified Lifted Index is the same as the Lifted Index except that the parcel is lifted from the highest wet bulb temperature in the lowest 300 mb of the atmosphere. The index was developed by Charles Doswell to better forecast thunderstorms on the High Plains. It was presented as a possible local enhancement to the lifted index, but was adopted for use by the NWS and military forecasters.
MLI = T_{ENVIR}(500) - T_{PARCEL}(500)
  parcel lifted from maximum wet bulb temperature in lowest 300 mb of atmosphere
Positive: No thunderstorms likely
0 to -2: Showers probable, thunderstorms possible
-3 to -5: Moderate indication of thunderstorms
<-6: Strong indication of severe thunderstorms

Another type of modified lifted index exists which raises a parcel to the -20 deg C isotherm instead of to the 500 mb level. The underlying concept is that a significant thunderstorm should have a cloud temperature of -20 deg C. The historical source of this index is not known.

■ Vertical Totals Index (VT)
The Vertical Totals Index is a measure of the lapse rate from about 5,000 to about 18,000 ft in the atmosphere. It makes no assumptions about parcel temperature. The VT Index was published by Robert Miller in 1967.
VTI = T(850) - T(500).
25 - Storms are unlikely
26 - Scattered thunderstorms
30 - Scattered thunderstorms, a few severe, isolated tornadoes
32 - Scattered to numerous thunderstorms, scattered to a few severe, a few tornadoes
34+ - Numerous thunderstorms, scattered severe storms, scattered tornadoes

■ Cross Totals Index (CT)
The Cross Totals Index relates the low-level moisture to the mid-level temperature, yielding an indirect estimate of lapse rate and convective instability. It was published by Robert Miller in 1967. The Surface Based Cross Totals Index (SCTI) is an alternative index that uses the surface instead of 850 mb.
CTI = T(850) - T(500).
<17 - Thunderstorms unlikely
18 to 19 - Isolated to few thunderstorms
20 to 21 - Scattered thunderstorms
22 to 23 - Scattered thunderstorms, isolated severe
24 to 25 - Scattered thunderstorms, few severe, isolated tornadoes
26 to 29 - Scattered to numerous thunderstorms, few to scattered severe, few tornadoes
30+ - Numerous thunderstorms, scattered severe, scattered tornadoes

■ Total Totals Index (TT)
The Total Totals Index attempts to integrate the lapse rate information of the Vertical Totals Index with the instability information in the Cross Totals Index. The index is sensitive to steep lapse rates, even if insufficient moisture is present. It was devised by Robert Miller in 1967.
TT = VT + CT
<43 - Thunderstorms unlikely
44-45 - Isolated or few thunderstorms
46-47 - Scattered thunderstorms
48-49 - Scattered thunderstorms, isolated

severe
- 50-51 - Scattered heavy thunderstorms, few severe, isolated tornadoes
- 52-55 - Scattered to numerous heavy thunderstorms, few to scattered severe, few tornadoes
- 56+ - Numerous heavy thunderstorms, scattered severe, scattered tornadoes

■ K Index (KI)

This parameter is a sum of lapse rate, 850 mb moisture, and humidity at 700 mb. Humidity at 700 mb is a significant contribution, though this is rare with Great Plains storm events or any event that depends on a mid-level cap. The index was published by J. J. George in 1960.
$KI = T(850) - T(500) + T_d(850) - D(700)$
- <15 - No thunderstorms (0%)
- 15-20 - Thunderstorms unlikely (<20%)
- 21-25 - Isolated thunderstorms (20-40%)
- 26-30 - Widely sct. thunderstorms (40-60%)
- 30-35 - Numerous thunderstorms (60-80%)
- 36-40 - Numerous thunderstorms (80-90%)
- 40+ - Definite thunderstorms (100%)

Indexes no longer in common use in the 2020s

Not in any specific order.

■ Showalter Stability Index (SSI)

The SSI lifts a parcel from 850 mb to 500 mb. It has the advantage of avoiding the problems inherent with shallow moisture situations, however this comes at the cost of ignoring boundary-layer characteristics. It does not work well in mountainous areas, and cannot be used when the 850 mb level is below ground level. The SSI was developed by Albert Showalter in 1947.
$SSI = T_{ENVIR}(500) - T_{PARCEL}(500)$
for a parcel lifted from 850 mb
- >+3: No thunderstorms likely
- +3 to +1: Showers probable, thunderstorms possible
- 0 to -3: Moderate indication of severe thunderstorms
- -4 to -6: Strong indication of severe thunderstorms
- <-6: Severe thunderstorms likely

■ Thompson Index (TI)

The Thompson Index is a combination of Lifted Index and K Index. It attempts to integrate elevated moisture into the index, using the 850 mb dewpoint and 700 mb humidity. Accordingly, it works best in tropical and mountainous locations. The historical origin of the index is not known.
$TI = K - LI$

Over Rockies
- <20: Thunderstorms unlikely
- 20-29: Thunderstorms
- 30-34: Thunderstorms approaching severe
- 35+: Severe thunderstorms

East of Rockies
- <25: Thunderstorms unlikely
- 25-34: Slight chance of thunderstorms
- 35-39: Few widely scattered thunderstorms approaching severe
- >40: Severe thunderstorms

■ Severe Weather Threat Index (SWEAT)

The SWEAT index uses a complex set of parameters. It was one of the first indices developed specifically to assess tornado potential. Important parameters are 850 mb dew points, lapse rates and parcel instability, wind speed at 850 mb and 500 mb, and directional shear between 850 mb and 500 mb. The SWEAT index was published in 1972 by Robert Miller. The index uses the Total Totals Index, which must be computed first.
$SWEAT = 12 \times T_d(850) + 20 \times (TT-49) + 2 \times FF(850) + FF(500) + 125 \times (\sin[DD(500) - DD(850)] + 0.2)$
- <300: Non-severe thunderstorms
- 300-400: Isolated moderate to heavy thunderstorms
- 400-500: Severe thunderstorms and tornadoes probable
- 500-800: Severe thunderstorms and tornadoes likely
- 800+: Possibly no severe weather (sheared convection)

■ KO Index

This index was developed by Swedish meteorologists and used heavily by the Deutsche Wetterdienst. It compares values of equivalent potential temperature at different levels. It was developed by T. Andersson, M. Andersson, and C. Jacobsson, and S. Nilsson.
$KO = 0.5 \times (\theta_e(500) + \theta_e(700)) - 0.5 \times (\theta_e(850) +$

$\theta_e(1000)$)
>6: No thunderstorms
2-6: Thunderstorms possible
<2: Severe thunderstorms possible

■ Boyden Index (BI)
This index, used in Europe, does not factor in moisture. It evaluates thickness and mid-level warmth. It was defined in 1963 by C. J. Boyden.
BI = Z(700) - Z(1000) - T(700) - 200
   where Z is height in dam
>95: Thunder possible

■ Bradbury Index (BRAD)
Also known as the Potential Wet-Bulb Index, this index is used in Europe. It is a measure of the potential instability between 850 and 500 mb. It was defined in 1977 by T. A. M. Bradbury.
BRAD = $\theta_w(500)$ - $\theta_w(850)$
<3: Thunderstorms possible

■ Rackliff Index (RI)
This index, used primarily in Europe during the 1950s, is a simple comparison of the 900 mb wet bulb temperature with the 500 mb temperature. It is believed to have been developed by Peter Rackliff during the 1940s.
RI = $\theta_w(900)$ - T_{500}
>30: Thunderstorms possible

■ Jefferson Index (JI)
A European stability index, the Jefferson Index was intended to be an improvement of the Rackliff Index. The change would make it less dependent on temperature. The version in use since the 1960s is a slight modification of G. J. Jefferson's 1963 definition.
JI = 1.6 x $\theta_w(850)$ - T(500) - 0.5 x (T(700) - $T_d(700)$) - 8
>30: Thunderstorms possible

■ S-Index (S)
This European index is a mix of the K Index and Vertical Totals Index. It was designed to be an optimized version of the Total Totals Index. The S-Index was developed by the German Military Geophysical Office.
S = KI - (T(500) + A)
   where A is 0 if the VT is greater than 25, 2 if the VT is between 22 and 25, and 6 if the VT is less than 22.

<39: No thunderstorms
41-45: Thunderstorms possible
>46: Thunderstorms likely

■ Yonetani Index (YON)
This index was developed by Japanese meteorologist Tsuneharu Yonetani in 1979 to forecast thunderstorms on the Kanto Plain. It provides a measure of conditional instability and low level moisture.
YON = $0.966\Gamma_L$ + $2.41(\Gamma_U-\Gamma_W)$ + 0.966γ - 15
   The final term is 16.5 instead of 15 if γ is less than or equal to 0.57. Γ is the layer lapse rate, with U representing 850-500 mb and L representing 900-850 mb, and W is the lapse rate at 850 mb. The term γ is the pressure weighted average of the relative humidity in the 900-850 mb layer, ranging from 0-1.
>0: Thunderstorms likely

■ Potential Instability Index (PII)
This index relates potential instability in the middle atmosphere with thickness. It was defined by A. J. Van Delden in 2001.
PII = ($\theta_e(700)$ - $\theta_e(500)$) / (Z(500) - Z(925))
>0: Thunderstorms likely

■ Deep Convective Index (DCI)
This index is a combination of parcel theta-e at 850 mb and lifted index. This attempts to further improve the lifted index. It was defined by W. R. Barlow in 1993.
DCI = T(850) + $T_d(850)$ - LI
10-20: Weak thunderstorms
20-30: Moderate thunderstorms
30+: Strong thunderstorms

Appendix 9. Miller's severe weather parameters

In 1972, Col Robert C. Miller published a list of parameters used by the Air Force Global Weather Central to identify thunderstorm risk areas. It was a basis for techniques used within the National Weather Service and worldwide for decades. Though slightly outdated and not a true ingredients-based approach, it is presented here for its informational value.

Rank	Parameter	Weak	Moderate	Strong
1	500 mb Vorticity	Neutral or Negative Advection	Contours cross vort pattern by <30 deg	Contours cross at more than 30 deg
2	Lifted Index Total Totals	-2 <50	-3 to -5 50 to 55	-6 >55
3	Mid-Level Jet Mid-Level Shear	<35 kt <15 kt per 90 nm	35-50 kt 15-30 kt per 90 nm	>50 kt >30 kt per 90 nm
4	Upper-Level Jet Upper-Level Shear	<55 kt <15 kt per 90 nm	55 to 85 kt 15-30 kt per 90 nm	>85 kt >30 kt per 90 nm
5	Low-level jet	<25 kt	25-34 kt	>34 kt
6	Low-level moisture	<8 g/kg	8-12 g/kg	>12 g/kg
7	850 mb max temp. field	E of moist ridge	over moist ridge	W of moist ridge
8	700 mb height no-change line	Winds cross line <20 deg	Winds cross line 20-40 deg	Winds cross line >40 deg
9	700 mb dry air intrusion	Not available or weak winds	Winds from dry to moist intrude at an angle of 10 to 40 deg are at least 15 kt	Winds intrude at an angle of 40 deg and are at least 25 kt
10	12 hr surface pressure falls	<1 mb	1-5 mb	5 mb
11	500 mb height chg	<30 m	30-60 m	>60 m
12	Height of wet-bulb zero above surface	Above 11,000 ft Below 5,000 ft	9,000-11,000 ft or 5,000-7,000 ft	7,000-9,000 ft
13	Surface pressure over threat area	>1010 mb	1010-1005 mb	<1005 mb
14	Surface dewpoint	<55 deg F	55-64 deg F	>64 deg F

Appendix 10. Satellite gallery

This section shows some of the major cloud forms as seen on typical high-resolution satellite imagery.

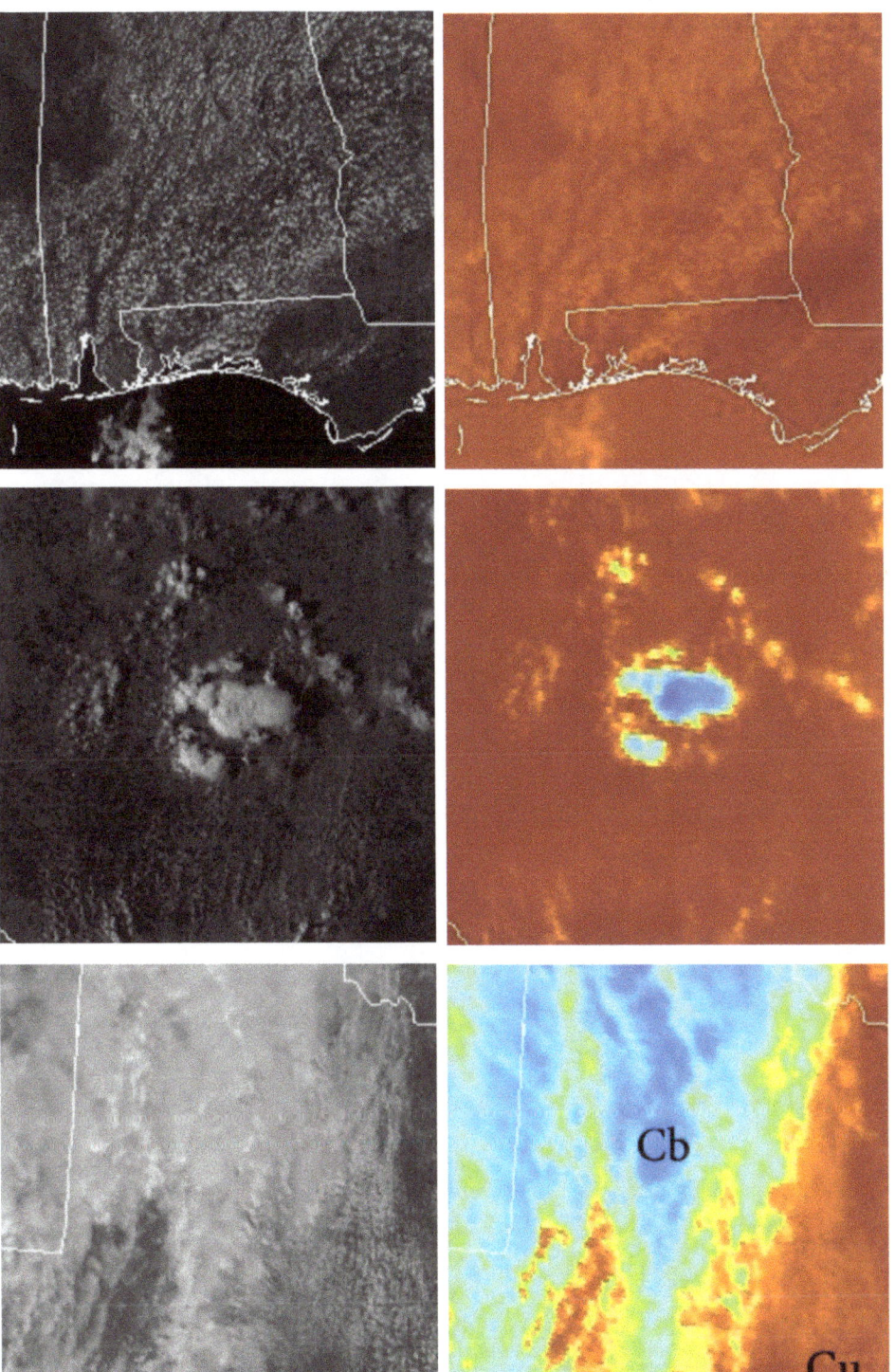

Cumulus usually appears in fields, as seen here, though it may prefer to develop along mountains. Cumulus fields have a random speckled appearance on visible imagery. The cumulus tops are not very cold, and on normal 4 km infrared imagery they tend to appear as a gray blur with speckled structure.

Isolated cumulonimbus is quite distinctive. Visible imagery shows very bright globular cloud masses which give way to a very bright circular or elongated anvil cloud. Infrared imagery shows small cloud elements with very cold temperatures; this expands over hours into a large, cold cirriform mass.

Embedded cumulonimbus is produced by significant synoptic-scale lift in marginal instability. Visible imagery shows multiple layers with mottled elements. Infrared shows organized spots of very cold tops. The majority of the cold cloud is actually cirriform anvil debris.

APPENDIX | 161

	VISIBLE	INFRARED
Stratocumulus appears as a bright white cloud with speckled edges indicative of cumulus. Infrared imagery shows a dull warm cloud, generally with soft edges. Some cirrus overlies this example, shown by the enhanced spots of yellow and green on the IR image.		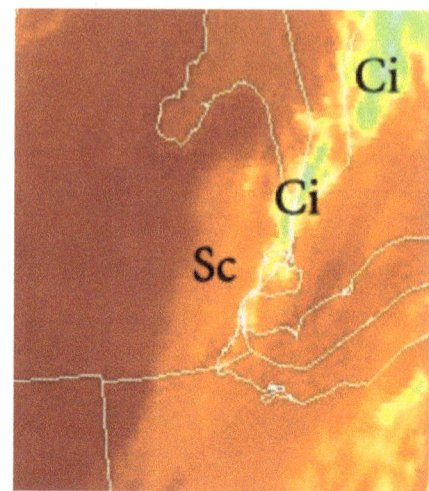
Stratus and fog looks on visible imagery like a featureless white cloud with edges that seem to be influenced by terrain; fog follows valley outlines. Infrared shows a very warm cloud that's dull, featureless gray; if warm lakes can be seen, then the cloud is probably fog and not stratus.	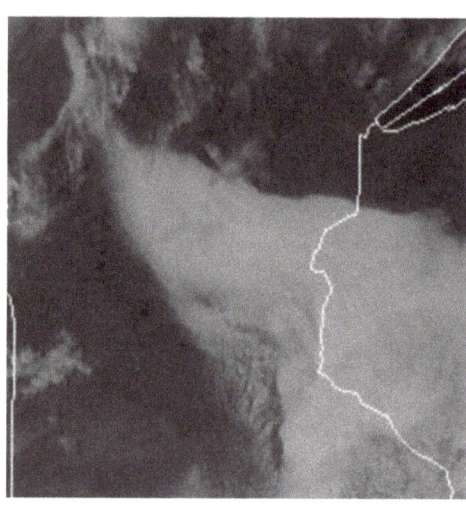	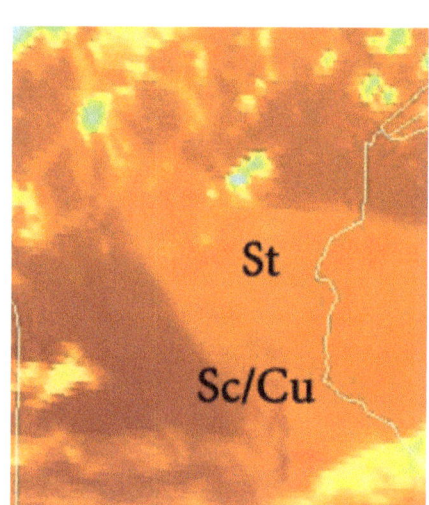
Nimbostratus, the classic rain cloud, does not occur by itself but is a amalgamation of very thick altostratus and cirrus layers. Visible imagery shows a large-scale overcast; infrared shows a moderately cold cloud with much the same. Numerous cold spots suggest more of a convective than stratiform situation.	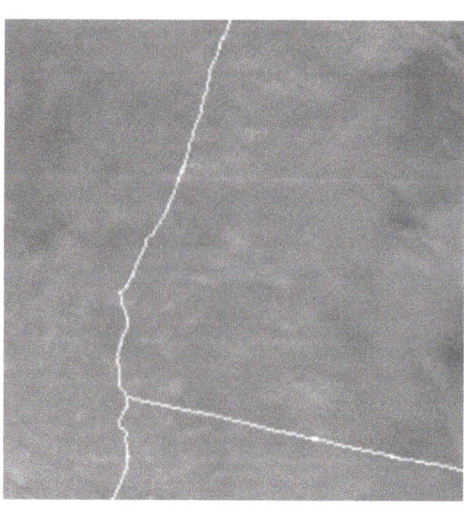	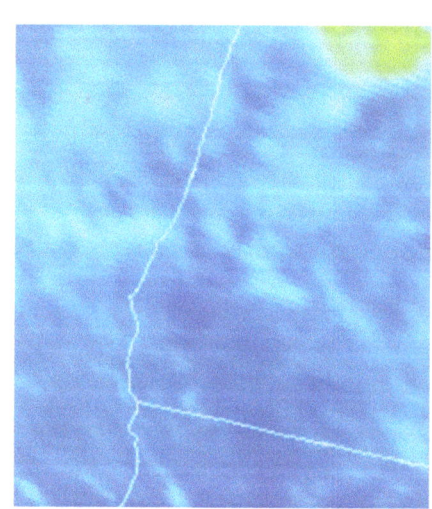

VISIBLE	INFRARED	
	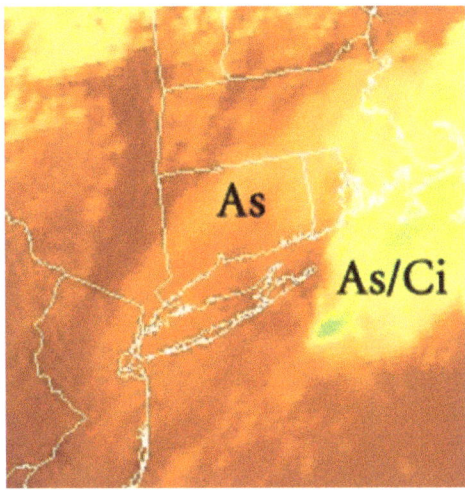	**Altostratus** is a fibrous bright cloud on visible imagery, rarely with sharp edges. Infrared imagery shows a marginally cool, featureless cloud. It usually occurs in conjunction with other layers. The example here appears to have breaks in the altostratus, with stratocumulus underneath.
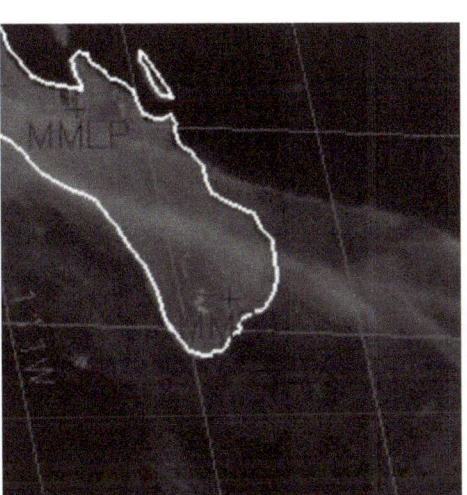	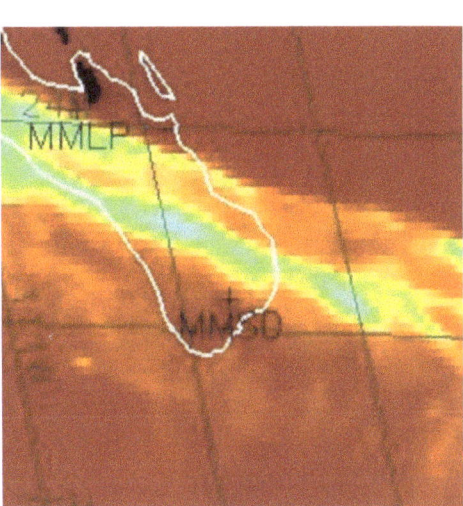	**"Fair weather cirrus"** has markedly different appearances on different channels. Visible imagery shows a veil-like layer with soft edges, which may have a patchy appearance. Infrared imagery shows a very cold cloud mass with soft edges and no concentrated cold spots that would be indicative of cumulonimbus underneath.
	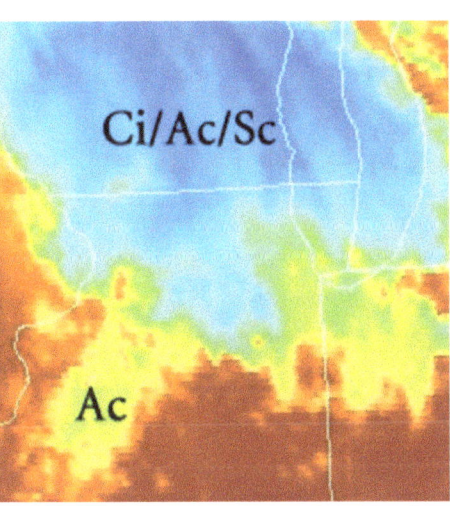	**Embedded cirrus** with other underlying layers appears bright on visible imagery but has the characteristic soft, fibrous appearance of cirrus. Infrared imagery shows a large, cold mass of cloud. This configuration is common with well-developed weather disturbances.

Appendix 11. Polarimetric precipitation detection

This sequence of charts was prepared by the NOAA Warning Decision Training Branch and provides an excellent overview of the base reflectivity signature (top of four products) and dual-polarization signature (bottom three products). Note that this only provides a guide to some of the fundamental precipitation types and does not take into account complex mixtures or what might be reaching the surface below the beam.

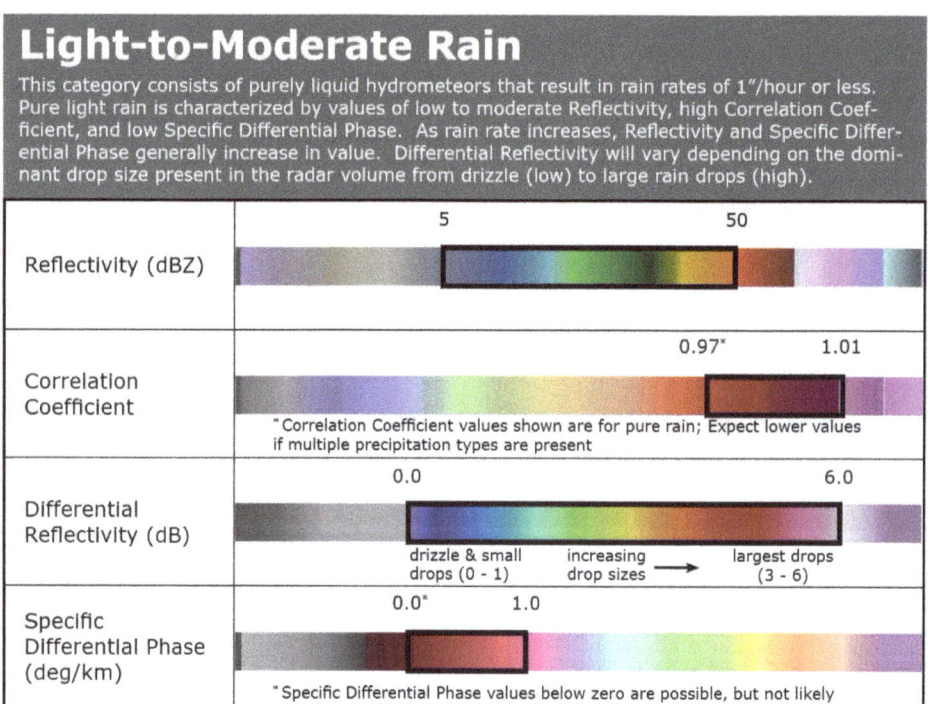

NOTES on light-moderate rain
R: The volume is filled with numerous small drops, reflecting moderate amounts of energy.
CC: The lack of mixed-phase precipitation and very uniform drops yields very high CCs.
ZDR: Large raindrops can become highly pancaked (oblate) while falling, yielding high ZDR.
KDP: Similar to reflectivity, KDP moves into higher ranges.

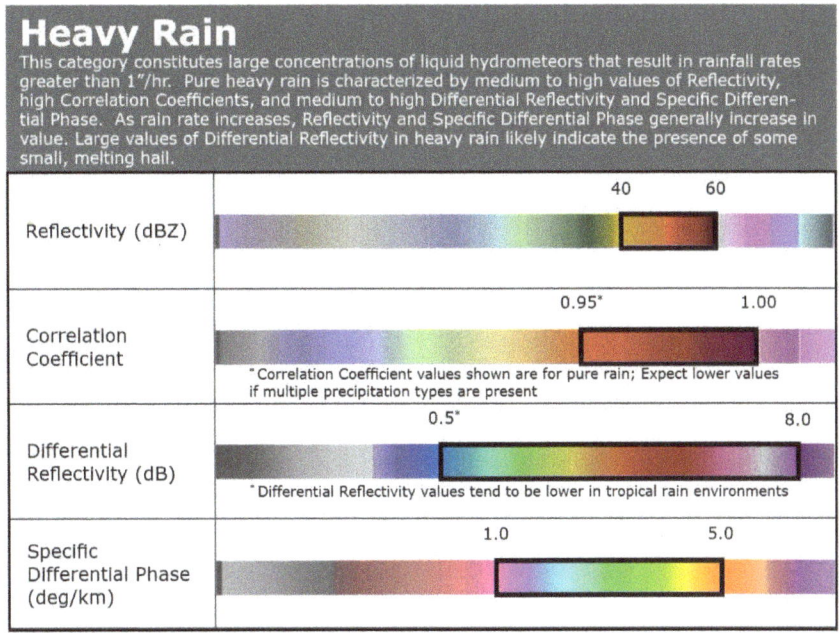

NOTES on heavy rain
R: The volume is filled with numerous, medium sized drops, reflecting significant energy.
CC: Fairly uniform with very high CCs, but mixed large/small drops lowers range slightly.
ZDR: Large raindrops can become highly pancaked (oblate) while falling, yielding high ZDR.
KDP: Heavy rain produces very high reflectivity and KDP.

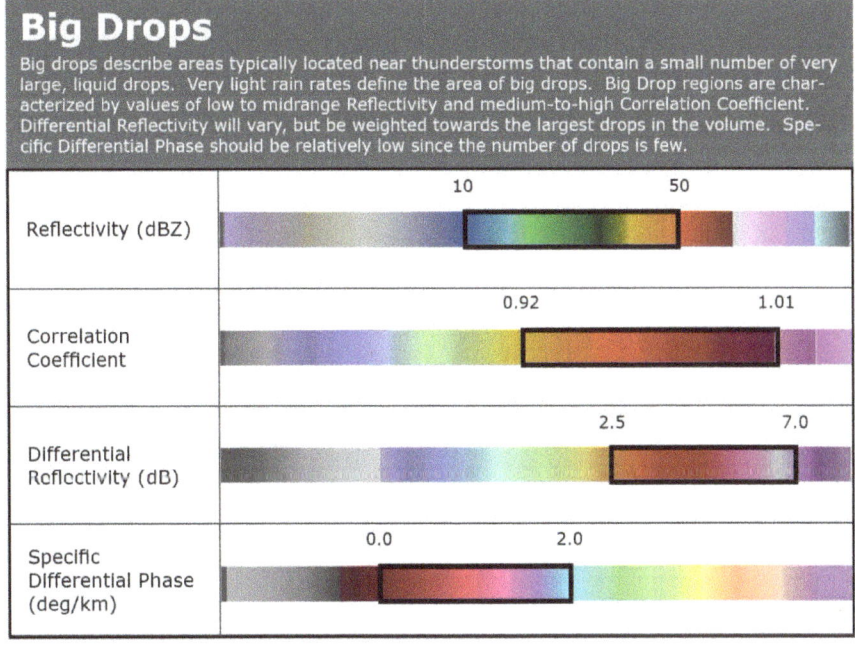

NOTES on big drops
R: Big drops are made up of only small numbers of drops but each is highly reflective.
CC: Radar discriminates differences in sizes/shapes of large drops, so CC range widens.
ZDR: Large raindrops can become highly pancaked (oblate) while falling, yielding high ZDR.
KDP: Similar to reflectivity, KDP moves into higher ranges.

APPENDIX | 165

Hail and Hail-Rain Mixtures

When hail is present, Reflectivity values are high. Correlation Coefficient is generally lower than with liquid precipitation alone, with extremely low values a possible indication of very large hail stones (i.e., larger than golf ball size hail). Dry, frozen hailstones usually have Differential Reflectivity and Specific Differential Phase values near zero. Smaller hail, when melting, is covered with water and has Differential Reflectivity and Specific Differential Phase values similar to heavy rain.

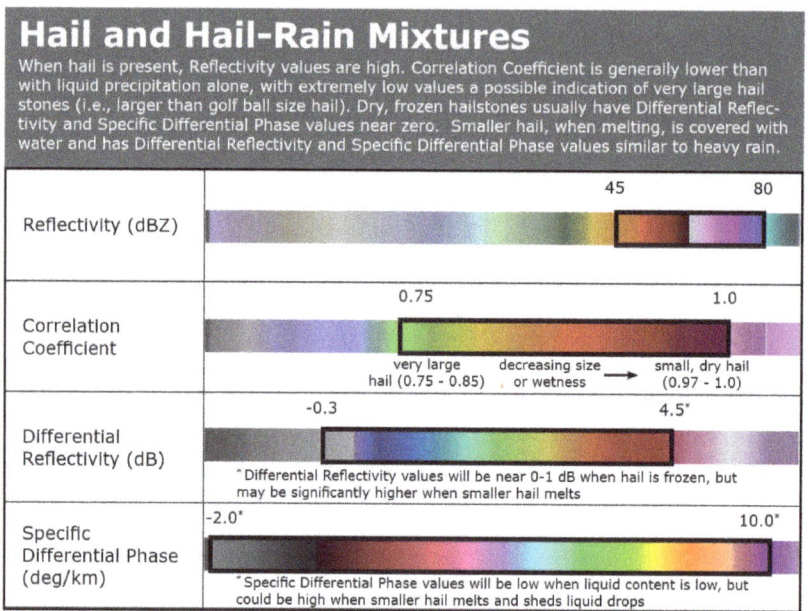

NOTES on hail
R: Hailstones are extremely good backscatterers, especially if wet in a hail-rain mixture.
CC: Hail-rain mixtures produce some of the lowest CC values of any meteorological target.
ZDR: Hailstones are spherical to prolate, yielding low ZDR, but rain biases ZDRs higher.
KDP: Wet hail produces extreme reflectivity and KDP. Dry hail is not detected well by KDP.

Graupel

Graupel will look similar to hail on a dual-polarization radar. Reflectivity values are lower and Correlation Coefficients are generally higher than with hail. Dry graupel will usually have Differential Reflectivity and Specific Differential Phase values near zero. As graupel melts, it will be covered with water and have Differential Reflectivity and Specific Differential Phase values similar to rain.

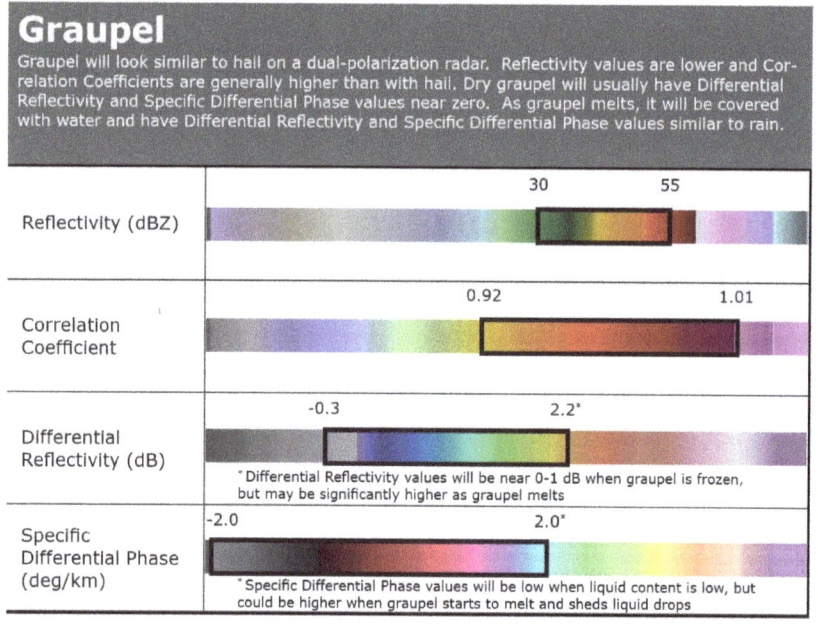

NOTES on graupel
R: Graupel particles appear similar to very small hail and will produce moderate R values.
CC: The relative uniformity of graupel is an important discriminator, yielding higher CCs.
ZDR: Graupel has a spherical signature and will normally have a low ZDR.
KDP: The solid nature of graupel means low ZDR values. Mixed phases means higher KDP.

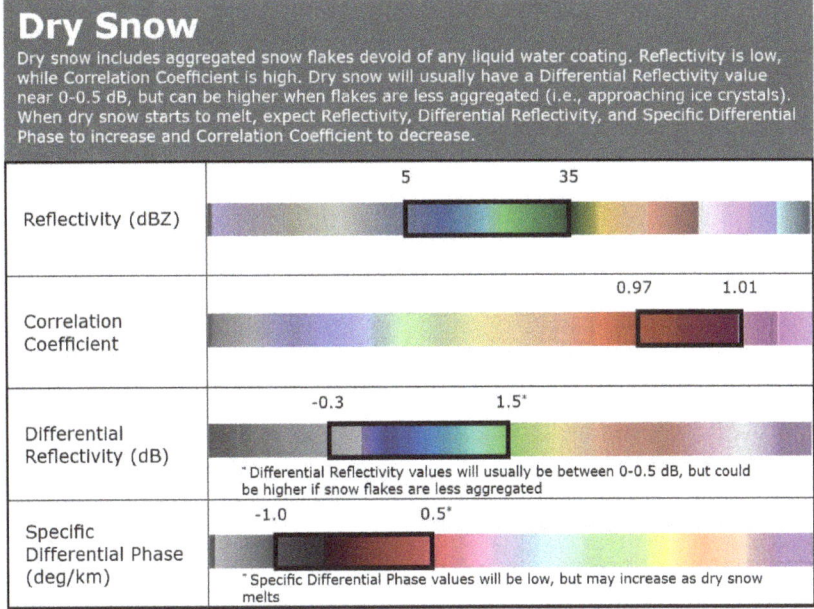

NOTES on dry snow

R: Dry snow produces very little backscatter, producing light intensities.
CC: Dry snow regions have exceptionally uniform sizes and shapes, yielding very high CC.
ZDR: Dry snow crystals fall with a vertical orientation, yielding low ZDR.
KDP: Reflectivity is weak and dry snow contains no liquid, so KDP is weak.

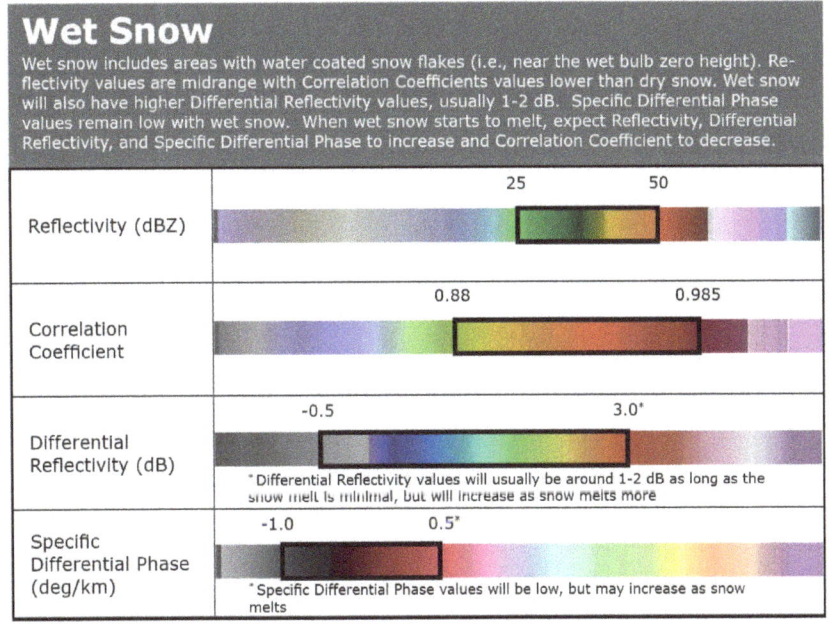

NOTES on wet snow

R: Wet snowflakes produces reflects moderate intensities back to the radar.
CC: The mixed phase regime of wet snow and droplets both lowers the CC range.
ZDR: Falling wet snow produces oblate (pancaked) signatures, resulting in high ZDR.
KDP: Wet snow has a fairly weak KDP signature.

Ice Crystals

Ice crystals are individual, non-aggregated frozen hydrometeors often found in drier areas of precipitation. Reflectivity is low, while Correlation Coefficient is high. Differential Reflectivity values are often high, but values depend on the density of the dominant crystals (i.e., density typically increases from needles to columns to plates). Likewise, Specific Differential Phase values will be low with dry snow.

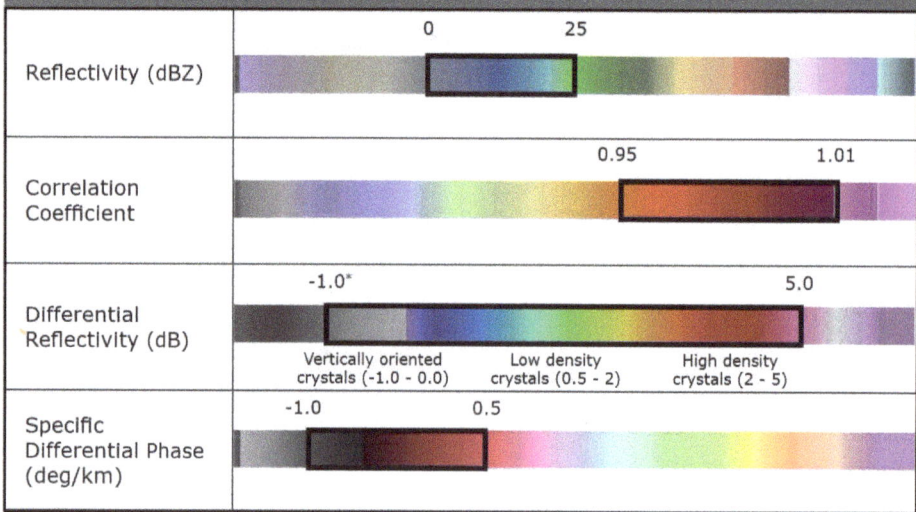

NOTES on ice crystals

R: The very small size of ice crystals produces very weak reflectivities.
CC: Very high homogeneity of the ice crystal field yields very high CC.
ZDR: Columns have low ZDR, while plates and dendrites have higher ZDR.
KDP: Weak backscatter results in weak KDP.

Appendix 12. WSR-88D product list

Contained herein is a complete list of **WSR-88D RPG** products as of May 2015. This can be helpful for identifying products to be viewed when only an abbreviation or message header is available.

PRODUCT NAME	RPG HDR	WMO HEADER		ELEVATION ANGLE
General Status Message	2/GSM	NXUS6i	GSM	—
Base Reflectivity – 124 nmi Range	19/R	SDUS5i	N0R	0.5
Base Reflectivity – 248 nmi Range	94/DR	SDUS5i	N0Q	0.5
	94/DR	SDUS5i	NAQ	0.9
	94/DR	SDUS2i	N1Q	1.3, 1.5
	94/DR	SDUS2i	NBQ	1.8
	94/DR	SDUS2i	N2Q	2.4, 2.5
	94/DR	SDUS2i	N3Q	3.1, 3.4, 3.5
Base Reflectivity – 248 nmi Range	20/R	SDUS7i	N0Z	0.5
Base Radial Velocity -124 nmi Range	27/V	SDUS5i	N0V	0.5
Base Radial Velocity -162 nmi Range	99/DV	SDUS5i	N0U	0.5
	99/DV	SDUS5i	NAU	0.9
	99/DV	SDUS2i	N1U	1.3, 1.5
	99/DV	SDUS2i	NBU	1.8
	99/DV	SDUS2i	N2U	2.4, 2.5
	99/DV	SDUS2i	N3U	3.1, 3.4, 3.5
Base Spectrum Width - 124 nmi Range	30/SW	SDUS6i	NSW	0.5
Digital Hybrid Scan Reflectivity	32/DHR	SDUS5i	DHR	—
Clutter Filter Control	34/CFC	SDUS6i	NC1	Segment 1
	34/CFC	SDUS6i	NC2	Segment 2
	34/CFC	SDUS6i	NC3	Segment 3
	34/CFC	SDUS6i	NC4	Segment 4
	34/CFC	SDUS6i	NC5	Segment 5
Composite Reflectivity (16 level) (124 nmi)	37/CR	SDUS5i	NCR	—
Composite Reflectivity (16 level) (248 nmi)	38/CR	SDUS6i	NCZ	—
Echo Tops	41/ET	SDUS7i	NET	—
Enhanced Echo Tops	135/EET	SDUS7i	EET	—
Velocity Azimuth Display Wind Profile	48/VWP	SDUS3i	NVW	—
Storm Relative Mean Radial Velocity	56/SRM	SDUS5i	N0S	0.5
	56/SRM	SDUS2i	N1S	1.3, 1.5
	56/SRM	SDUS2i	N2S	2.4, 2.5
	56/SRM	SDUS3i	N3S	3.1, 3.4, 3.5
Vertical Integrated Liquid	57/VIL	SDUS5i	NVL	—
Digital Vertical Integrated Liquid	134/DVL	SDUS5i	DVL	—
Storm Tracking Information*	58/STI	SDUS3i	NST	—
Hail Index*	59/HI	SDUS6i	NHI	—
Tornadic Vortex Signature*	61/TVS	SDUS6i	NTV	—
Storm Structure*	62/SS	SDUS6i	NSS	—
Layer Composite ReflectivityMiddle Level	66/LRM	SDUS6i	NML	—
Layer Composite Reflectivity AP Removed	67/APR	SDUS6i	NLA	—
Radar Coded Message	74/RCM	SDUS4i	RCM	—
Free Text Message	75/FTM	NOUS6i	FTM	—

Product	Code	Type	ID	Elevations
Surface Rainfall Accumulation (1 hr)	78/OHP	SDUS3i	N1P	—
Surface Rainfall Accumulation (3 hr)	79/THP	SDUS6i	N3P	—
Surface Rainfall Accumulation (Storm)	80/STP	SDUS5i	NTP	—
Digital Precipitation Array	81/DPA	SDUS5i	DPA	—
Supplemental Precipitation Data*	82/SPD	SDUS6i	SPD	—
Layer Composite Reflectivity High Level	90/LRM	SDUS6i	NHL	—
Digital Storm Total Precipitation	138/DSP	SDUS5i	DSP	—
Mesocyclone*	141/MD	SDUS3i	NMD	—
Archive III Status Product	152/ASP	SDUS4i	RSL	—
Digital Differential Reflectivity	159/DZD	SDUS8i	N0X	0.5
	159/DZD	SDUS8i	NAX	0.9
	159/DZD	SDUS8i	N1X	1.3, 1.5
	159/DZD	SDUS8i	NBX	1.8
	159/DZD	SDUS8i	N2X	2.4, 2.5
	159/DZD	SDUS8i	N3X	3.1, 3.4, 3.5
Digital Correlation Coefficient	161/DCC	SDUS8i	N0C	0.5
	161/DCC	SDUS8i	NAC	0.9
	161/DCC	SDUS8i	N1C	1.3, 1.5
	161/DCC	SDUS8i	NBC	1.8
	161/DCC	SDUS8i	N2C	2.4, 2.5
	161/DCC	SDUS8i	N3C	3.1, 3.4, 3.5
Digital Specific Differential Phase	163/DKD	SDUS8i	N0K	0.5
	163/DKD	SDUS8i	NAK	0.9
	163/DKD	SDUS8i	N1K	1.3, 1.5
	163/DKD	SDUS8i	NBK	1.8
	163/DKD	SDUS8i	N2K	2.4, 2.5
	163/DKD	SDUS8i	N3K	3.1, 3.4, 3.5
Digital Hydrometeor Classification	165/DHC	SDUS8i	N0H	0.5
	165/DHC	SDUS8i	NAH	0.9
	165/DHC	SDUS8i	N1H	1.3, 1.5
	165/DHC	SDUS8i	NBH	1.8
	165/DHC	SDUS8i	N2H	2.4, 2.5
	165/DHC	SDUS8i	N3H	3.1, 3.4, 3.5
Melting Layer	166/ML	SDUS8i	N0M	0.5
	166/ML	SDUS8i	NAM	0.9
	166/ML	SDUS8i	N1M	1.3, 1.5
	166/ML	SDUS8i	NBM	1.8
	166/ML	SDUS8i	N2M	2.4, 2.5
	166/ML	SDUS8i	N3M	3.1, 3.4, 3.5
Digital Inst. Precip. Rate(in/hr)	176/DPR	SDUS8i	DPR	—
Hybrid Scan Hydrometeor Classification	177/HHC	SDUS8i	HHC	—
One Hour Accum	169/OHA	SDUS8i	OHA	—
Dig. Accum Array (unbiased)	170/DAA	SDUS8i	DAA	—
Storm Total Accum	171/STA	SDUS3i	PTA	—
Dig. Storm Total Accum	172/DSA	SDUS8i	DTA	—
Dig. User-Selectable Accum:3hr/hrly	173/DUA	SDUS8i	DU3	—
Dig. User-Selectable Accum:24hr/12Z	173/DUA	SDUS8i	DU6	—
Dig. One Hour Difference Accum	174/DOD	SDUS8i	DOD	—
Dig. Storm Total Difference Accum	175/DSD	SDUS8i	DSD	—

* Product is not transmitted when radar is operating in clear air mode

Glossary

This section provides a summary of acronyms that may be encountered in journals, case studies, and forecasting discussions.

ablation Depletion of snow and ice by melting and evaporation.
Ac Altocumulus (q.v.)
AC 1. Convective Outlook bulletin; from the Family of Services data stream header. 2. Altocumulus
ACARS Aircraft Communications and Reporting System. A block of data transmitted by aircraft that often contains weather data.
ACCAS Altocumulus Castellanus. Altocumulus which forms in a convectively unstable layer.
accessory cloud A cloud which is dependent on a larger cloud system for development.
ACSL Altocumulus standing lenticular (q.v.)
acre-foot The amount of water required to cover one acre to one foot of depth. This equals 326,851 gallons or 43,560 cubic feet.
adiabatic The change in temperature without a transfer of heat. It may be caused by compression or expansion.
advection Horizontal movement of air that causes changes in the physical properties of air at a specific location.
advection fog Fog that forms as warmer, moist air moves over a cold surface. The air is forced to condense as it loses heat to conduction.
advisory In the United States, a weather bulletin which is less serious than a warning.
AFD Area Forecast Discussion (q.v.)
AFGWC Air Force Global Weather Central, which became obsolete on 15 October 1997. It is now known as AFWA.
AFOS Automation of Field Operations and Services (discontinued). The backbone computer system of NWS offices; developed in 1976 and fielded in 1979; retired between 1996 and 1999.
AFWA Air Force Weather Agency (q.v.)
AGL Above Ground Level.
Air Force Weather Agency (AFWA) Weather component of the U.S. Air Force. Activated 15 October 1997, it is a combination of Air Weather Service Headquarters and the Air Force Global Weather Center.
albedo The portion of incoming radiation which is reflected by a surface.
air mass A body of air which contains relatively uniform properties of temperature and moisture.
AJ Arctic jet (q.v.)
algorithm A computer program designed to solve a specific problem. Often used in WSR-88D radars.
aliasing A process in which a radar return has a frequency too high to be analyzed within the given sampling interval but at a frequency less than the Nyquist interval.
ALSTG Altimeter setting (q.v.)
altimeter setting The pressure at which an altimeter must be set so that it reads the correct elevation.
altocumulus Mid-level clouds composed primarily of water or supercooled water. The base is traditionally at a height between 6,500 and 23,000 ft AGL (26,000 ft in the tropics and 13,000 ft at the poles).
Altocumulus Standing Lenticular (ACSL) Clouds formed at the tips of vertical waves in the wake of mountain ranges.
altostratus A bluish veil or layer of clouds having a fibrous appearance. The outline of the sun may show dimly as if through frosted glass. The base is traditionally at a height between 6,500 and 23,000 ft AGL (26,000 ft in the tropics and 13,000 ft at the poles).
anafront Also called active front. A cold front in which there is a tendency for air to ascend the frontal surface. Generally associated with lift and weather behind the front. (cf. katafront)
anemometer A device that measures wind speed.
ANL Analysis
anomalous propagation Unexpected radio wave propagation that occurs due to non-standard atmospheric conditions. Usually refers to ducting of the beam to the ground, returning ground clutter.
anticyclone An area of high pressure, around which the wind blows clockwise (counterclockwise in the Southern Hemisphere).
anvil The spreading top of a cumulonimbus cloud.
AOA At or above
AOB At or below
AP Anomalous propagation (q.v.)
arctic air Air which has its roots over the snow-covered region of northern Canada, the polar basin, and northern Siberia.
arctic jet Baroclinic jet which develops in association with the polar vortex.

Area Forecast Discussion (AFD) A discussion of the meteorological thinking used in the creation of a zone forecast. (NWS)

ARINC Aeronautical Radio, Incorporated. Company based in Annapolis, MD responsible for ACARS (q.v.)

ARTCC Air Route Traffic Control Center

As Altostratus (q.v.) Also AS.

ASCII American Standard Code for Information Interchange

ASOS Automated Surface Observing System. The network in place across the United States which have provided automated meteorological reports since 1992.

ATTM At this time

AVA Anticyclonic vorticity advection

AVN 1. The NCEP Aviation model, also known as the global spectral model, comprising the United States' primary global weather model; replaced by the GFS model (q.v.) 2. Aviation.

AWC Aviation Weather Center

AWIPS Advanced Weather Interactive Processing System

AWOS Automated Weather Observing System

back door cold front A cold front moving westward or southwestward.

backing Referring to a change in wind direction that is counterclockwise, with respect to either height or time. Contrast with veering.

backscatter Power that returns to the radar dish after striking a target.

baroclinic zone An area in which a horizontal temperature gradient exists. Rapid weather changes may occur in such zones.

barotropic More properly referred to as equivalent barotropic, this term refers to a weather system which has weak or insignificant temperature contrasts

barotropic zone An area in which a significant horizontal temperature gradient does not exist. Rapid weather changes are not as likely as in a baroclinic zone.

Base Reflectivity (BR) A simple reflectivity product as obtained from any elevation of a radar scan (not necessarily the lowest one).

base velocity A simple velocity product as obtained from any elevation of a radar scan (not necessarily the lowest one).

beam width In radar meteorology, the width within which the power density is at least half that of the axis of the beam (i.e. within 3 dB)

blizzard A winter storm which produces, for at least 3 hours, both winds gusting to 35 mph and falling/drifting/blowing snow reducing visibility to less than 1/4 mile.

block A long wave pattern, usually revealed on 200/250/300 mb charts, in which the long waves are neither progressing nor retrogressing. Often refers to the responsible feature, such as an omega or rex block (q.v.)

boundary layer (BL, PBL) The layer in contact with the ground in which friction is significant. This is usually the lowest few thousand feet of the atmosphere but may vary greatly with weather pattern, season, and insolation.

broken Partial coverage of the sky by a layer of more than half (5 to 7 oktas). (cf. clear, few, scattered, and overcast)

BR Base Reflectivity (q.v.)

BRN Bulk Richardson Number (q.v.)

BUFR Binary Universal Format for Data Representation

Bulk Richardson Number (BRN) The ratio of CAPE to vertical wind shear. It has been found that values of less than 45 support supercellular structures, while greater than 45 favors multicells. However it is not as good of a predictor as its component terms are.

BWER Bounded Weak Echo Region

CAA Cold air advection (q.v.)

cap A layer of warm air aloft that acts as an inversion and suppresses convective development. It may be measured by the Convective Inhibition Index, or CINH (q.v.)

CAPE Convective Available Potential Energy (q.v.)

CB, Cb Cumulonimbus (q.v.)

CC, Cc Cirrocumulus (q.v.)

CCL Convective Condensation Level (q.v.)

CI, Ci Cirrus (q.v.)

CIN Convective inhibition

CISK Convective instability of the second kind

cirrocumulus (Cc) A layer of high, fibrous clouds with convective cells. The cloud is made up entirely of ice crystals. Its bases are traditionally as low as 16,000 ft (20,000 ft in the tropics; 10,000 ft in polar regions).

cirrostratus (Cs) A thin layer of high, fibrous clouds without detail and often appearing as a sheet covering the sky. It is composed entirely of ice crystals. Its bases are traditionally as low as 16,000 ft (20,000 ft in the tropics; 10,000 ft in polar regions).

cirrus (Ci) A layer of high, fibrous clouds composed entirely of ice crystals. Its bases are traditionally as low as 16,000 ft (20,000 ft in the tropics; 10,000 ft in polar regions).

clear Complete absence of cloud. (cf. few, scattered, broken, and overcast)

cloud height The height of a cloud's base, usually rounded to the nearest hundred feet (thousand feet above 10,000 ft).

cold front The leading edge of an air mass that is replacing a warmer air mass.

Composite Reflectivity (CR) A WSR-88D radar product that displays the maximum reflectivity observed in a grid box at a given location.

confluence A pattern in which wind flows inward into a common axis. It is not the same as convergence. (cf. difluence, convergence, divergence)

convection The transport of heat and moisture by the vertical movement of air in an unstable atmosphere. This may cause cumuliform clouds and thunderstorms.

Convective Available Potential Energy (CAPE) The vertically integrated buoyancy of a rising air parcel. Measured in j/kg.

Convective Condensation Level (CCL) The height at which a parcel of air, if heated sufficiently from below, will rise adiabatically until saturation begins.

Convective Inhibition (CIN) A measure of negative buoyancy that prevents a rising parcel from reaching its Level of Free Convection, or LFC (q.v.). It is measured in j/kg.

convective temperature The theoretical surface temperature for a given atmospheric profile that must be reached to start the formation of convective clouds.

convergence A wind pattern in which more air is entering than leaving, either through speed convergence or confluence. (cf. divergence, diffluence, confluence)

CONUS Continental United States

Coordinated Universal Time (UTC) See Universal Coordinated Time.

COOSAC Committee on Operations, Standards, and Conventions

Coriolis effect The effect caused by the Earth's rotation which deflects parcels to the right (left in the Southern Hemisphere).

COTR Contract Office Technical Representative

couplet Adjacent maxima of radial velocities of opposite signs.

CPC Climate Prediction Center

CR Composite Reflectivity (q.v.)

cross section A diagram of the atmosphere in which horizontal distance is expressed on the X-axis and height on the Y-axis.

Cross Totals index (CT) An expression of instability, equalling $Td_{850}-T_{500}$. Values of greater than 18-30 are considered significant.

Cs, CS Cirrostratus (q.v.)

CSI Conditional symmetric instability

CT Cross Totals index (q.v.)

CU, Cu Cumulus (q.v.)

cumulonimbus (Cb) A large, cauliflower-shaped cloud whose upper portions are usually fibrous. Often associated with precipitation and thunder.

cumulus (Cu) Low, heaplike clouds that are associated with convective weather. The three "categories" of cumulus are typically fair-weather, moderate, and towering. Further cumulus development will evolve into cumulonimbus.

CVA Cyclonic vorticity advection

cyclogenesis The intensification of a low-pressure system.

cyclone An area of low pressure with a closed circulation. The wind flow rotates counterclockwise (clockwise in the Southern Hemisphere).

dBZ Decibels of reflectivity factor.

decoupling The intensification of the contrast between the boundary layer and the free atmosphere, which strengthens winds above and weakens winds below. Tends to occur at night.

DELMARVA Delaware-Maryland-Virginia

derecho A widespread and fast-moving convective wind event.

difluence Alternate spelling of diffluence (q.v.)

diffluence A pattern in which wind flows outward from a common axis. It is not the same as divergence. (cf. confluence, convergence, divergence)

diurnal 1. Occurring on a daily basis. 2. Occurring during the day. (cf. nocturnal)

divergence A wind pattern in which more air is leaving than entering, either through speed divergence or diffluence. (cf. convergence, diffluence, confluence)

DOCBLOCK Computer program documentation block

dryline A boundary which separates dry, warm continental air from moist, warm oceanic air. It is most common in the Great Plains, the Sahel, India/Bangladesh, Australia, and China.

dynamics A term that generally refers to forces produced by air out of geostrophic balance which in turn produces vertical motion.

easterly wave A disturbance embedded in the trade winds that moves east to west.

EBDIC Extended Binary-Coded Decimal Interchange

Code
ECMWF European Centers for Medium Range Weather Forecasting
EHI Energy Helicity Index (q.v.)
EL Equilibrium Level (q.v.)
EMC Environmental Modelling Center (q.v.)
Energy Helicity Index (EHI) An index that is a product of shear and instability, and is defined as CAPE x SRH / 160,000, where CAPE is j kg^{-1} and SRH is in m^{2}s^{-2}.
Environmental Modelling Center (EMC) A center of NCEP that is focused on improving numerical modelling technologies.
Equilibrium Level (EL) The height, sometimes within the stratosphere, at which a rising parcel's temperature becomes equal to that of the environment. Upward momentum is sharply lost beyond this point.
Eta Eta model
European model The ECMWF global spectral model.
exit region The region downstream from a jet max. The poleward side typically is associated with divergence aloft and upward motion; the equatorward side with convergence aloft and downward motion.
FA Area forecast
FAA Federal Aviation Administration
Family of Services (FOS) The public connection to National Weather Service data which was established in 1983.
FD Winds and temperatures aloft forecast
FEW Partial coverage by cloud material of a quarter or less (1 to 2 oktas). (cf. clear, scattered, broken, overcast)
FFG Flash flood guidance
FNL Final production run for a given cycle
FNMOC Fleet Numerical Oceanography Center
FNOC Fleet Numerical Oceanographic Center (obsolete; replaced by FNMOC)
FOS Family Of Services (q.v.)
FT Terminal forecast (obsolete; now TAF)
FTP File Transfer Protocol
FTS Federal Telecommunications System
GBL Global production run for a given cycle
GDAS Global Data Assimilation System production run for a given cycle
GDM Graphic Display Model
geostrophic wind The imaginary wind that would result from a balance of both pressure gradient force and the Coriolis effect.
GES Guess
GFS Global Forecast System (q.v.)
Global Forecast System (GFS) The most sophisticated global spectral model currently used by the United States. It incorporates both the AVN and MRF models, whose names have been "retired".
GMT Greenwich Mean Time
GOES Geostationary Operational Environmental Satellite. The United States' network of geostationary weather satellites poised at the Equator above the Western Hemisphere continuously since 1974.
GRIB Gridded Binary data
GTS Global Telecommunications System
HADS Hydrometeorological Automated Data System
HIC Hydrologist In Charge
HMT Hydrometeorological Technician
hodograph A polar coordinate graph showing the wind profile of the atmosphere at a given point, with respect to ground-relative azimuth and speed.
HPC Hydrometeorological Prediction Center (q.v.)
hurricane A warm-core tropical system that has sustained surface winds exceeding 63 kt.
Hydrometeorological Prediction Center (HPC) A center of NCEP which is responsible for centralized forecasting functions of the National Weather Service.
ICAO International Civil Aviation Organization
IMSL International Mathematical and Statistical Library
instability An atmospheric state in which warm air is able to continue rising and accelerating.
inversion An increase in temperature with height, comprising a stable layer in the atmosphere. Vertical motion through the inversion is suppressed.
INVOF In vicinity of
IR Infrared
isallobar A line of equal atmospheric pressure change.
isentrope A line of equal potential temperature.
isentropic lift Lift produced by motion of air along surfaces of constant potential temperature which slope upward relative to the parcel's motion. This typically occurs when the parcel is traversing from warmer to colder air below.
isentropic subsidence Sinking motion produced by motion of air along surfaces of constant potential temperature which slope downward relative to the parcel's motion. This typically occurs when the parcel is traversing from colder to warmer air below.
isobar A line of equal pressure.
isochrone A line of equal time.
ISPAN Information Stream Project for AWIPS/NOAAPORT
j kg^{-1} Joules per kilogram (CAPE, q.v.)

jet max A region of maximum winds within a jet stream. Also jet streak, speed max.

jet streak A region of maximum winds within a jet stream. Also jet max, speed max.

JIF Job Implementation Form

JMA Japan Meteorological Agency

JSC Johnson Spaceflight Center

JSPRO Joint Systems Program Office for NEXRAD

katafront Also called inactive front. A cold front in which there is a tendency for air to descend the frontal surface. Generally associated with subsidence behind the front and weather ahead of the front. (cf. anafront)

K-Index (KI) A measure of the thunderstorm potential based on vertical temperature lapse rate, moisture content of the lower atmosphere, and the vertical extent of the moist layer. Equals $T_{850}-T_{500}+Td_{850}-DD_{700}$ where DD equals dewpoint depression. Values above 20-35 are significant.

KI K-Index (q.v.)

knot A measure of velocity, nautical miles per hour, equal to 1.15 statute miles per hour.

lapse rate The change in temperature with height. Normally is 6.5 Celsius degrees per km.

LAWRS Limited Aviation Weather Reporting Station (usually a control tower)

LCN Loosely Coupled Network

LI Lifted Index (q.v.)

LCL Lifted Condensation Level (q.v.)

LEWP Line echo wave pattern

LFC Level of free convection

LFM Limited-area Fine Mesh model, a numerical model that was used by NMC (NCEP) from 1971 to 1996.

LFQ Left-front quadrant of a jet streak. In the Northern Hemisphere this is usually associated with upward motion.

LRQ Left-rear quadrant of a jet streak. In the Northern Hemisphere this is usually associated with downward motion.

Lifted Condensation Level (LCL) The height at which a parcel of air will become saturated if lifted adiabatically.

Lifted Index (LI) The temperature difference between a lifted parcel and that of its environment at 500 mb. This is a single-level expression of instability. It equals T_E-T_P where E is the environment and P is the parcel. Negative values are unstable, and below -5 are significant.

LLJ Low Level Jet (q.v.)

Low Level Jet (LLJ) An elongated area of strong winds, generally below 10,000 ft MSL, which may occur in advance of extratropical lows. It is significant in transporting heat and moisture poleward, reinforcing baroclinicity and destabilizing the atmosphere.

long wave A large-scale wave in the upper atmosphere, either a trough or a ridge. There are usually four or five long waves around a hemisphere.

$m^2\ s^{-2}$ Meters squared per second squared

MAR Modernization and Associated Restructuring Program

MAX Maximum

Maximum Parcel Level (MPL) The highest attainable level a thunderstorm updraft can reach, where all further upward velocity of a parcel is lost. Factors in overshooting tops.

mb Millibars

MCC Mesoscale convective complex

MCD Mesoscale discussion

MCIDAS Man-Computer Interactive Data Access System

MCS Mesoscale Convective System

MDR Manually Digitized Radar (now obsolete)

mesocyclone A low pressure area which is the embodiment of a rotating thunderstorm; it usually measures 1 to 5 miles in diameter. It is a misnomer because it is not a mesoscale system.

mesolow A mesoscale low-pressure area. Not to be confused with mesocyclone.

mesohigh A mesoscale high-pressure area, sometimes associated with stagnating thunderstorm outflow air.

mesoscale Referring to weather systems with scales of about 50 to 500 miles, or 1 to 24 hours.

METAR Meteorological Aviation Report

MIC Meteorologist In Charge

MLCAPE Mean Layer CAPE. Calculated using a parcel that contains mean temperature and mixing ratio of a layer, typically 100 mb deep.

MOA Memorandum of Agreement

Model Output Statistics (MOS) A statistical forecasting model, usually calculated city-by-city.

monsoon A seasonal shift in wind direction.

MOS Model Output Statistics (q.v.)

MOU Memorandum of Understanding

MPL Maximum Parcel Level (q.v.)

MPC Marine Prediction Center

MRF Medium Range Forecast model (obsolete; replaced by GFS)

MSL (above) Mean Sea Level

MSLP Mean Sea Level Pressure

MUCAPE Most Unstable CAPE. CAPE calculated from a parcel that provides the most unstable CAPE possible.

NASA National Aeronautics and Space Administration

National Centers for Environmental Prediction (NCEP) Was NMC (National Meteorological Center) from 1958-1995. An agency falling under NOAA which provides guidance and products to the National Weather Service. It is comprised of nine centers: Aviation Weather Center (AWC); Climate Prediction Center (CPC); Environmental Modelling Center (EMC); Hydrometeorological Prediction Center (HPC); NCEP Central Operations (NCO); Ocean Prediction Center (OPC); Space Environmental Center (SEC); Storm Prediction Center (SPC); and Tropical Prediction Center (TPC).

National Climatic Data Center (NCDC) The United States government agency responsible for archival of meteorological data.

National Oceanic and Atmospheric Administration (NOAA) The United States government agency falling under the Department of Commerce, which is responsible for all civilian programs engaged in work with the atmosphere, oceans, and lakes.

National Weather Service (NWS) A branch of NOAA responsible for all United States public forecasting.

NCCF NOAA Central Computer Facility

NCDC National Climatic Data Center (q.v.)

NCEP National Centers for Environmental Prediction (q.v.)

NCO NCEP Central Operations

negative tilt Description of an upper-level trough whose axis is tilted to the west with increasing latitude. It is often associated with strengthening dynamics.

NESDIS National Environmental Satellite Data and Information Service

NEXRAD Next Generation Weather Radar (WSR-88D)

NEXUS Next Generation Upper-Air System

NGM Nested Grid Model

NHC National Hurricane Center

NIDS NEXRAD Information Dissemination Service

nimbostratus An amorphous cloud thick enough to completely obscure the sun, with its base almost indistinguishable and typically obscured by precipitation. Does not produce showers or thunder. Abbreviated Ns.

NMC National Meteorological Center (obsolete; now NCEP)

NMFS National Marine Fisheries Service

NOAA National Oceanic and Atmospheric Administration (q.v.)

NOS National Ocean Survey

NOTAM Notice to Airmen

nowcast A forecast of about six hours or less. Also called a short-term forecast.

NS, Ns Nimbostratus (q.v.)

NSSFC National Severe Storms Forecast Center

NVA Negative vorticity advection. Advection of negative vorticity into a region.

NWS National Weather Service (q.v.)

NWSTG National Weather Service Telecommunications Gateway

occlusion The convergence of three air masses, in which the least dense is displaced aloft and the remaining two are demarcated by an occluded front. Typically occurs when a cold front "catches up" to a warm front.

OFOAR Office of Oceanic and Atmospheric Research

OHP One-Hour Precipitation, as used in weather radar estimates.

OI Optimum Interpolation method

okta An eighth of sky cover.

omega block An upper-level pattern in which a high pressure (height) area intensifies to a very high amplitude, resembling the greek letter omega. It "locks in" the long wave pattern.

ON Office Note

outflow boundary The leading edge of outflow from a thunderstorm downdraft. It may persist hours or days after the dissipation of the storm.

overcast A cloud layer completely covering the sky (8 oktas of cover). (cf. clear, few, scattered, and broken)

overrunning An oversimplification of the process of isentropic lift (q.v.).

PE Primitive Equation model

PFJ Polar front jet (q.v.)

PIREP Pilot Report

polar front jet (PFJ) The jet that is associated with the gradient between polar and tropical air masses. (cf. arctic jet, subtropical jet, low-level jet)

polar vortex A large cold-core low aloft that typically is found over northern Hudson Bay in North America during the winter months. Occluding polar front systems are usually absorbed into the polar vortex.

POP Probability of Precipitation (q.v.)

positive area The area formed on a sounding between an environmental temperature line and a warmer parcel temperature line. Its area is roughly proportional to CAPE.

positive-tilt Description of an upper-level trough whose axis is tilted to the east with increasing latitude. It is often associated with weakening dynamics.

potential temperature The temperature which a parcel would have if brought to a common level, by standard convention 1000 mb.

pressure gradient The change in pressure over a given distance.

Probability of Precipitation (POP) A quantity that describes the likelihood of a measurable amount of precipitation at any given location in a forecast area. The NWS expressions are 20% for slight chance, 30-50% for a chance, and 60-70% for likely.

PROD Production (for operational jobs)

profiler Also wind profiler. A radio detection device designed to measure wind direction and speed vertically in the troposphere above a given point.

PVA Positive vorticity advection. Equal to CVA in the Northern Hemisphere and AVA in the Southern Hemisphere (q.v.)

Q vector A horizontal vector representing the rate of change of the horizontal potential temperature gradient. Convergence or divergence of the vectors are associated with forcing for vertical motion.

QG Quasi-geostrophic

QPF Quantitative Precipitation Forecast

RAOB Radiosonde observation

radial velocity The component of motion along an axis extending from a radar unit. The NEXRAD base velocity product depicts radial velocity.

RAFS Regional Analysis and Forecast System (NGM)

range folding A process by which a radar echo returns after another pulse has been transmitted, creating an echo that might be incorrectly distanced by the radar unit.

RAREP Radar Report

RCM Radar Coded Message. An automated product of the WSR-88D unit which provides a summary of the echoes and signatures from a given radar.

rex block A blocking pattern in the upper levels in which a closed high is located poleward of a closed low. The long-wave pattern tends to "lock up".

RFQ Right-front quadrant of a jet streak. In the Northern Hemisphere this is usually associated with downward motion.

RGL Regional Model

ridge An elongated area of high pressure or heights.

RRQ Right-rear quadrant of a jet streak. In the Northern Hemisphere this is usually associated with upward motion.

RUC Rapid Update Cycle model

RUNHIST Run History

SBCAPE Surface based CAPE; resulting from a parcel that is lifted from the surface with no other modifications.

SC, Sc Stratocumulus (q.v.)

scattered Partial coverage of a cloud layer, covering more than a quarter to half of the sky, of 2 to 4 oktas. (cf. clear, few, broken, and overcast)

SD 1. Radar Report (now obsolete) 2. Storm Data, a publication of NCDC.

SDM Senior Duty Meteorologist

SEC Space Environment Center

Showalter Stability Index (SSI) The difference in temperature between the environment at 500 mb and a parcel lifted from 850 mb, expressed as $T_{500}-T_{850}$. A negative value corresponds to instability. Lifted Index and CAPE are usually preferred over the SSI.

SIGMET Significant Weather bulletin for pilots

SOO Science and Operations Officer

sounding A plot of temperature and dewpoint above a given station with respect to temperature (X-axis) and height (Y-axis). A thermodynamic diagram, usually the SKEW-T log P, is typically used.

SPC Storm Prediction Center (q.v.)

spectrum width The variance in velocity of scatterers within a given volume of air.

speed max A region of maximum winds within a jet stream. Also jet max, jet streak.

SPENES NESDIS satellite precipitation estimate

SRH Storm-relative helicity

SSI Showalter Stability Index (q.v.)

ST, St Stratus (q.v.)

STJ Subtropical jet (q.v.)

STK Storage Technology

Storm Prediction Center (SPC) A branch of NCEP, located in Norman, Oklahoma, which is responsible for providing short-term forecast guidance for convective storms.

stratocumulus A relatively flat, low cloud with little vertical development. It has distinct globular masses or rolls.

stratus (St) A low, sheetlike cloud which may either occur alone, or with precipitation (in which case it is referred to as scud, fractus, or stratus of bad weather).

subsidence Sinking motion.

sub-synoptic Mesoscale.

subtropical jet (STJ) An upper-level jet stream that is usually found between 20 and 30 deg of latitude and is associated with thermal differences within the subtropical high. (cf. arctic jet, polar front jet, and low-level jet)

SWODY1 Severe Weather Outlook - Day 1

SWODY2 Severe Weather Outlook - Day 2

synoptic-scale Spanning a distance scale of over 500 miles or a time scale of days.

TAF Terminal Aerodrome Forecast

TCU Towering Cumulus

TD Tropical Depression

teleconnection A strong statistical relationship between weather in different parts of the globe.

theta-e Equivalent potential temperature

Total Totals Index (TTI) A sum of the Cross Totals and Vertical Totals indices. It is equal to $T_{850}-T_{500}+Td_{850}-T_{500}$. A value of greater than 44-56 is considered significant.

TPB Technical Procedures Bulletin

TPC Tropical Prediction Center (q.v.)

Tropical Prediction Center (TPC) A branch of NCEP responsible for tropical weather forecasting, including hurricanes.

tropical storm A warm-core storm with a maximum sustained surface wind of 34-63 kt.

tropopause The point between the troposphere and stratosphere at which a positive tropospheric lapse rate becomes neutral or negative.

trough An elongated area of low pressure or heights.

TS Tropical Storm

TTI Total Totals Index (q.v.)

typhoon A tropical storm of hurricane strength in the Western Pacific basin.

UA Pilot Report; or upper air.

UCAR University Corporation for Atmospheric Research

UCL UNICOS Control Language (shell script)

UKMO United Kingdom Met Office

UKMET United Kingdom Met Office

ULJ Upper level jet

UPS Uninterruptable Power Supply

UTC Universal Coordinated Time

UVV Upward vertical velocity

VAD Velocity Azimuth Display. A plot of radial velocity (Y-axis) with respect to azimuth (X-axis) by a weather radar for a given level. It is used as a basis for construction of VWP diagrams (q.v.).

VAFTAD Volcanic Ash Forecast Transport and Dispersion

VC Vicinity

veering Referring to a clockwise change in the wind direction, with respect to either height or time. Contrast with backing.

vertical stack The tendency for a weather system, usually a closed low or high, to have the same location aloft as at the surface. This typically indicates a lack of baroclinicity and suggests a warm-core or cold-core structure.

Vertical Totals index (VT) An expression of the low to mid-level lapse rate, as given by $T_{850}-T_{500}$. A value of 26 or more is considered significant.

VIL Vertically Integrated Liquid

volume scan The complete scan of a weather radar for all assigned elevations. When a volume scan is complete, the radar is able to generate all possible products (with the exception of products that require a history of an echo). The WSR-88D completes a volume scan in 5 to 10 minutes.

vort max The highest vorticity in a given region.

vorticity The rotation in a volume of air, made up of shear and curvature.

VSB Visible

VIS Visible

VWP VAD Wind Profile. A plot of winds with height above a given station, as determined by a weather radar. The profile is displayed with height as the Y-coordinate and time as the X-coordinate.

VT Vertical Totals index (q.v.)

WAA Warm air advection

WAFS World Area Forecast System

WBZ Wet Bulb Zero (q.v.)

Wet Bulb Zero (WBZ) The height at which the wet bulb temperature drops below freezing, expressed as height above ground level (AGL). It is a measure of depth through which a hailstone will melt. Values of less than 10,000 ft are associated with large hail, given enough instability.

WFO Weather Forecast Office

WMSC Weather Message Switching Center

WMO World Meteorological Organization

WS Significant Weather bulletin (SIGMET) for pilots

WSFO Weather Service Forecast Office

WSO Weather Service Office

WST Convective SIGMET for pilots

WW Weather watch (thunderstorm or tornado)

WWB World Weather Building

Z Zulu Time (Greenwich Mean Time)

ZFP Zone Forecast Product (q.v.)

Zone Forecast Product (ZFP) A NWS bulletin that provides a clear, chronological statement of the weather conditions in a county or a given set of counties for the general public.

Suggested references & further reading

Presented here is a list of source materials for this book and suggestions for further reading. This is not intended to be a complete survey of the literature. Items included are those that may be of most interest to operational forecasters. Internet URLs and ISBNs are provided where possible to assist the reader.

General forecasting

Ahrens, C. Donald (2018). *Meteorology Today: An Introduction to Weather, Climate, and the Environment.* Cengage Learning, ISBN 1337616664. 656 pp.

Barry, Roger and Chorley, Richard J. *Atmosphere, Weather and Climate* (2010). Routledge, ISBN 9780415465700. 516 pp.

Bluestein, Howard B. (1992). *Synoptic-dynamic Meteorology in Midlatitudes: Observations and theory of weather systems (Volume 2).* Taylor & Francis, ISBN 9780195062687. 594 pp.

Carlson, Toby N. (1991). *Mid-Latitude Weather Systems.* Routledge, ISBN 0415109302. 507 pp.

Cole, Franklyn W. (1980). *Introduction to Meteorology.* John Wiley & Sons, ISBN 0471047058. 505 pp.

Gedzelman, Stanley D. (1980). *The Science and Wonders of the Atmosphere.* John Wiley & Sons, ISBN 0471029726. 535 pp.

Lackmann, Gary (2011). *Midlatitude Synoptic Forecasting.* American Meteorological Society, ISBN 9781878220103. 360 pp.

Markowski, Paul M. & Richardson, Yvette P. (2010). *Mesoscale Meteorology in Midlatitudes.* Wiley-Blackwell, ISBN 9780470742136. 430 pp.

Moran, Joseph M. (1994). *Meteorology: The Atmosphere and the Science of Weather.* Macmillan, ISBN 0023833416. 517 pp.

National Weather Service (1993). *Forecasters Handbook No. 1.* 340 pp.

Stull, Roland B. (1995). *Meteorology Today for Scientists and Engineers.* West Publishing Co., ISBN 0314064710. 385 pp.

Vasquez, Tim (2021). *Weather Analysis & Forecasting Handbook.* Weather Graphics Technologies, ISBN 9780996942348. 340 pp.

Chart analysis

Doswell, Charles A. III (1986). The human element in weather forecasting. *Nat. Wea. Dig.*, 11, 6-18. <www.cimms.ou.edu/~doswell/human/Human.html>

Djuric, Dusan (1994). *Weather Analysis.* Prentice-Hall, ISBN 0135011493. 304 pp.

Miller, Robert C. (1972). *Notes on Analysis and Severe-Storm Forecasting Procedures of the Air Force Global Weather Central.* AWS Technical Report 2000 (Rev), Air Weather Service, Scott AFB. 190 pp.

National Weather Service (1993). *Graphical Guidance.* National Weather Service, Washington. 169 pp.

Reed, R. J., & Sanders, F. (1953). An investigation of the development of a mid-tropospheric frontal zone and its associated vorticity field. *J. of Atmos. Sci.*, 10(5), 338-349.

Sanders, Frederick & Doswell, Charles A. III (1992). A Case for Detailed Surface Analysis. *Bull. of the Amer. Met. Soc.*, 76, 505-521.

Young, G. S. & Fritsch, J. M. (1989). A Proposal for General Conventions in Analyses of Mesoscale Boundaries. *Bull. of the Amer. Met. Soc.*, 70: 1412-1513.

Satellite imagery

GOES ABI Quick Guides (Online resource). Cooperative Institute for Meteorological Satellite Studies. <http://cimss.ssec.wisc.edu/

goes/GOESR_QuickGuides.html>

National Environmental Satellite, Data, and Information Service (1983). *The GOES Users Guide*, NESDIS, Washington. 164 pp.

Schmit, T. J., Griffith, P., Gunshor, M. M., Daniels, J. M., Goodman, S. J., & Lebair, W. J. (2017). A closer look at the ABI on the GOES-R series. *Bull. of the Amer. Met. Soc.*, 98(4), 681-698.

Radar products

Office of the Federal Coordinator for Meteorology (2021). *Federal Meteorological Handbook #11: Doppler Radar Meteorological Observations*. OFCM, Washington. <https://www.icams-portal.gov/resources/ofcm/fmh/allfmh2.htm>

Miller, Sam (2020). *Applied Radar Meteorology*. Published online. <https://core.ac.uk/download/pdf/327105104.pdf>

Rauber, Robert & Nesbitt, Stephen (2018). *Radar Meteorology: A First Course*. Wiley, ISBN 9781118432624. 496 pp.

Vasquez, Tim (2013). *Weather Radar Handbook*. Weather Graphics, ISBN 9780996942317. 166 pp.

Dynamical models

Carr, Frederick H. (1988). *Introduction to Numerican Weather Prediction Models at the National Meteorological Center*. 63 pp.

Innes, Peter & Dorling, Steve (2012). *Operational Weather Forecasting*. Wiley, ISBN 9781118447635. 248 pp.

National Meteorological Center (1987). *Section 2.2.1: The NMC Production Suite*. NMC Handbook., National Meteorological Center, Washington.

United Kingdom Met Office (2003). NWP Gazette, quarterly. <library.metoffice.gov.uk>

University Corporation for Atmospheric Research (2003). Operational Models Matrix: Characteristics of Operational NWP Products. UCAR Homepage. <meted.ucar.edu/nwp/pcu2>

Warner, Thomas Tompkins (2011). *Numerical Weather and Climate Prediction*. Cambridge University Press, ISBN 9780521513890. 526 pp.

Text products

Office of the Federal Coordinator for Meteorology (1998). *Federal Meteorological Handbook #1: Surface Weather Observations and Reports* (FCM-H1-1998). OFCM, Washington. <www.ofcm.gov/fmh-1/fmh1.htm>

Office of the Federal Coordinator for Meteorology (1997). *Federal Meteorological Handbook #3: Rawinsonde and Pibal Observations* (FCM-H3-1997). OFCM, Washington. <www.ofcm.gov/fmh3/text/default.htm>

Office of the Federal Coordinator for Meteorology (1998). *Federal Meteorological Handbook #12: United States Meteorological Codes and Coding Practices* (FCM-H12-1998). OFCM, Washington. <www.ofcm.gov/fmh12/frontpage.htm>

World Meteorological Organization (1988). *Manual on Codes*. WMO Publication No. 306, World Meteorological Organization, Geneva

Index

Symbols

4DVAR 122
200 mb 22
250 mb 22
300 mb 22
500 mb 20
540 dam line 26
700 mb 18
850 mb 14, 16
1000-500 mb thickness 26
1000-700 mb thickness 26
1000-850 mb thickness 26

A

absolute vorticity 30
advection 26, 30
analysis 10
anticyclonic flow 22
anticyclonic rotation 72
anticyclonic vorticity advection 30
ascent 2, 28
AVA 30

B

baroclinicity 20
baroclinic zones 11
base reflectivity 68
beam width 88
big drops 82
Boyden Index 159
Bradbury Index 159
bright band 68
BRN 156
BRN Shear 156
Bulk Richardson Number 156

C

cap 18
CAPE 42, 156
CC 78
chaff 84
CINH 156
clear air mode 66
cloud code groups 151
cold advection 30
conditional probability 126
Convective Availability of Potential Energy 42
Convective Available Potential Energy 156
Convective Inhibition 156
convergence 3, 72
Coordinated Universal Time 4
correlation coefficient 78
couplet 72
CPR 126
Cross Totals Index 157
CT 157
CVA 3, 30
cyclonic flow 22
cyclonic rotation 72
cyclonic vorticity advection 3, 30

D

dam 26
Deep Convective Index 159
dekameters 26
descent 28
descriptors 149
diagnosis 10
differential phase 80
differential reflectivity 76
divergence 3, 72
drylines 11
dynamics 2
dynamic tropopause 38

E

Echo Tops 88
ECMWF 122
EFR 124
EHI 156
elevated mixed layer 18
Elevated TVS 96
elevation gaps 88
EML 18
Energy-Helicity Index 156
enhanced imagery 56
ensemble 126
ensemble mean 126
EPS 126
Eta model 131
ETVS 96
European Centre for Medium-Range Weather Forecasts 122

F

forcing 2
four-cell concept 2
Free Text Message 100
frontal systems 16
frontogenesis 34
frontolysis 34
fronts 11, 12
FTM 100
further reading 181

G

gate-to-gate shear 72
geostationary satellites 50
geostrophic balance 2
GEPS 126
global 112
Global Environmental Multi-scale 124
Global Spectral Model 116
GMS 50
GOES 50
graupel 82

H

Hail Detection Algorithm 92
Hail Index 92
hand analysis 10
HCA 82
HDA 92
HHC 82
High Resolution (HRES) 122
High Resolution Rapid Refresh (HRRR) 120
hook echo 68
HREF 126
HRRR 120
Hydrometeor Classification Algorithm 82

I

ICAO regions 148
infrared imagery 56
Integrated Forecast System (IFS) 122

inversion 18
isentropic analysis 28
isentropic lift 3
isobars 12
isopleths 10

J

Jefferson Index 159
jet maxes 22
jet streaks 22

K

KDP 80
KI 158
K Index 158
KO Index 158

L

lapse rate 42
left front quadrant 22
left rear quadrant 22
LFQ 22
LI 42, 157
lift 3
Lifted Index 42, 157
Limited-area Fine Mesh (LFM) 130
LLJ 16
long waves 22
low-level jet 16
LRQ 22

M

MAPS 132
Maximum Expected Hail Size 92
maximum unambiguous velocity 72
mean 126
MEHS 92
member 126
mesocyclone 94
Mesocyclone Detection Algorithm 94
mesoscale 112
Mesoscale Analysis and Prediction System 132
METAR 138
METEOSAT 50

Met Office 124
microbars 32
MLI 157
MM5 132
MN 126
model biases 113
Modified Lifted Index 157
moisture 18

N

NAM 118
National Severe Storms Forecast Center 104
Navy Operational Global Atmospheric Prediction Sys 125
Nested Grid Model 131
NGM 131
NOGAPS 125
NSSFC 104
numerical weather prediction 112
Nyquist co-interval 72

O

oblate 76
obstructions to vision 149

P

paintball 126
pattern vectors 96
PB 126
PBL 16
Pennsylvania State University 132
planetary boundary layer 16
POES 51
POH 92
polar orbiters 51
POSH 92
Potential Instability Index 159
potential temperature 28
PR 126
precipitation 42
precipitation mode 66
present weather 150
probability 126
Probability Of Hail 92
Probability Of Severe Hail 92
prolate 76

PSU 132

R

Rackliff Index 159
Radar Data Acquisition 66
Radar Products Generator 66
Radio Acoustic Sounding System 44
radiosonde data 142
radiosonde plot 147
rain-snow transition 26
range folding 72
RAP 120
Rapid Refresh Forecast System (RRFS) 118, 120
Rapid Refresh (RAP) 120
Rapid Update Cycle (RUC) 120, 132
RASS 44
RDA 66
regional 112
relative humidity 18
REPS 126
RFE 124
RFQ 22
rho 78
right front quadrant 22
right rear quadrant 22
rotation 72
RPG 66
RRQ 22
RUC 132

S

SCIT 92
SD 126
SEF 124
self development 20
Severe Weather Threat Index 158
SHARPpy 128
short waves 20
Showalter Stability Index 42, 158
S-Index 159
SP 126
spaghetti diagram 126
specific differential phase 80
SREF 126
SRH 156

SRM 72
SSI 42, 158
ST 126
stamps 126
standard deviation 126
Storm Cell Identification and Tracking 92
storm centroid 90
storm component 90
storm precipitation total 84
Storm Relative Helicity 156
storm relative motion 72
storm segments 90
storm total precipitation 84
Storm Tracking Information 90
STP 84
stratosphere 22
surface chart 12
surface plot 146
SWEAT 158
SYNOP 136

T

TAF 140
TBSS 74
Television InfraRed Observation Satellite 50
TEMP format 142
Terminal Aerodrome Forecast 140
thermal advection 26
thermodynamic diagram 42
thickness 26, 46
Thompson Index 158
three-body scatter spike 74
TI 158
TIROS 50
Tornado Detection Algorithm 96
tornadoes 72, 74
Total Totals Index 157
tropopause 22
  dynamic 38
troposphere 22
trough 11
TT 157

U

UCAR 132

UKMET 124
uncorrelated shear 94
Unified Model System 124
University Corporation for Atmospheric Research 132
upper air plot 147
upslope flow 3

V

VAD 98
VAD Wind Profile 98
Velocity Azimuth Display 98
Vertically Integrated Liquid 86
Vertical Totals Index 157
vertical velocity 3
VIL 86
VT 157
VWP 98

W

WAA 3
warm advection 30
warm air advection 3
wave number transition 113
weather 149, 150
Weather Research and Forecasting 118
wind profiler 44
wind shift 20
WMO regions 148
WRF 118
WSR-57 66
WSR-74C 66

Y

Yonetani Index 159

Z

ZDR 76

Printed in the USA
CPSIA information can be obtained
at www.ICGtesting.com
JSHW070734130923
47983JS00011B/5